P9-DUK-570

3 1404 00769 2806

DEC 15

WITHDRAWN

JUN 2 7 2024

PROPERTY OF
DAVID O. McKAY LIBRARY
BYU-IDAHO
REXBURG ID 83460-0405

A HISTORY OF NERVE FUNCTIONS

Recent developments have extended our knowledge of the basic functions of nerves – notably, demonstration of the mechanism within nerve fibers that transports a wide range of essential materials. To understand how this discovery occurred, it is necessary to examine its history. The story begins in ancient Greece when nerves were conceived of as channels through which animal spirits carried sensory impressions to the brain. As science developed, the discoveries of various physical and chemical agents supplanted the agency of animal spirits until the molecular machinery of transport was recognized. In this fascinating and complete history, Sidney Ochs begins with a chronological look at this path of discovery, followed in the second half by a thematic approach, wherein the author describes the electrical nature of the nerve impulse, fiber form, and its changes in degeneration and regeneration, reflexes, learning, memory, and other higher functions in which transport participates. *A History of Nerve Functions* will serve as an invaluable resource for historians of neuroscience and medicine, philosophers of science and medicine, as well as for neuroscientists.

Sidney Ochs is Professor Emeritus of Cellular and Integrative Physiology at the University of Indiana School of Medicine. He has been a pioneer in research in the field of axoplasmic transport in nerves, publishing the first monograph on the subject. He has contributed more than 300 publications on various aspects of the peripheral and central nervous systems, including a textbook on neurophysiology. He founded the *Journal of Neurobiology* and was a Regional Organizer in the establishment of the Society for Neuroscience, as well as later acting as a Councilor of the Society.

A HISTORY OF NERVE FUNCTIONS

From Animal Spirits to Molecular Mechanisms

SIDNEY OCHS

Indiana University

CAMBRIDGE
UNIVERSITY PRESS

PUBLISHED BY THE PRESS SYNDICATE OF THE UNIVERSITY OF CAMBRIDGE
The Pitt Building, Trumpington Street, Cambridge, United Kingdom

CAMBRIDGE UNIVERSITY PRESS
The Edinburgh Building, Cambridge CB2 2RU, UK
40 West 20th Street, New York, NY 10011-4211, USA
477 Williamstown Road, Port Melbourne, VIC 3207, Australia
Ruiz de Alarcón 13, 28014 Madrid, Spain
Dock House, The Waterfront, Cape Town 8001, South Africa

http://www.cambridge.org

© Sidney Ochs 2004

This book is in copyright. Subject to statutory exception
and to the provisions of relevant collective licensing agreements,
no reproduction of any part may take place without
the written permission of Cambridge University Press.

First published 2004

Printed in the United States of America

Typefaces ITC Stone Serif 9/12 pt. and ITC Symbol   *System* LaTeX 2_ε [TB]

A catalog record for this book is available from the British Library.

Ochs, Sidney.
   A history of nerve functions : from animal spirits to molecular
mechanisms / Sidney Ochs.
     p. cm.
   Includes bibliographical references and index.
   ISBN 0-521-24742-X
   1. Axonal transport–History.   2. Nerves–History.   3. Neurology–History.   I. Title.
QP363.O273   2003
573.8'5'09 – dc21   2003055050

ISBN 0 521 24742 X hardback

CONTENTS

PREFACE

Within the last half of the twentieth century, two fundamental properties of nerve were established: in midcentury, the ionic nature of the propagated action potential; and, later in the century, the process in the fibers known as *axonal flow, axoplasmic transport, axonal transport, neuroplasmic transport,* and so on. By means of the transport mechanism, essential components synthesized in the nerve cell bodies are carried out within the relatively long length of nerve fibers to maintain their viability and function. Components transported include the ion channels and ion pumps needed to maintain membrane potentials all along the length of the fibers, metabolic and structural components supporting the form and viability of the fibers, and substances providing for reception at sensory terminals and neurotransmitters at motor terminals. This is indeed a protean mechanism, fundamental for an understanding of modern neuroscience and a rational basis for interpretation of neuropathies and eventually their therapy.

Although the discovery of the properties and molecular nature of the transport mechanism and related topics is a major theme, this account is not restricted to the last half century. The concept can be traced back to its earliest beginnings in the sixth and fifth centuries B.C., respectively, when philosophy and science had their origins in ancient Greece. Nerves were then conceived of as channels carrying sensory impressions by animal spirits to the brain where consciousness awareness and reasoned judgment were located, and from it willed commands were carried by nerves to actuate the muscles. (Notable exceptions to the concept that the brain played this role were Aristotle and the Stoics, who viewed the heart as serving those higher functions.) From ancient Greece, as step-by-step physiological and anatomical sciences evolved and new physical principles and chemical substances were discovered over the centuries, various agents were proposed to replace animal spirits without prevailing, until, in the nineteenth century, the electrical nature of the nerve impulse was established and, later in the century, the elementary unit of the nervous system, the neuron, was recognized.

In the last half of the twentieth century, electron microscopy revealed the fine structure of the nerve fiber and isotope tracers, and biochemical techniques were used to characterize the movement of proteins and other materials in them and reveal the molecular nature of the transport mechanism. The same processes discovered in peripheral nerves were also found to take place in the neurons of the central nervous system, where transport in the brain was seen as underlying neuronal changes related to higher behavior, to learning and memory. Thus, a close connection can be traced from the earliest conceptions of nerve as a channel for an agency responsible for sensation carried to and motor responses from to the brain.

Whereas this history makes a case for the similarity of the earliest concept of nerve to its present understanding, a Whiggish view of history, the interpretation that the past simply evolves in a direct progressive path to the present is not taken. Our story includes false steps, strong personal oppositions, and periods of stagnation or even regression, these setbacks overcome with new thinking, often provided by the importation of concepts and techniques from other sciences. Some account of the cultural background out of which the science developed has been touched on, where philosophical and religious teachings have acted to further or hinder the progress of discovery. Liberal use was made of the best available scholarship to give English translations of some of the primary sources to convey what was thought at the time. This, it is hoped, will help avoid anachronisms, the temptation to judge the work of our predecessors on the basis of present information and standards. I have retained the older phraseology; and for some obsolete words, I have supplied present-day equivalents in square brackets. Square brackets were also used where elisions or additions were made to further the sense of a quoted passage.

A chronological order has in general been followed for the first six chapters until, by the late eighteenth century, the known properties of nerve and the nervous system had grown to the point that keeping to a strict chronological order became cumbersome. From the seventh chapter on, a thematic course was followed, with some aspects of the subject carried forward to our day; the electrical nature of the nerve impulse, the form of the nerve fiber, its degeneration and regeneration, the characteristics and molecular mechanisms proposed for axonal transport, agents interfering with transport and their relationship to neuropathology, reflex responses, the relationship of brain structures and transport to higher behavior, to learning and memory. Some implications of those latter subjects are dealt with in a more speculative manner in the last chapter, as part of the postscript.

Rather than an attempt to chronicle all aspects of nerve, the main purpose is to give an account of those concepts thought to be most essential that are related to transport. Even so, the large and ever-increasing body of literature has made the problem of selection, particularly in dealing with the more recent literature, acute. I am only too aware that, to keep within the bounds of a single volume, some aspects and worthwhile contributions could not be included. My hope is that omissions in this respect may be overcome by recourse to the copious references supplied.

Although, as far as is known to me, this is the first book-length treatment of the history of nerve in which axonal transport figures in a major way, some recent histories of neuroscience have dealt with some aspects of transport or related topics. Clarke and O'Malley have given translations of a number of important historical writings on nerve, brain, and spinal cord.[1] The book by Clarke and Jacyna, dealing with fundamental developments in the history of the nervous system during the first half of the nineteenth century, contains important sections relative to transport.[2] Liddell discusses the early microscopic studies of the nerve fiber.[3] Some of the earlier history of transport was briefly touched on in my book, which gives a general exposition of axoplasmic transport up to 1982.[4] Spillane's book on nerve has a more extensive clinical orientation.[5] The histories of neuroscience by Brazier[6] and those by Finger[7] provide useful background information. These deal for the most part with the central nervous system. Important reviews dealing directly with portions of the history of transport have been given by Clarke,[8] Rothschuh,[9] and Billings[10] and in collected volumes on axoplasmic transport.[11]

The writing of this book was shaped by studies of peripheral nerve, spinal cord, and brain properties carried out over a period of more than 50 years. My interactions in those studies with colleagues, post-doctoral fellows, and students was not limited to only the investigative work at hand, but they acted as a stimulus to further understand the historical basis of what we were involved with. I give my heartfelt thanks to all who shared their studies with me. I would like to express my gratitude to Ralph Waldo Gerard who took me on as a doctoral student at the University of Chicago, and to Anthonie van Harreveld, with whom I served as a post-doctoral Fellow at the California Institute of Technology, who treated me as a colleague and friend. Thanks must also go to the librarians at the Ruth Lilly Library of the Indiana University School of Medicine who have been unfailing in their help with difficult to obtain materials. I wish to express my thanks to Dr. Katrina Halliday, my editor at Cambridge University Press, for her support and to those other editors at Cambridge who through the years did not lose faith that this book would eventuate. Last, but not least, this book would not have been possible without the support of my wife Bess and our children Rachel, Raymond, and Susie who each in their own way has helped in the course of making this book.

[1] (Clarke and O'Malley, 1968). [2] (Clarke and Jacyna, 1987).
[3] (Liddell, 1960). [4] (Ochs, 1982). [5] (Spillane, 1981).
[6] (Brazier, 1984) and (Brazier, 1988). [7] (Finger, 1994) and (Finger, 2000).
[8] (Clarke, 1968) and (Clarke, 1978). [9] (Rothschuh, 1958). [10] (Billings, 1971).
[11] (Weiss, 1982), and (Iqbal, 1986).

1

INTRODUCTION: GREEK SCIENCE AND THE RECOGNITION OF NERVE AS A CHANNEL

Before the dawn of civilization, primitive man believed, as does primitive man today, in animism, magic, and supernatural forces to account for events in the world he experienced.[1] The powers of nature are seen when, after the death of vegetation in winter, its rebirth occurs in spring. Storms with their lightning and thunder, wild animals, and the unpredictable and often turbulent behavior of man in relation to man were powers anthropomorphized through the action of spirits who were either beneficent or malevolent. The emotions felt within himself, man projected to other men, to other living beings, and even to inanimate objects moved by unseen forces.

With the rise of Greek philosophy and science, another view of nature and man arose: the belief that the cosmos and man are ruled by impersonal laws, that the gods do not take a providential interest in the affairs of man. As scientific knowledge evolved and the structures and functions of the various body organs became recognized, the nerves were singled out as having an integral relation to sensation and body movements. In some of the earliest accounts of nerve, they were thought of as channels carrying a spiritual influence to the brain in which consciousness and willed motor control over the body was located.

THE EARLY CONCEPTION OF NERVE CONFOUNDED WITH TENDONS
The artifacts and cave drawings left by prehistoric man attest to his powers of observation. In the course of hunting or warfare, he would have seen muscles become lax and limbs made useless by the severing of large tendons. But the tendon was not clearly identified. It could be muscle tendons (such as those of the hamstrings) or a major nerve (such as the sciatic). In the archaic Greek civilization represented in Homer's epic *The Iliad*,[2] the word *neuron*

---

[1] (Frazer, 1922).
[2] c. 8th century B.C., probably first written down in the sixth century.

was applied both to tendons and nerves, and analogized to a bowstring taut with tension applied to it or lax when severed. The standard dictionary definitions in use to this day attest to the order of its ancient derivation. Under the dictionary heading of "nerve," we find the first meanings given are "sinew" and "tendon," with the idea of putting forward the utmost exertion, as to "strain every tendon."[3] The later meaning given is that it is "cord-like or filamentous tissue connecting parts of the nervous system and organs of the body." It was defined in 1606 as, "A fibre or bundle of fibres arising from the brain, spinal cord, or other ganglionic organ, capable of stimulation by various means, and serving to convey impulses (especially of sensation and motion) between the brain, etc., and some other part of the body."[4] A number of other meanings given for its modern use relate to its original meaning as a tendon, namely as subserving vitality, force, physical strength, fortitude, vigor, endurance, and also curiously its inverse, oversensitiveness and nervous weakness.[5]

In the *Old Testament*, reference is made to "the sinew present in the thigh" with some confusion with blood vessels, as in the translation "the principal vein of the leg which is in the thigh, commonly the sciatic nerve" and, to the "sinew of the thigh-vein or thigh-nerve." That the nerve was implicated is indicated when, in the struggle of Jacob with a strange man (Angel or God?), Jacob was "touched on the hollow of his thigh, which lamed him."[6] This is why it is said that, "the Israelites to this day do not eat the sinew of the nerve that runs in the hollow of the thigh." From the era of the Patriarchs, eating the flesh of that part of the leg (the rump) containing this nerve (the sciatic) was forbidden, and this proscription was maintained as part of Kosher law. However, the flesh may be eaten if the nerve is first removed

---

[3] (Webster's, 1966).

[4] (Oxford, 1944). In earlier dictionaries devoted to medicine (Castelli, 1761), the latin term *nervus* derived from the Greek *neuron* while described in relation to tendons and ligaments, the bulk of the definition dealt with its then modern role as channels conveying animal spirits to support sensation and motion.

[5] (Oxford, 1971). Samuel Johnson gave as the first meaning for nerve an organ of sensation, the second the use by poets for sinew and tendon, and the third for force and strength (Johnson, 1827). It is interesting that in our time only by the eighth definition do we find reference to nerve as a fiber or bundle of fibers arising from the brain and acting to convey impulses of sensation or motion. Another relation of the term nerve to tendons comes from the early experiences with stringed instruments. The strings, when stretched, give rise to a higher note on plucking them. Thus, an individual, if overly sensitive or nervous, is said to be high strung.

[6] (Bible, 1970), Genesis 32:31–32. This is variously translated. The New English Bible has it that a dislocation at the hip occurred. Translating the passage from the Latin Vulgate, "He touched his (Jacob's) *nervum femoris* (femoral nerve)." But, the femoral nerve supplies the anterior surface of the thigh, and the reference is clearly to the hollow of the thigh. Continuing, the passage reads, "immediately (the nerve or leg) became feeble" (reading *emascuit*-withered or feeble).

in the process known as *porging*.[7] The custom of porging appears to have been carried out in an old, isolated Jewish enclave in China by the Jews who lived in Kaifeng (K'ai-feng), the old Capital city of the province of Honan,[8] where the Chinese referred to them as the *T'iao-chin chiao*, the people who "pick out the tendons."[9]

An analogy to the biblical injunction against eating the rump because of the sciatic nerve within it was the practice carried out among some North American Indian tribes of regularly cutting out and throwing away the thigh muscles containing the nerve.[10] The reason given by the Cherokee Indians was that the "tendon," when cut, retracts, with the muscles becoming lax; and they did not wish to expose themselves to the danger of also becoming weakened if they were to eat it. The notion is clearly based on sympathetic magic.[11] The struggle of Jacob in Genesis may very well also have had a similar origin in sympathetic magic that was later given a mythic interpretation.

PRIMITIVE ANIMISTIC BELIEF IN SPIRITS ANIMATING THINGS, AS WELL AS LIVING BEINGS

An insight into the thought of the ancient man was given by the studies of aboriginal peoples in Polynesia and elsewhere in remote corners of the world by explorers, evangelizers, and anthropologists. An extraordinary opportunity to directly study the thinking of a primitive man came about when in 1911 an Indian, Ishi, emerged from the foothills of a remote mountain in northern California. He was the last of an isolated Stone Age tribe that had completely died out but for him.[12] Ishi was patient, cheerful, good-natured, with the capacity to learn equal to that of modern man. He

---

[7] (Klein, 1979).

[8] Jews entered China via the Silk Road as traders in a number of places in China, perhaps as early as during the Han dynasty (206 B.C.–220 A.D.) (White, 1966), p. 52, though the earliest tangible evidence of their presence is around 718 with a settled Jewish community in Kaifeng more definitely attested to by a synagogue built in 1163.

[9] (White, 1966) ref. to (Lépine, 1894), Part 1, p. 51, Part 2, note 18, p. 24 and p. 110. The term "T'iao" refers to jumping, "chin" to the tendon or nerve, the jumping nerve, and "chiao" to a hollow, wherein by digging the sciatic nerve could be removed from the rump and thigh muscles.

[10] (Gaster, 1969), pp. 210–211. [11] Ibid., pp. 211–212.

[12] (Kroeber, 1994), pp. 23, 78. Hungry, sick, and alone, Ishi came into the hands of two California anthropologists: T. T. Waterman and Alfred L. Kroeber. With the assistance of an Indian from a bordering Indian tribe who had a dialect close to that of the Indian and the linguist Edward Sapir, he was able to communicate the story of his life. Ichi was a Stone Age man who had stepped out of the remote shadows of the past. He lived on for four plus years of his life quartered in the museum transmitting his language and culture. He demonstrated how he made obsidian knives and arrowheads, bows and arrows. He also fished and hunted. While housed in a hospital, Ishi witnessed a number of surgeries. These did not

lived at ease with the supernatural and the mystical, which for him were so pervasive in all aspects of his life, feeling no need to differentiate mystical truth from directly observed evidential or "material" truth, the supernatural from the natural.

Anthropologists have described the primitive's belief that the world is ruled by spirits, with the power to do him harm or good.[13] To the primitive mind, there are no accidents. Everything has a cause. A falling rock or a swaying tree was directed by some hidden force.[14] Misfortunes and diseases are attributed to malevolent agencies.[15] These include taboo violation, disease-object intrusion, spirit intrusion by sorcery, and soul loss. When human ghosts feel lonesome or if disrespect has been shown to its body after death, they can cause disease. The power of the wind was of importance, and the meaning given to it by American Navajo Indians gives some insight into what prehistoric man may likely have understood of the world. The chants of the singers transmitted the traditional sacred lore in which the term *Niłch'i*, meant Wind, Air, or Atmosphere, which was conceived by the Navajo to be endowed with powers.[16] Suffusing all of nature, *Holy Wind* gives life, thought, speech, and the power of motion to all living things, and serves as the means of communication among all elements in the living world. By the winds of the East, West, North and South, man is given direction to life, to movement, thinking, and action to carry out his plans; these external winds are the same as "the wind standing within us."[17] The great force of the wind is shown by its power to knock down large trees. The wind was also thought by the ancients to have procreative power. Mares were reported to become impregnated by facing their hind quarters into the wind; and, at a relatively later time, the Roman writer Varro (116–127 B.C.) affirmed it as certain truth "that about Lisbon some mares conceived by the wind, at a certain season, as hens conceive what is called a 'wind egg.'"[18]

trouble him but for the induction of anesthesia. He thought that during sleep, the soul leaving the body might have difficulty in returning.

[13] A short discussion of animism given by Clodd included the various terms given for the concept around the world (Clodd, 1905). See also (Vogel, 1970).

[14] (Frazer, 1922).   [15] (Rivers, 1924).

[16] (McNeley, 1981), p. 1. By carefully cross-checking various oral versions, McNeley was able to determine what the Indians generally understood by the concept of "wind." The relative isolation of the Navajo subjects on the reservation, uncontaminated with modern Western notions and his personal participation in Navajo culture makes his account an important document. The concepts he described could very well represent a deposition of earlier thought reaching back to the Stone Age.

[17] Ibid., p. 16.

[18] (Oxford, 1971), p. 3786. The wind-egg is an imperfect or soft-shelled egg that is unproductive. This term was used by Plato in the dialogue Theaetetus for an argument or a concept that was unproductive (Cornford, 1935), p. 163.

The belief in powers that control the world and man was the basis of magical practices. If followed in exact accordance with the laws of those powers, a desired end – such as the cure of a disease – could be brought about. An important distinction was made by Fraser between the practice of magic and religion.[19] The magician acts to bring about the desired end by following his belief in the immutable laws of nature as he understood them. In this sense, his thought process is analogous to that of the scientist – the difference being that, for the scientist, those laws and the understanding of them can evolve and be replaced by others. Religion, on the other hand, is the belief that superhuman agencies, the gods, take a personal interest in man. The function of the priest is to intervene to bring about a desired end by placating the gods or to make them turn from their foreordained detrimental purposes. In the course of the development of man's thought, the offices of magician and priest were combined. The priest could, as in ancient Egypt, not only placate, but also compel the highest of gods to do his bidding.

To primitive man, the dramatic transformation of a living man or animal, warm and moving, into a cold immobile form when dead, appeared to be due to the loss of an internal "spirit" or "soul" essential for life. The very term *spirit* is related to breath, which confers life. The ancient Egyptians describe how Isis, by breathing on the dismembered body of Osiris, brought him back to life. In the *Old Testament*, the Hebrew word for spirit, *neshuma*, encompasses the concept of breath and wind. It is written in Genesis 2:7 that man formed from clay (or dust) was inert until "God breathed into his nostrils the breath of life and he became a living soul." In the book Ezekiel 36:2–11, the prophet was put down in a plain full of bones. Prophesy to these dry bones he was told and hear the word of the Lord:

> I will put breath into you and you shall live. There was a rustling sound and the bones fitted themselves together. Sinews appeared, flesh appeared and skin covered them but there was no life. Prophesy to the wind to come from every quarter that they might come to life. Breath came into them. They came to life, and rose to their feet a mighty host.

In the Egyptian Coptic rite, the priest immediately after baptism breathes on the face of the infant saying, "Receive the Holy Ghost"; and their priests are ordained by the reigning bishop breathing on the new prelate's face. And, by inhaling a dying breath capture its power.

Pneuma, the ancient Greek term cognate with the Latin *spirit*, was thought to suffuse the cosmos, with wind being the expression of its presence and power. To the Vedic Indians, the cognate term *prana* was used to indicate

---

[19] (Frazer, 1922), pp. 56–69.

the presence of breath-giving life:

> The cosmic wind that blows in the atmosphere motivates and regulates the normal course of things or the cosmic order in the same way that the breath in living beings motivates life. Thus, wind is the breath of the cosmic person and the dead person's spirit (*atman*) goes to the wind.

This *pneumatological* belief is widespread. It is the basis of the practice of Yoga in India[20] and in Chinese and Japanese medicine where the term *qi* is cognate with the Greek term pneuma and its Latin equivalent spirit (*spiritus*).[21]

The word *soul* used in the *New Testament* is the translation of the Hebrew word for spirit that embraces a complex of ideas extending the idea of spirit. The soul can have appetites, hunger, and thirst. It is the seat of emotions, desires, and can experience sorrow or joy. It can show volition and think; it is the self-conscious manifestation of life. The Homeric Greeks pictured the soul as thin and impalpable, a shadow haunting a ghostly realm, a replica of the human body that resembles its former self.[22] According to Tylor, "nothing but dreams and visions could have ever put into men's minds such an idea as that of souls being ethereal images of bodies."[23] To the primitive mind, the dream is no less real than the waking state, and the dreamer may accuse another of a crime he saw committed in a dream.[24]

ANCIENT BELIEF IN POWERS RESIDING IN THE HEAD
INDICATED BY TREPANATION

The association of important spiritual powers residing in the brain is seen by practices such as those of the head hunting cannibals of Dutch New Guinea and adjacent islands who took their enemies' heads to gain the powers felt to reside within them. In the 27 skulls found by Lord Moyne in an abandoned village on the Bloemen River region in Dutch New Guinea, enlargements were made around foramen magnum at the bottom of the skulls "that shows the effect of a deliberate and successful removal of a large part of its base to facilitate the extraction of the contained brain, for use either for food or for ritual purposes."[25] Cave quotes Frazier who gives examples of the eating of the brain of a vanquished warrior to "acquire thereby his virtues

---

[20] (Zysk, 1995).   [21] (Kuriyama, 1995).

[22] This is strikingly described in the *Iliad*, when Achilles falls asleep and Patroclus' ghost visits him in a dream, looking exactly as he had while alive. He asks to be "buried without further ado so that I can win admittance to the Underworld." Achilles "stretched out both arms, but the phantom evaded his clutch, turning to vapor and sinking through the earth with a shriek." Achilles leaped up, horror-stricken: "Then it is true!" he exclaimed. "There are spirits of the dead in Hades' kingdom; active minds, though unsubstantial and lifeless! Yet, how marvelously Patroclus' ghost resembled his living self when it stood lamenting and pleading with me" (Homer, 1997).

[23] (Tylor, 1871).   [24] Ibid. and see also (Clodd, 1905).   [25] (Cave, 1937), p. 15.

and qualities."[26] Among primitives in the Celebes, it was to acquire the victim's bravery, among those in the Philippines it was to obtain courage, and among those in the New Guinea territory it was to procure strength. In Southern Guinea, "the decayed brain of a wise man is consumed in the belief that his wisdom passes to the consumer. . . ."[27] Some additional ritualistic reasons for the practice of taking the brains other than that of vanquished warriors is indicated by the presence among Lord Moyne's collection of the skulls of children 4–8 years of age that had similarly been broken into at the base. In recent years, transmission of the disease called *kuru*, a form of Creutzfeldt-Jakob disease affecting the brain, was traced to the practice of eating brains.[28]

The loss of consciousness and prostration or death following a blow to the head and convulsions must not have been an uncommon observation made by prehistoric and primitive man. This could well have suggested the idea that the head plays an especially powerful role in controlling conscious life, as well as controlling the lively vigor of the body leading to the ancient practice of *trepanation*, the opening of holes in the skull of a living subject.[29] The first example of trepanned skulls that had been made in ancient times was discovered in France in 1834. Subsequently, such skulls were found elsewhere in Europe, North Africa, parts of Asia, and America, with greatest frequency in the highlands of Peru and Bolivia.[30] That these openings had incurred in life was shown by the healed edges of the holes. Also testifying to the subjects outliving the operation are skulls found with two, three, or more holes in them (Figure 1.1). Perhaps epileptic attacks, chorea, insanity, or some other aberrant behavior could have led to the thought that an indwelling evil spirit was responsible and had to be let out. That the basis for the operation was magical is indicated by the portion of bone removed from the trepanated skull, the *rondel*, being used as an amulet. With the dawn of the science of medicine, Hippocrates describes trepanation as a surgical procedure in the treatment of head injuries. It is of interest that trepanation, outside of its use in surgical interventions, was still carried out in various remote places in Europe and the Americas on into the nineteenth century and exists today as a bizarre cult practice performed in supposedly normal people.[31]

---

[26] (Frazer, 1922), pp. 576–578. The sympathetic magic practice of eating various parts of slain valiant enemies is to absorb their virtues: strength, bravery or to thereby acquire skills, intelligence, and wisdom.

[27] Ibid., p. 16.   [28] (Gajdusek, 1977).

[29] (Guiard, 1930). See also (Lisowski, 1967) and (Margetts, 1967), pp. 673–701.

[30] (Hrdlička, 1939).

[31] A site on the Internet advances the use of trepanation for rejuvenation, supposedly by reversing the falloff of brain blood volume and metabolism associated with aging and offers instructions for self-trepanation!

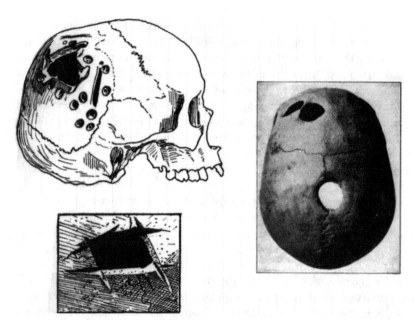

1.1. Examples of trepanation. Different types of openings made into the skulls with signs of healing seen at their edges. The skull on the right reveals five round holes. An example of a square hole is shown at the bottom left. From (Guiard, 1930), Plates 1 and 8, Figures 1 and 2.

The rise of Greek philosophy and science

It was the genius of the ancient Greek thinkers of the sixth and fifth centuries B.C. that replaced animistic belief in spiritual powers, magic, and the supernatural anthropomorphic religion expressed by Homer and Hesiod (c. ninth and eighth centuries B.C.) with philosophy and science.[32] The gods were no longer thinly veiled anthropomorphic powers controlling human events. They were replaced by physical forces and entities operating by inexorable laws. The impetus was the search for the *arche*, the principle underlying things and the basic "stuff" of which all things are made. In the Platonic dialogue *Timaeus*, at the onset all was chaos out of which the Demiurge (a lesser god) made the world. This was presented as a fable, what we may call a hypothesis, one that could undergo modification. In contrast, in the religious account given in Genesis, God created the world from nothing,

[32] It should not be thought that the philosophers struggled only against the mythic gods of Homer and Hesiod. The more archaic, darker, animistic beliefs in the supernatural, mysticism, magic, fetishism, and irrational practices and rituals were endemic in the population (Harrison, 1962), (Dodds, 1951), and (Macchiori, 1930). These irrational beliefs remain with us today, not only in primitive societies, but also in somewhat altered forms in the most highly civilized nations of the world.

ex nihil (by divine fiat),[33] and continues to intervene in the affairs of men – a providential God. In the Greek philosophical accounts, the whole of *physis* (nature) of the *Cosmos* that encompasses all – the earth, sea, and sky with its heavenly bodies of the sun, moon, and stars and the earth and its living creatures – is ruled by impersonal fixed law.[34] Some of the earliest of these thinkers, the *physiologists*,[35] appeared in the important commercial Greek city of Miletus at the western shore of Anatolia, now Turkey, facing the Aegean Sea.[36]

Among them was *Thales of Miletus* (c. 640–546 B.C.), who taught that water was the fundamental substance, the concept likely inspired by its necessity for life. For *Anaximenes* (fl. 546 B.C.), the fundamental substance was *aer* (air), which includes pneuma (breath or spirit), the term later extended to include the *psyche* (soul) present in animate beings.[37] In us the soul enables life: "Just as our soul being air, holds us together, so do breath and air encompass the whole world."[38] The term then further extended to include the concept of the mind. Philosophers appeared in other city states of the Greek world. *Pythagoras of Samos* (c. 580–489 B.C.) founded a mystical philosophical brotherhood at Croton, a Greek colony located at the foot of Italy. He taught that the world could be understood on rational principles, on mathematical laws, and considered the world to be made of four elements: earth, water, air, and fire. His pupil *Empedocles* (490–430 B.C.) advanced the concept that, under the opposing forces of "love" and "strife," the four fundamental elements of earth, air, fire, and water become transformed into one thing or another.[39] *Anaxagoras* (500–428 B.C.) viewed the cosmos as being formed by *nous* (mind or reason) from an infinite number of different elements. These come together on the basis of "like coming together with like," just as "fluid" merging with other "fluid" forms the oceans, the various parts of the body are formed the same way with bone formed from

---

[33] (Burnet, 1948), (Zeller, 1931).

[34] The world order was called the *Cosmos* in the fifth century B.C., but likely could have been referred to as such earlier, in the sixth century B.C. (Hussey, 1972), p. 18.

[35] The term "physiologist" was at first used for those who study *physis*, nature. The term was then extended to include additional meanings, making it necessary to assess the use of the term used at a given time (Peters, 1967). The term *physiologist* is now used for those who study animate nature, with the term *physicist* reserved for those who study inanimate nature.

[36] Miletus at the time was the greatest metropolis in the East. It was on the western coast of modern Turkey and was rich from its commerce and trade both inland with the Persian Empire and via the sea with the port cities of Greece and other peoples around the Mediterranean Sea.

[37] (Peters, 1967).

[38] Kirck & Raven, 1971). See Chapter IV, and (Hankinson, 1998), and (Russell, 1945), p. 28.

[39] (Freeman, 1957).

elementary bone, flesh from elementary flesh, and so on. As Greek thought developed, the special role played by air, the term originally given as spirit, was extended to soul as the animating principle responsible for life and then extended to encompass the mind by nous, the cognitive (*noetic*) function, whereby things, their number, harmony, and their relations to one another could be understood.[40] Nous was not only deemed a property of the mind of man, but it was also revealed by the cosmos with the regularity of the movement of the stars, sun, and moon in the heavens expressing by the regularity of their progression in the skies harmony and order, the presence of intelligence. This was the *anima mundi*, the immortal animated world, the cosmos that included the air, the sky, and the heavens; the *macrocosm* that is related to man, the *microcosm*. The divine air of the cosmos, from which all things take their origin and are held together by its invisible organizing principle, is taken in by respiration to animate the body, hold it together, and regulate changes within it.[41]

The concept that inspired air carried in the carotids confers consciousness and thought in the brain was taught by *Diogenes of Apollonia* (412?–323 B.C.). He took air to be the source of all things, identifying the warm air within the body (apparently carried by the arteries) to mix within the brain where the soul is located. Perception arises there with the psyche acting as the mediator between sensation and cognition.[42] This relationship of the intelligence of the cosmos brought in by air to man, the microcosm, was clearly expressed when he wrote:

> And it seems to me that that which has Intelligence is that which is called Air by mankind; and further, that by this, all creatures are guided, and that it rules everything; for this in itself seems to me to be God and to reach everywhere and to arrange everything and to be in everything. And there is nothing which has no share of it; but the share of each thing is not the same as that of any other, but on the contrary there are many forms both of the Air itself and of other Intelligence; for it is manifold in form: hotter and colder and dryer and wetter and more stationary or having a swifter motion; and there are many other differences inherent in it and infinite (*forms*) of savor and color. Also in all animals the Soul is the same thing, (namely) Air, warmer than that outside in which we are, but much colder than that near the sun.[43]

CONCEPTION OF NERVES AS CHANNELS FOR SENSATION AND MOTION

With *Alcmaeon of Croton* (fl. fifth century B.C.), a pupil of Pythagorus, the differentiation of science from philosophy was widened. He is credited with carrying out animal dissections, which led to his proposing that special

[40] (Peters, 1967).   [41] (Rusche, 1933).   [42] (Solmsen, 1961).
[43] (Freeman, 1957), p. 88.

channels in the body exist for sensory reception.[44] His idea of sensory channels appears to have risen from his observation of the empty blood vessel
seen in sections of the optic nerves. By extension, he thought channels
invisible to the naked eye were also present in other nerves to convey auditory, olfactory, gustatory, and somatic sensory impressions from the ears,
nose, tongue, and skin to the brain to there give rise to the perceptions of
objects. An argument he gave to support the specificity of the various sensory channels was that damage of the optic nerve causes blindness without
involving the other senses, similarly damage to the ear results in deafness
alone, and so on. Furthermore, such specific sensory deficits can occur without a perturbation of general sensibility and intellect, or of motor powers.[45]

Fundamental to Alcmaeon's concept of the brain was his study of the
development of bird eggs in which he found the brain to be the first rec-
ognizable body structure to develop. He regarded the brain as the terminus
for the various sensory inputs, with each sense possessing a territory of its
own and the brain acting as the chief organ of sensation. The senses meet
there in a common site wherein the ruling or directing faculty of the soul,
the *hegemonikon*, the seat of the intellect, was located.[46] Alcmaeon differentiated consciousness and intelligence from sensation, with man having
intelligence, which is more penetrating, greater, and vaster than that of the
animals who only have sensations. It was apparently to Alcmaeon's concepts that *Socrates* (470–399 B.C.) alluded to in the Platonic dialogue *Phaedo*
when he said:

> When I was young I had been inflamed by a desire to know that which is called the
> history of nature because I have found great and divine science which has taught
> the causes of each thing, that which has made it to be born and that which has
> made it to die... to know if animals have come to be born as some of us pretend
> when heat and cold have conceived some type of corruption. If it is the blood
> which makes thought, if it is the air, or if it is fire, or if it is not any of these things,
> but only of the brain which is the motor of our senses, of our vision, of hearing, or
> smelling. If the senses result in memory and imagination and if the memory and
> imagination after a time of rest is born into science.[47]

PLATONIC TEACHING ON THE BRAIN
Plato (427–347 B.C.), in his dialogue *Phaedo*, said that the function of the
brain was to "furnish the sensations of hearing, sight, and smelling, from

---

[44] (Stratton, 1917), and (Codellas, 1932).   [45] (Souques, 1936).
[46] (Peters, 1967), p. 78. *Hegemonikon* embodies the same concept later designated as
the *sensus commune*, the site where the senses combine to give rise to an object
apprehended by several senses and perceived as a single entity. Although Alcmaeon
placed this in the brain, it was believed by the Stoics and Aristotle to reside in the
heart. See section below.
[47] (Plato, 1937), pp. 480–481.

which memory and judgment are born, and from whose sensations, once established, wisdom is also born." In the dialogue *Timaeus*, Plato said that sight was given by the god to see the stars, sun, and sky. The sight of them, of "day and night, of months and revolving years, of equinox and solstice, has caused the invention of number and bestowed on us the notion of time and the study of the nature of the world; whence we have derived all philosophy...the god invented and gave us vision in order that we might observe the circuits of intelligence in the heavens and profit by them for the revolutions in our own thought which are akin to them."[48]

Plato gives his view of the *tripartate* soul with the immortal *intellectual* part located in the brain, wherein experience is reflected in the mind and wherein the pure idea of things is contained; the *emotional*, spirited part, is located in the heart and a *sensual* or *desirous* (concupiscent) part in the liver. In his mythic model, the body was constructed under the direction of a lesser god, a demiurge:

> The bones and flesh, and other similar parts of us, were made as follows. The first principle of all of them was the generation of the *marrow*. For the bonds of life which unite the soul with the body are made fast there, and they are the root and foundation of the human race. The marrow is created out of other materials: God took such of the triangles [the primitive basis of materials] which were adapted by their perfection to produce fire and water, and air and earth [the elements] and mingling them in due proportion made the marrow out of them to be a universal seed of the whole race of mankind; and in this seed he then planted and enclosed the souls, and in the original distribution gave to the marrow as many and various forms as the different kinds of souls were hereafter to receive. That which, like a field, was to receive the divine seed, he made round every way, and called that portion of the marrow, *brain*, intending that, when an animal was perfected, the vessel containing this should be the head; that which was intended to contain the remaining and mortal part of the soul he distributed into figures at once round and elongated [the spinal cord], and he called them all by the name "marrow"; and to these, as to anchors, fastening the bonds of the whole soul, he proceeded to fashion around them the entire framework of our body, constructing for the marrow first of all, a complete covering of bone.[49]

Plato goes on to describe the formation of the bony globe of the head with its narrow opening at its base, the foramen magnum, by which the marrow in the neck and the spinal cord along the back enters it. The vertebrae

---

[48] (Cornford, 1937), pp. 157–158. According to C. S. Lewis, the dialogue *Timaeus* was the only book of Plato known in earlier Medieval times. See (Luscombe, 1997) p. 23. It has been called the first physiology text. The importance with which it was held is indicated by Raphael's painting, "The School of Athens," hanging in the Vatican where Plato accompanied by Aristotle is shown holding his book with its Greek title *Timeo*.

[49] (Plato, 1920), Volume 2, pp. 51–52. The demiurge (gk. demiurgos) is the craftsman god who carries out the plans of the higher god.

were placed by the demiurge "under one another like pivots, beginning at the head and extending through the whole of the trunk." Other bones containing marrow, the "bone marrow," were sharply differentiated from the marrow of the brain and spinal cord. The sacral portion of the spinal marrow was of particular interest, in that it constitutes the "universal seed stuff" (semen) of mortals that finds its way to the genitalia for procreation. The route would be through channels leading from the lower part of the spinal cord into the penis, where it is discharged into the female, as was pictured by Leonardo da Vinci in his "coitus figure."[50] This connection of the spinal cord with semen could have come about from the physical similarity of semen and the material of the spinal cord. Keele records that this concept of the spinal cord substance forming the semen was tested by Alcmaeon, who supposedly measured the dimensions of the spinal cord in animals before and after coitus.[51] The relation of the semen to bodily health was later adumbrated by Tissot in the nineteenth century.[52]

Plato taught that the highest intellectual soul is immortal and does not perish. It either returns to its original home in the stars or undergoes a transmigration into other persons or animals, the concept deriving from Pythagoras. The still older primitive concept was that some animating spirit was responsible for consciousness, a soul that was thought of as an *eidolon*, an unsubstantiated wraith, that would be dissipated on death when it became separated from the body. The proof Plato gave in the dialogue *Meno* was a slave who, when questioned, was able to give the solution to a difficult problem in geometry. This, Plato said, was not taught to him but was "recollected" as it were in a dream. Such knowledge must have been acquired before birth, the theory of *anamnesis*. "If then, the truth of things is always within our soul, the soul must then be immortal."

Another central tenet of the Platonic philosophy was that of "forms," of unseen intelligible realities. They have intercourse with the body through the senses and the soul through reflection. The distinction between the unseen intelligible forms and the visible objects of the senses is that the latter changes (those of the body, the "becoming"), whereas the forms are

[50] (Keele, 1957), p. 27.
[51] Ibid., p. 28. How this could have been carried out with the primitive techniques then available stretches the imagination. Although it could have been attempted, it most likely was a thought experiment.
[52] This was the theoretical basis for the proscription against excessive venery and onanism as the basis of disease, a widespread belief in the nineteenth century. Tissot drew on a number of authorities, starting with Hippocrates and listing among them Boerhaave who viewed seminal fluid as necessary for life and connected its loss to signs of nervous disorders. He wrote: "We do not know if animal spirit and the 'genital liquid' are the same thing but ... the two things have a great analogy and the loss of one or the other produces the same ills" (Tissot, 1840), pp. 486–487.

real and unchanging (the soul or "being"). The world of the real was said in the later dialogue *Sophist* to consist not wholly of unchangeable forms, but also to contain life, soul, intelligence, and such changes as they imply.[53]

The teachings of Plato had a widespread influence on philosophy, and along with its later elaboration known as *Neoplatonism*, teachings that have lasted to this day through its effect on the development of Christianity.

HIPPOCRATIC TEACHING ON THE BRAIN

The concept of diseases as a natural occurrence, rather than an intervention of a malign supernatural agent, is attributed to *Hippocrates of Cos* (460–375 B.C.), the Father of Medicine.[54] This represented a fundamental advance over the magico-theological beliefs that were embedded in the ancient civilizations of Egypt and Mesopotamia. In their medical practice, incantations and exhortations were used to induce the baleful indwelling spirits to leave, along with disgusting medicines containing fecal material used as further inducement to encourage their expulsion. In the book *On the Sacred Disease*, Hippocrates declared epilepsy to have a natural cause and not to be considered a judgment of the gods or the intrusion of malevolent spirits.[55] He gave a physiological explanation of the disease: Pneuma (spirit) drawn into the lungs on inspiration is distributed in the blood throughout the body by the heart and blood vessels to keep the body and its organs living. If passage of blood into a limb is prevented, as by a ligature, the abstraction of the vital principle from the blood cannot take place and as a consequence the limb withers and dies. The brain similarly abstracts the vital principle from the blood supplied to it. Interruption of blood flow to the brain, as from a blow to the head, results in a failure to extract the vital principle from the blood. As a result, sensibility is altered and loss of consciousness ensues. An excess of pneuma in the brain, from an obstruction preventing its normal outflow, accounts for the convulsions seen in epilepsy. The frothy phlegm seen on the lips of the epileptic during a convulsion was taken as evidence of the excess of pneuma forced by the increased pressure to flow into one motor channel or another, leading from the brain to various parts of the body to cause the convulsions.

The pneuma drawn into the body on inspiration passes from the lungs to the heart, where it acts to nourish its internal fire and cool the blood. From the heart, pneuma is carried by the carotid vessels to the brain, the seat and agent of the soul. The name given to the carotids, the "sleep-producing

---

[53] See the dialogue *Sophist* and Cornford's comments on it (Cornford, 1935), pp. 241–248.

[54] It is not certain which of the Hippocratic writings are due to him directly or to others in his school. For example, Polybus, the son-in-law of Hippocrates, was likely the writer of "On the Nature of Man" in the Hippocratic Corpus, (Hippocrates, 1849), p. 34.

[55] (Hippocrates, 1923) and (Keele, 1957).

vessels," could have originated from the experience of wrestlers and soldiers who found that application of pressure to the neck could bring about a rapid syncope.[56] Hyrtl traced the thought entailed by the term. He noted that Aristotle wrote that the result of compressing the veins (carotids) was to cause a collapse, loss of sensibility, and closure of the eyes, suggesting an induction of sleep. As Aristotle viewed the heart as the site of consciousness, the effect of compression was to prevent the flow of pneuma to the heart rather than to the brain. *Rufus of Ephesus* (fl. 150) later also wrote that the compression of the carotids causes deep sleep and aphonia. Throughout the Middle Ages, it was believed that compression of the carotids caused a pathological sleep with stupor, lethargy, and apoplexy. This was put to an experimental test in 1551 by Matthaus Curtius, who ligated the carotids in living animals. Apoplexy did not appear, and he concluded that the older observations of their relation to sleep and stupor were in error.[57]

A section in the *Hippocratic Corpus* deals with the use of trepanation for the treatment of injuries to the head. But whereas primitive man practiced trepanation motivated by his belief in supernatural or magical principles, in the Hippocratic writings it appears as a surgical treatment to relieve the effects of head injury and prevent untoward consequences of damage to the underlying brain. It is of interest that the author shows some awareness of the crossed tracts leading from the brain when he relates that an injury to one side of the head causes paralysis on the opposite side of the body.[58] The spinal cord was considered to be an extension of the brain with injury to it causing paralysis, sensory disorders, and difficulty with defecation and urination.[59]

THE ATOMISM OF THE EPICUREANS

A major branch of classical Greek philosophical thought was *Atomism*, the view that the universe is composed of an infinity of small indivisible and

---

[56] (Hyrtl, 1970), pp. 93–94. This we know could elicit the carotid sinus reflex causing blood pressure to fall and fainting.

[57] Ibid., p. 94. However, my own experiences are in accord with the earlier descriptions. When the carotids are firmly compressed in the neck of a standing human volunteer, collapse with a loss of consciousness occurs in a matter of seconds. In some cases, this was accompanied by a brief but alarming convulsion. This procedure is not to be undertaken lightly. The lack of symptoms found by Curtius was likely due to a sufficient supply of blood supplying the brain of the animals through the vertebral arteries and also their recumbent position in his experiments.

[58] The crossing of the tracts was later rediscovered by Francois Pourfour du Petit in 1710. See (Kruger, 1963), and (McHenry, 1969), p. 68.

[59] The striking effect of injury to the cord seen by the ancients is indicated by the bas-relief from the palace of Assurbanipal which shows a lioness struck by an arrow in the spinal cord dragging her paralyzed hindlimbs (McHenry, 1969), Figure 2.

indestructible particles, *atoms*,[60] which undergo a random and everlasting motion within an infinite space, the *void*. The philosophy was first advanced by *Leucippus* (c. 480 B.C.), of whose writings only fragments remain, the teaching better known through his disciple *Democritus of Abdera* (c. 460–370 B.C.).[61] The doctrine was further promoted by *Epicurus* (341–270 B.C.) and his school of philosophy known as *Epicureanism*. The Atomistic philosophy is more fully known to us from the Latin poem written by *Lucretius* (96?–55 B.C.).[62] In a letter to Herodotus, Epicurus gave the fundamental thesis of the philosophy – namely that atoms and the void are infinite and that "nothing comes into being out of what is nonexistent."[63] The atoms differ in their shape, size, and weight, combining in various ways to form the things of the world. Atoms and things have their effect on other atoms and things by mechanical means, by impact or pressure, and by their aggregation things, the source of motion being inherent in the atoms themselves. Things are known to us through the senses by their secondary properties of color, smell, and taste and not through the primary properties that atoms possess: shape, weight, and size. Atoms do not change when they combine in various ways to form the composite bodies that are appreciated by the senses.

Sensation is our only means of knowing things. Sight is from the outline of things, exceedingly thin films emitted from solid bodies, *simulacra* to which the name *images* are also given.[64] These have the same shape as the object, but of "exceeding thinness" and move with "enormous" velocity to enter our eyes and then via the optic nerves to the mind where they can be recognized. Sound is a blow that displaces atoms to enter the ear and give rise to the sensation of hearing. Smell results from fine bodies conveyed from the object that are of the proper sort to excite the organ of smell. Perception of these sensations is due to a change of the soul by influences that are similar to it, on the principle that like affects like. The soul that perceives the sensory body bears the same relation to its matter as the perfume of the rose to the flower, or light and heat to ignited fuel.

A central part of atomistic philosophy has to do with its view of life and the mind.[65] Soul (psyche) consists of very small, round atoms that give life

---

[60] This has been translated from the Greek term *atomoi* meaning uncutable. Though modern physics recognizes that in actuality atoms are made up of still other entities, the term atomism was generally taken to represent the concept of ultimately indivisible particles responsible for the unique properties of the elements.

[61] Note should be taken that our Democritus is to be distinguished from as many as four persons known to have had that name in ancient times (Gaskin, 1995).

[62] (Lucretius, 1951).   [63] (Gaskin, 1995), pp. 13–29.

[64] (Gaskin, 1995), p. 16 ff and (Diogenes Laertius, 1966), Volume 2, X 45–55. For an interesting overview of the atomists and the other philosophical schools with respect to sensation and perception, see (Hamlyn, 1961).

[65] (Guthrie, 1965), p. 430 ff.

to the lifeless body. It is closely associated with heat, this likely inferred from the difference between the cold corpse and a living warm body. The soul is believed to consist of fire-like atoms that are finer and smoother and hence more mobile than others. Drawn into the body by respiration, these finer atoms penetrate to warm it and bring it to life; with the soul atoms constituting the noblest and most divine part of man. Life is maintained by breathing, as the very tiny mobile soul atoms tend to be continually lost from the body. The air taken in by respiration containing the mind and soul particles restores that loss.[66] When respiration ceases, death occurs. As the soul of an individual is mortal and dispersed on death, the fear of torment after death for sins done while living is unreasonable and can be dispensed with. This was the message preached by Epicurus and Lucretius. By their rejection of the immortality of the soul, they and other atomists earned the enmity of the Stoic philosophers and the implacable hatred of Christian theologians for whom the immortality of the soul became the cornerstone of their faith (Chapter 3).

The soul atoms are dispersed in the body such that atoms of soul and body alternate, the soul imparting its motive power to the more ponderous atoms distributed throughout the body.[67] The mind, the thinking part of the soul, is an exception to this general distribution. Democritus placed it in the head (as did Alcmaeon, Anaxagoras, Diogenes, and Plato), whereas Epicurus and Aristotle located it in the heart. The relation of sensation to thought was based on mechanistic principles in the atomistic account. Sensations reach the mind through channels, and then the brain or heart, where the soul atoms are so closely packed that the sensations cannot pass without moving them aside, thus giving rise to the sensation we call thought.

STOICISM

The great competing philosophy to Epicureanism was *Stoicism*, the philosophy founded by *Zeno of Cyprus* (fl. 335 B.C.) and enlarged on by *Chrysippus of Cnidus* (fl. 340 B.C.), among others. Unlike Epicureanism, Stoicism underwent many modifications by its adherents. In general, the fundamental difference between it and Epicureanism was that, for the Stoics, there was no void with indivisible atoms in the cosmos. The world was a plenum, a medium in which matter moved, much as fish swimming in water or wine merges with water. The Stoics accepted the theory of four elements, with a special role given to air (aer) and fire with pneuma (spirit) present everywhere pervading all things. As was later said by Galen (Chapter 3) commenting on the Stoics, for them; "the pneuma-like substance causes cohesion making the matter-like substances [water and earth] cohesive; and so they say air and fire make cohesive and earth and water are made

---

[66] Ibid., p. 434. [67] Ibid., p. 432.

cohesive."[68] Entities are unified and held together by pneuma, which acts as a dynamic continuum pervading all things. In the living organism, it holds all its parts together. This expresses the view of the universe as the macrocosm, a living being pervaded by vital spirit (pneuma), life/soul (psyche), and reason (nous) with the living beings, the microcosm, similarly pervaded.[69] The necessity for breathing to remain alive can readily be understood from this interrelationship.

The philosophical difference between the atomists and the stoics is fundamental. For the Atomists, change is the key to the world; chance brings about all the different things into being by the interactions between the atoms. There is no place for a providential god. All is in flux, a view expressed a century earlier than Leucippus by *Heraclitus* (556–460 B.C.). This was expressed by Heraclitus in his famous statement that "one cannot bathe twice in the same river."[70] For the Stoic, the world is ordered and is itself a god that follows a plan that unfolds, one in which, however, providential acts occur by the intervention of the gods in human affairs – a view that the atomists and the Epicureans denied. The contrasting positions of the two philosophies with their differing conception of the gods were given by *Cicero* (106–43 B.C.) in the form of an engaging dialogue between a Stoic and an Epicurean, along with a criticism of both positions by a Platonist.[71]

ARISTOTLE: THE HEART AS THE SEAT OF THE SOUL

Aristotle (384–322 B.C.), the great philosopher and preeminent scientist of Greek antiquity, wrote a renowned book on the philosophy of the mind, *De Anima*.[72] Besides his own exposition of its nature, he also commented on the views of still older Greek philosophers, preserving what would otherwise be lost to us. His book was used in the teaching of psychology until the rise of modern science (Chapter 4). Some difficulty in interpreting Aristotle's thought lies in the terminology he used to express his concepts, terms that through time changed by taking on extended meanings. Of these, psyche is of the greatest importance for our consideration of the mind and the nervous system, the term is variously translated as life-giving breath, a ghost, vital principle, soul.[73] Aristotle, in dealing with the views of his precursors, thought they meant by the term the source of movement and perception. For Homer, psyche was a ghost that gave life to the body, its absence seen in the inertness of the dead corpse. After Homer, the term psyche absorbed the meaning of *thymos*, which describes the psychic totality of man, whereas *soma* was used to describe the body. Thymos includes not only mind, soul, feeling, and temper, but also eager desire, avidity, (carnal) appetite, and lust –

---

[68] (Lloyd, 1973), p. 28.   [69] Ibid., p. 29.   [70] (Dewitt, 1967), p. 265.
[71] (Cicero, 1972).   [72] (Ross, 1923) and (Aristotle, 1968).
[73] (Peters, 1967), pp. 166–176.

in general terms emotional states. Its location to the chest is shown by the agitation that can be felt there in emotional states. The term thymos may perhaps be connected with the thymus gland, which is not too distant from it.[74] The term used for the psyche, the "soul," does not only mean the animating principle giving life to the body, but also it includes the mind, which is considered to be the highest development of life.

For Plato, ideal "forms"exist that determine what "matter" will be.[75] The forms have a real existence and act on material to give it its identity. The soul can be considered a form that acts to make the body what it is. It was considered as having three parts, with the highest (the rational) responsible for the mind. Below it is the "spirited" part, the source of action. The lowest is the appetitive, responsible for appetites, concupiscence, – carnal desires. The highest part of the soul having to do with reason, consciousness, thought, and judgment imposes its control over the lower parts of the soul. It is immaterial, entering the body at birth and on death leaves it for another existence in another human being or an animal, the transmigration teaching of the Pythagoreans.

Aristotle, in his philosophy considered four sorts of causes to bring about a thing:

- the *material cause* – the material substance of which things are made
- the *efficient cause* – that effecting the change by which things are made
- the *formal cause* – that which gives the essence or the nature of a thing
- the *final cause* – the end to which the thing is formed

Matter has the potentiality of becoming an animated being, and form is that which brings it into actuality; the material is the body, its animation brought about by the soul. With this view, we could assume that the soul is mortal – that as the body dies, so does the soul. As one writer stated: "Aristotle is clear on the subject of personal immortality. Since the soul is the formal and final cause of body, it cannot survive the dissolution of the union with that body, except perhaps, as part of the species."[76] However, it is not certain that this was Aristotle's final view. Some passages of his manifold writings suggest an immortal part of the soul. What he thought on this subject (as opposed to Plato's views) was a major scholastic debate throughout the Middle Ages (Chapter 3),[77] and it remains a matter of philosophical controversy to this day.[78]

Although accepting Plato's tripartate view of the soul and distinguishing pneuma (spiritus or air) from psyche (soul), Aristotle placed nous (mind),

[74] (Hyrtl, 1970), p. 546.
[75] (Durling, 1993), pp. 237–239, and (Lewes, 1864), pp. 221–229.
[76] (Peters, 1967), p. 172, in ref. to De Anima II, 415b.   [77] (Pegis, 1934).
[78] (Cherniss, 1962).

the highest faculty of reason, in the heart rather than the brain. The soul on his account is the prime agent for motion, and pneuma is its instrument (organ). Nous, the energizing and ordering (reasoning) principle and guide of development, consists of a finer matter, somewhat like an "ether" permeating all things, most vividly so in animated beings. Psyche, the principle of motion and perception, is taken in with the breath, to exist in the body as an *innate pneuma*. It is a hot, foamy, humid substance that in the heart provides the sensitive and moving link to the physical organs of the body.

As evidence that the soul is located in the heart, Aristotle found it hot to the touch in contrast to the brain. This was significant, for heat was considered a prime characteristic of life in higher animal forms.[79] The function of the brain was relegated to tempering the heat of the heart brought to it by the blood carried in the carotids. His identification of the heart as the source of motion and heat and the chief organ of the soul was likely because of his embryological studies. In studying the development of bird eggs, he saw the heart to be the first organ formed (contra Alcmaeon). It first appears as a leaping red speck, the *saliens punctum*, which then goes on to develop into the heart of the mature animal. Aristotle considered the rapid beating and quivering motion of the mammalian heart, and the movement of its valves to be brought about by their tendon-like chorda tendinea, which he confused with nerves, tendons, and ligaments. The vessels were thought to end as solid sinews, as tendons connected with muscular action:

> The sinews of animals [have their point of origin in] the heart; for the heart has sinews within itself in the largest of its three chambers [instead of the four we recognize], and the aorta is a sinew-like vein; in fact, at its extremity it is actually a sinew, for it is there no longer hollow, and is stretched like the sinews where they terminate at the jointing of the bones.... In the ham [back of thigh], or the part of the frame brought into full play in the effort of leaping, is an important system of sinews; and another sinew, a double one, is that called 'the tendon' [sciatic nerve?], and others are those brought into play when a great effort of physical strength is required; that is to say, the [contraction seen in] episthotonos or [that of the] back-stay, and the shoulder-sinews.[80]

CHANNELS FOR SENSATION AND PERCEPTION

Theophrastus (370–287 B.C.), the successor of Aristotle as head of the Lyceum, the school Aristotle founded, wrote the most extensive treatise on the senses that has come down to us from antiquity. It contains, along with his own views, those of his predecessors that would otherwise have been lost.[81]

---

[79] (Mendelsohn, 1964).   [80] (Aristotle, 1910), Book III, 515[a].
[81] (Stratton, 1917).

He describes the organs responsible for hearing, smell, taste, touch, and pleasure and pain. Taking into account what had been written on vision, Theophrastus thought that it was achieved by means of the eye's internal fire passing out through the pupil. This is an *extromission* theory of vision, the reverse of what we think of as the entry of visual information into the eye, the *intromission* theory.[82] Evidence given for the extromission theory is the impression of light experienced when the side of the eyeball is pressed on by a finger.[83] In the extromission theory, a stream (of internal fire) passes out of the eye; the light of day surrounds it and by coalescing with the rays from the external object in the visual field the perception of objects is realized. The Stoics considered that optical pneuma (a mixture of air and fire) flows from the seat of consciousness (the hegemonikon, the ruling part of the brain) down to the eyes and moves out from it to excite the air just adjacent to the eyes putting it under tension. When the air is illuminated, contact is made with the visible object. In the Stoic view, the simultaneous action of the optical pneuma and sunlight on air transforms it into an instrument of the soul, making it an extension of the body for perception.[84]

For the other senses, Theophrastus refers to Empedocles who:

> has a common method of treating of all the senses: he says that perception occurs because something fits into the passages of the particular (sense organ). For this reason the senses cannot discern one another's objects, he holds, because the passages of some (of the sense-organs) are too wide for the object, and those of others are too narrow. And consequently some (of these objects) hold their course through without contact, while others are quite unable to enter.[85]

Differing atom structures account for the different taste sensations: An astringent taste is sensed because the atoms of substances have many angles, whereas bitter atoms, though considered to be smooth and round, have their surfaces furnished with hooks and as a consequence are sticky and viscous. The various sensations are transmitted through channels to the *sensorium* to different parts of the brain, where they are perceived on the basis of the principle that "like knows like." The nature of the sensory channels, however, was not spelled out by Theophrastus. He could possibly have thought of them as blood vessels.[86] It is when we come to the Alexandrians in the third century B.C. that the nature of nerves is clearly differentiated from blood vessels and tendons.

---

[82] (Lindberg, 1976).
[83] This can be observed by exerting a relatively light pressure to the side of the eyeball through the closed eyelids. Additional evidence is the appearance of a light shining out from the tapetum of animal eyes, as seen in cat eyes at dusk.
[84] (Lindberg, 1976), p. 9.   [85] (Stratton, 1917).   [86] (Solmsen, 1961).

THE ALEXANDRIANS: THE DIFFERENTIATION OF NERVES
FROM TENDONS

Alexandria, the Egyptian port city on the Mediterranean founded by Alexander in 332 B.C., contained a celebrated institution of learning that attracted scholars from all over the Greek and Roman world. Its museum contained a large library holding the greatest collection of scientific and literary treasures of the ancient world, and it contained facilities for carrying out anatomical studies on humans (elsewhere banned) and for doing animal research.[87] The great contributions of the Alexandrians *Herophilus of Chalcedon* (c. 335–280 B.C.) and *Erasistratus of Chios* (310–250 B.C.) on the anatomy and physiology of the nervous system[88] are known to us mainly through the writings of Galen (Chapter 2). They clearly showed the origin of nerves from the brain and spinal cord, differentiated nerves from tendons, and even held that different nerves are responsible for sensation and motor control.[89]

Of the two Alexandrians, Erasistratus was considered to be more concerned with function and thus has been judged to be the first physiologist.[90] He distinguished three main channels in the body: arteries, veins, and nerves. The veins carry blood manufactured in the liver. They contain the *natural spirit* that nourishes the body. The arteries, on the other hand, he held to be free of blood. They carry the *vital spirit*, which is present in the atmospheric air that – following its inspiration into the lungs – is carried by the pulmonary vein to the left ventricle of the heart. From there, it is distributed by the arteries throughout the body to support the viability of the body organs.[91] Carried to the brain by the carotid arteries, the vital spirit is elaborated there into *pneuma psychikon*, the *animal spirits*, that carries out the higher functions of the brain. In the cerebrum, the animal spirits were said to be compounded in the large anterior (1st and 2nd) ventricles, which then moves back down into the next lower (3rd) ventricle, and then into the hindmost (4th) ventricle, which "some say is the most important of all the ventricles throughout the whole of the brain."[92] From there, animal spirits move down into the spinal cord and then out into the various nerves distributed throughout the body subserving sensation and motor movements. The flow of animal spirits in the nerves was considered to take place via hollow channels within them.

---

[87] (Parsons, 1952). Works from the Alexandrian library later aided the West in its recovery from the Dark Ages.

[88] See (Dobson, 1925) and (von Staden, 1989) for Herophilus, and (Dobson, 1927) for Erasistratus.

[89] (Solmsen, 1961).   [90] (Dobson, 1927).   [91] (Wilson, 1959).

[92] (Longrigg, 1993), pp. 86, 87. Damage to this region can rapidly cause death. Modern studies have shown this to occur through interference with the medulla and the autonomic centers in it rather than to the cerebellum per se (Neuburger, 1981), (Ochs, 1983).

The central tenet in Erasistratus' schematization, that pneuma and not blood is contained in the arteries, appears to have been founded on the anatomical technique he used in his animal studies. Following Aristotle's practice to better show the blood vessels, animals were first starved to reduce obscuring fat and then killed by strangulation.[93] This has the effect of causing the arterioles in the lungs to constrict, whereas the heart – continuing to beat on for a while before failing – causes a runoff of blood from the arteries into the veins. This leaves the veins of the body engorged with blood and the arteries empty, the basis for Erasistratus' concept that the arteries serve to carry pneuma. The very term *arteria* was derived from the Greek term for air. The trachea, which was obviously seen as a channel for air, was called the *rough artery*.[94] Erasistratus' identification of pneuma with vitality categorizes him as a *pneumatologist*,[95] whereas his fellow Alexandrian, Herophilus, was a *humoralist* who taught that human temperament was the result of a balance of body humors. As will be seen in the following chapter, both views were incorporated into Galen's physiology, wherein both spirits and humors are carried by the blood.

[93] (Wilson, 1959).
[94] This is the term still in use for the trachea in modern Greek medicine.
[95] (Allbutt, 1921).

2

GALEN'S PHYSIOLOGY OF THE NERVOUS SYSTEM

After the contributions of the Alexandrians Herophilus and Erasistratus in the fourth and third centuries B.C. (Chapter 1), the next great advance in anatomy and experimental physiology of the nervous system was made in the second century A.D. by *Galen of Pergamon* (c. 129–216).[1] His extensive writings on medical practice, science, logic, philosophy, and his studies of anatomy and animal experimentation, a large portion of that work on the nervous system, made him a towering figure in the history of medicine. He was likened in importance to Hippocrates, with his influence extending into the nineteenth century.[2] In addition to a fairly large part of his writings on the anatomy and physiology of the nervous system, his discussion of significant portions of the writing of his predecessors, of Erasistratus and others on the nervous system, has preserved what would otherwise be lost to us.[3]

GALENIC PHYSICS IN RELATION TO HIS PHYSIOLOGICAL THEORY
In his physical science, Galen accepted the Pythagorian-Empedoclean-Platonic accounts that matter is composed of four elements; fire, water, air, and earth with their qualities of hot, moist, cold, and dry, respectively (Chapter 1). The elements do not exist as such in the body, but are

---

[1] Dates in our era are given without the ascription A.D. Earlier estimates of Galen's life span were given as 130-200. More recent scholarly studies have opted for the later dates (Ballester-Garcia,2000) I p. 6.

[2] The dominating influence of Galen on medical science and practice over the centuries can be found in Temkin's authoratative summarization (Temkin, 1973).

[3] Ambiguities and contradictions among his voluminous writings have been noted. Some are explainable as reactions to an idea or to an author he was opposing at a given time [i.e., a debating tract (Wilson, 1959)]. Knowledge of the temporal ordering of his writings would also, if known, give an indication of changes in his views over the course of his long career that could account for some of the contradictions that have been noted.

characterized as blood, yellow bile, black bile and phlegm. His humoral theory was based on the theory of Hippocrates and Aristotle,[4] in which blood is believed to contain the "hot and moist," yellow bile the "hot and dry," black bile the "dry and cold," and phlegm the "moist and cold." The humor "blood" that predominates in the veins is taken by extension to refer to all the contents in the vessel.[5] The relative amount of the four humors determines the state of health: In the healthy state, they are in balance. An excess of one humor or other can change the temperament of an individual, making those with an excess of blood *sanguine*, with yellow bile *choleric*, with black bile *melancholic*, and with phlegm *phlegmatic*. A greater unbalance of humors causes disease.

Of the elements entering into the humors, that of earth, fluid, and water are supplied by drink and the consumption of food, whereas air is taken in by respiration. As the ancient philosophers believed, matter becomes alive through its organization by the fine spirit that pervades the cosmos. Taken in by respiration, it becomes the organizing principle or soul of living creatures (Chapter 1). With Plato, Galen[6] recognized a tripartite division of the spirit (soul) in man: the *Spiritus naturalis* (*pneuma naturalis*) residing in the liver that is responsible for blood formation and general metabolism; the *Spiritus vitalis* (*pneuma zooticon*) residing in the heart that is responsible for emotions, the pulse, and the movement of blood distributing heat and giving life to the body; and the *Spiritus animalis* (*pneuma psychicon*) that is responsible for sensory appreciation and animation, its highest part that of reason located in the brain.[7]

As for whether the soul is immortal or not, Galen was guarded. In his book *On the Soul's Dependence on the Body*,[8] he is unconvinced of the soul's immortality.[9] If the reasoning faculty is a form of the soul, namely a mixture within the brain, it must be mortal. If it is immortal why should it depart when the brain undergoes excessive cooling, heating, drying, or moistening (physical elements of which it is composed)?... Why does a great loss of blood, drinking of hemlock, or a raging fever cause such separation?[10] And why does "a build up of yellow bile in the brain lead to a derangement [delerium] or a buildup of black black bile to melancholy... phlegm lethargic complaints... impairment of memory and understanding... [and] wine relieve us of sadness and low spirits." And, in addition to these difficulties

---

[4] (Temkin, 1973), p. 18.   [5] Ibid., p. 17.

[6] Chapter 1, (De Lacy, 1972) and (De Lacy, 1978).

[7] The role of natural spirit in Galen's schematization appears to be minor (Temkin, 1951) and (Temkin, 1973). It does enter importantly into his treatise on the formation of the embryo (Galen, 1997), pp. 183. The embryo draws its sustenance from the mother through blood vessels in the placenta and Galen makes an analogy of the embryo's vessels to plants that send their roots into the soil for their growth.

[8] (Galen, 1997), pp 150–176.   [9] Ibid., p. 152.   [10] Ibid., p. 153 ff.

he cannot see how the soul, if it is of a non-bodily substance, can exist in isolation (on the supposition that it can remain after the body's death), and can extend throughout the whole body to give it its form. Temkin pointed out[11] that, after prolonged study, the historian Charles Daremberg decided that Galen had taken a materialistic position and so also did the historian Garcia Ballester.[12] Galen's most consistent position seems to be that the soul functions represent the temperament of the body (i.e., the composition of its component elements). *Nemesius, Bishop of Emesa* (fl. 380–390), who later made use of much of Galen's writings in his book *On the Nature of Man*,[13] could not help but be suspicious of the materialistic position implied by Galen's view that the soul is the temperament of the body. The consequence of that view would be that the soul is mortal.[14] Conflicting opinion over what Galen believed regarding the soul's immortality was expressed throughout the Middle Ages (Chapter 3).

THE PRESENCE OF BLOOD IN THE ARTERIES

According to Erasistratus (Chapter 1), inspired air when taken into the body passes to the left ventricle of the heart where, as *vital spirit*, it is distributed within the arteries to the body and brain. The vital spirit carried to the brain is there transformed into *animal spirit*, which passes from the ventricles of the brain down into the spinal cord, and then into the sensory and motor nerve channels of the body. The basic tenet of Erasistratus that vital spirit and not blood is carried in the arteries seems to be contradicted by the common observation that the blood flows from an artery when cut. Erasistratus accounted for this by postulating that the arteries and veins are connected at their distal terminations. When an artery is cut, a vacuum is created and blood rushes in from the veins to replace the vital spirit. This position was challenged experimentally by Galen. In a living animal, he ligated the aorta at two places, removed the tied-off segment, and submerged it in a container of water. When he punctured it, instead of gas bubbles emerging, as would be expected on Erasistratus' theory, blood immediately flowed from it. In another experiment, he pricked the artery of an animal and found that the "the blood spurts out at once... [while] the pneuma should be evacuated not immediately, but over a considerable time. For how could blood escape [so quickly] from the wound, since Erasistratus himself says it is the arteries lying furthest away [at their terminations] that first by the transfusion [of blood from the veins takes place into the arteries]?"[15] And, the pneuma would have to exit in a single mass, and all of it move out before blood leaves, according to Erasistratus. The fact that the animal lives on after blood appears when all the vital spirit would have been vented makes

---

[11] (Temkin, 1973), p. 44.   [12] Ibid., pp. 44–45.   [13] (Nemesius, 1955).
[14] Ibid., p. 82.   [15] (Galen, 1984), pp. 149–153.

Erasistratus's theory untenable. Galen therefore concluded that the arteries contain blood with vital spirit in it and not only spirit as Erasistratus believed.

TRANSFORMATION OF SPIRITS

The veins containing natural spirits in its blood nourishes the tissues, whereas the blood carried in the arteries contains vital spirit essential for keeping the tissues in their living state. Ligating or cutting the vessels to a limb depriving it of either spirit causes the tissue to die. Galen, as did Erasistratus, considered that the two vascular systems are connected at their distal extremities so that under certain circumstances, as when an artery is cut, blood can pass between the two sets of vessels. Here, Galenic physiology stopped tantalizingly short of the concept of a circulation that was later established by Harvey (Chapter 4).

In Galen's physiology (Figure 2.1), food taken into the stomach undergoes a "*coction*" (digestion). The chyle produced is carried to the liver and from it venous blood containing *natural spirit* is produced. Carried by the vein descending from the liver, this blood nourishes the tissues of the body. The ascending vein carries venous blood into the right ventricle of the heart. There, blood passes through pores in the interventricular septum of the heart to the left chamber, where natural spirit becomes transformed into *vital spirit* under the influence of the inspired air brought into the ventricle from the lung by the *arteria venalis*, the "pulmonary vein."[16] The arterial blood containing vital spirit is carried by the aorta and its arterial branches to the body, where it confers vitality on its various tissues. Ascending to the brain via the carotid arteries, it enters a complex of vessels at its base, the *rete mirabili*.[17] There, the vessels divide and subdivide to form the network that delays the blood long enough for the transformation of its vital spirit into *animal spirits*. The vessels from the network penetrate the brain; and, in the ventricles, the passage into animal spirit is further aided by the entrance of air by channels from the olfactory bulbs.[18] The animal spirits also enter the ventricles from the choroid plexuses:

It is reasonable that this pneuma is [also] produced at the ventricles of the brain, and that for this reason no small number of arteries and veins terminate there forming the choroid plexuses... and that the pneuma is as I said the first instrument of the soul and you would expect even more that this pneuma is produced

---

[16] Excess vapors and waste fumes from the heart passes out through the pulmonary vein back to the lungs, where they are exhaled as noxious air. The spent air breathed out was analogized to the sooty vapor emitted from a burning oil lamp.

[17] This network of vessels is found at the base of the brain in some animals, but not in man.

[18] (Galen, 1968), p. 47.

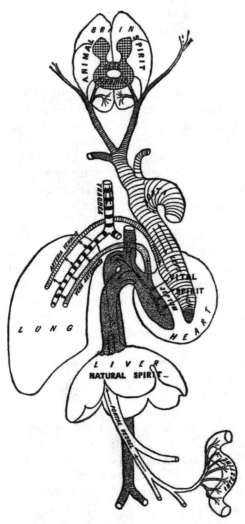

2.1. Diagram showing Galenic physiology. Natural spirit generated in the liver is carried by venous blood through the interventricular septum of the heart to the left ventricle where, under the influence of air carried to the left ventricle of the heart by the arteria venalis, it is transformed into vital spirit. Carried to the brain via the carotids, vital spirit is transformed by the rete mirabili at the base of the brain (shown hatched) into animal spirits. VENT, ventricle. From (Singer, 1925), Figure 30.

when the vessels especially the arteries, breathe it out into the ventricles of the brain.[19]

The animal spirits pass from the anterior ventricles (1st and 2nd) into the middle (3rd) ventricle through an aqueduct and from it into the posterior

_____

[19] (De Lacy, 1978), pp. 444–445.

(4th) ventricle, where it exerts motor functions.[20] Animal spirits then move down into the spinal cord, which was considered to be an extension of the brain, and from it out into the hollow nerve channels where it subserves sensation and control of the musculature.

DIFFERENCES IN THE VENTRICLES IN RELATION TO FUNCTION

Galen championed the brain as the seat of sensation, motion, and mentation rather than the heart, as Aristotle did. In discussing the ventricles of the brain, Galen showed his detailed knowledge of their anatomy and function:

> Now first we shall consider the ventricles of the brain, the size and position of each of them, their shape, their communications with one another, their number, and then the parts that are superimposed upon or adjacent to them. The two anterior ventricles perform the inspiration, expiration, and exsufflation of the brain. Elsewhere we have demonstrated these facts. We have also demonstrated that they prepare and elaborate for it the *psychic pneuma* [*animal spirits*].[21]

Galen describes the communicating passage between the 3rd and 4th ventricles and its role:

> When all the parts described here have been properly revealed, you will see the third ventricle between the two first and the fourth in the posterior. For you will observe the passages on which the pineal gland rests, extending to the middle ventricle in such a way that two [ducts] of some size appear in the opening. One stretches back to the cerebellum, and if you extend a probe or spatula through it you will realize that it ends in the posterior ventricle.... The aforesaid passage [later termed the aqueduct of Sylvius] which extends from the middle [3rd] to the posterior [4th] ventricle and is situated between these nates [inferior quadrigeminal bodies].... A certain small part of the brain, which has an outline similar to the shape of the worm found in wood [vermis of the cerebellum], lies upon it. Notice how the posterior [4th] ventricle is exposed when [the vermis] are reflected forwards, and, when it is moved backwards, how most of it is concealed and only that part appears which Herophilus compared to the groove of a reed pen [calamus scriptorius] on the floor of the fourth ventricle.[22]

The flow of animal spirits between the ventricles was assumed to be controlled by "the vermiform process [vermis inferior]... [which] forms a sort of lid for it [over the ventricle]."[23]

That the pneuma flows down from the brain into the spinal cord, which Galen considered to be a continuation of the brain, was demonstrated by

---

[20] (Singer, 1956), pp. 235–236. In his Figure 26, a bristle in the aqueduct is pictured. This passage was later named, with mistaken priority, the aqueduct of Sylvius (Galen, 1962), pp. 1–6.

[21] (Clarke and O'Malley, 1968), p. 711.   [22] Ibid., pp. 712–713.

[23] (Galen, 1962), p. 1. The concept of its control function was later used by Descartes, with the pineal gland serving that role (Chapter 5).

transecting the cord in animals. His directions were to do this:

> so as to leave no part of the spinal medulla undivided. After the incision, in all
> the nerves which lie below the place where the transection has been made, both
> the two potentialities are lost, I mean the capacity of sensation and the capacity
> of movement, and also all the bodily parts of the animal in which they are dis-
> tributed become insensitive and motionless, a result that is inevitable, clear and
> intelligible.[24]

The loss of sensory and voluntary motor control below the transection follows as expected by the cord receiving these faculties from the brain. Transection of the cord at the highest level of the cervical cord caused paralysis
of all four extremities, whereas a transverse section at a lower level affects
only the legs. A hemisection has a more restricted effect. "Transverse section of either the right or left part of the cord causes paralysis below the
sections only on that side directly under the cut: the right side when the
right side of the cord is cut, the left when it is cut."[25] Galen analogized his
experimental studies to the effects seen in patients who have undergone a
stroke.[26] The loss of motion seen on only one side in patients that suffered
what he called *hemiplegy* (hemiplegia), he related to his animal experiments
in which a hemisection of the cord similarly caused the limbs on only that
side to be affected. He concluded from these studies that the motor nerves
draw their energy from the spinal cord and the cord from the brain.

VENTRICULAR INTERVENTIONS
Animal spirits did not only pass through the ventricles. They are present in
the substance of the brain:

> The pneuma psychikon is not only contained in the cavities of the brain, as I de-
> scribed elsewhere, but it is distributed throughout the entire substance of the brain.
> Most of the pneuma diffuses into the posterior part of the brain, the *parencephalon*
> [the cerebellum]. Nerves originating from it spread into the spinal cord and then
> throughout the body.[27]

A compression of or incision made into the different ventricles showed
variations in the outcome:

> After you have laid open the brain, and divested it of the dura mater, you can
> first of all press down upon the brain on each one of its four ventricles, and ob-
> serve what derangements have afflicted the animal. Should the brain be com-
> pressed on both the two anterior ventricles, then the degree of stupor which
> overcomes the animal is slight. Should it be compressed on the middle ventri-
> cle, then the stupor of the animal is heavier. And when one presses down upon
> that ventricle which is found in the part of the brain lying at the nape of the neck

[24] (Galen, 1962), Ibid., pp. 22–23.   [25] (Souques, 1936), p. 209.
[26] (Siegel, 1968), pp. 304–307.   [27] Ibid., p. 193.

[4th ventricle], then the animal falls into a very heavy and pronounced stupor. This is what happens also when you cut into these ventricles, the animal does not revert to its natural condition as it does when you press upon them.[28]

These effects, that Galen supposed were due to an interruption of the function of the 4th ventricle, were more likely to have been the result of damage to the underlying brain stem in which the vital functions were later found to be controlled. Even as late as the nineteenth century, a similar surgical approach was being used to assess functional loss with uncertainty still remaining as to whether damage to the cerebellum or to the underlying brain stem was producing symptoms.[29] Galen found a difference in the results obtained when pressing on the ventricles or making incisions into them. Following cuts:

> it does sometimes do this [revert to the precut state] if the incision should become united. This return to the normal condition follows more easily and more quickly, should the incision be made upon the two anterior ventricles. But if the incision encounters the middle ventricle, then the return to normal comes to pass less easily and speedily. And if the incision should have been imposed upon the fourth, that is, the posterior ventricle, then the animal seldom returns to its natural condition; [unless done quickly and then you can] see how the animal blinks with its eyes, especially when you bring some object near to the eyes, even when you have exposed to view the posterior ventricle.[30]

The differences in the return of function seen after cutting into or by applying pressure to each of the ventricles allowed him to infer something of the nature of the soul (e.g., that it can have its residence in the brain substance):

> If the soul is incorporeal, the pneuma is so to speak in its first home; or if the soul is corporeal, this very thing is the soul. But when presently, after the ventricles have been [cut and] closed up, the animal regains sensation and motion, it is no longer possible to accept either alternative. It is better then to assume that the soul dwells in the actual body of the brain, whatever its substance may be . . . and that the soul's first instrument for all the sensations of the animal and for its voluntary motions as well be as this pneuma; and therefore when the pneuma has escaped, and until it is collected again, it does not deprive the animal of its life but renders it incapable of sensation and motion. Yet if the pneuma were itself the substance of the soul, the animal would immediately die along with the escape of the pneuma.[31]

Galen, as did the Alexandrians, clearly distinguished nerves from tendons and ligaments, and knew that a muscle's action is dependent on its innervation. He demonstrated this in the vivisected pig. Ligatures were passed under the laryngeal nerves and when they were tightly pulled on, the squeals of

---

[28] (Galen, 1962), pp. 18–19.   [29] (Neuburger, 1981), and (Ochs, 1983).
[30] (Galen, 1962), pp. 18–19.   [31] (De Lacy, 1978), pp. 443–445.

the pig promptly stopped and were quickly resumed when the ligatures were relaxed. Thus, the laryngeal muscles act in this respect no differently than did other ligatured muscles. How Galen was led to investigate the laryngeal nerve is of interest in showing his experimentalist approach.[32] He wrote that earlier writers spoke of the carotids as the "arteries of stupor" or "stupefying arteries" because they viewed stupor to be caused by compression of them cutting off the flow of vital spirit to the brain (Chapter 1). Galen thought they were unaware of the nerves lying alongside the arteries. He was then led to examine the nerves in the neck of the pig, which showed that the function of the laryngeal nerves is to control the laryngeal muscles that are responsible for the voice.

In addition to the voluntary control over the muscles by the motor nerves giving rise to "genuine" contractions, Galen described the heart and other muscle-like structures – such as the stomach and uterus – that exhibit autonomous movements that are not under the control of the will.[33] In this, Galen foreshadowed the awareness of the involuntary (autonomic) nervous system (Chapter 6).

GALEN'S VIEW OF VISION

Galen did not support an interoceptive view of vision, one in which "thin films" emanating from the surface of objects pass into the eye to be recognized. He supported the exteroceptive concept of vision favored by Theophrastus and the Stoics (Chapter 1). In their view, *pneuma* from the eye flows out to its surface, where with sufficient illumination present objects are recognized.[34] This position was supported by his observation that when one eye is closed the pupil of the other eye becomes enlarged. This he took to be due to an increased downflow of pneuma into it. He added as evidence that, "you may make trial of this by artificial means for if you inflate [an isolated eye], the grape-like tunic [the iris of the eye] from within, you will see the aperture dilate."[35]

He was at a loss to account for how the image of a large structure, such as a mountain, could shrink to enter the pupil.[36] And he had difficulty with the fact that the images from an object would have to reach a large multitude of observers simultaneously. The difficulty was resolved when a

---

[32] (Galen, 1962), pp. 211–212.   [33] (Siegel, 1968), pp. 41–44.

[34] This recognition would not be carried out by a physical instrumentality, like that of a walking stick, as the Stoics would have it (De Lacy, 1978), p. 475. A stick would give information as to hardness or softness, whereas vision entails the perception of color, size, and position.

[35] (Galen, 1968), volume 2, book 10, p. 476.

[36] (De Lacy, 1978), Book 7, pp. 453–455. St. Augustine also had difficulty in conceiving this (Chapter 3).

better understanding of optics and the function of the eye in vision was disclosed in the Middle Ages (Chapter 3).

GALEN'S ALTERNATE CONCEPTION OF HARD AND SOFT NERVES

While Galen espoused the role of pneuma flowing in nerve channels, the difference he found between the softer sensory and harder motor nerves led him to advance an alternative model of the agency of nerve action:

> With regard to the ventricles of the brain, when they were [cut into and] emptied of pneuma the whole animal lost the power of sensation, and for that reason we said that the pneuma was useful for the sensations and motions of the parts; should we therefore similarly suppose that there is also a certain pneuma in each nerve? And should we suppose that this pneuma is something local and native to the nerves, being struck by the pneuma that comes like a messenger from the source (of power), or is there no native pneuma in them, and does it rather flow in from the brain at that moment when we choose to move the member? I have no ready answer. Let these two alternatives be investigated in common, and in addition to them a third, the qualitative change of continuous parts, a view that seems to me to be hinted at by those who say that the flow is by virtue of some power without substance. The transmission of qualities to continuous bodies by alteration they call a flow [of] power, as when in the surrounding air some transmission of quality sets out from the light of the sun and reaches every part of the air while [the] actual substance of the sun remains in its place. . . . It is not possible to state readily whether power flow in this way from the brain to the limbs through the nerves, or the substance of the pneuma reaches all the way to perceiving and moving members, or it falls upon the nerves for a certain distance so as to alter them violently, and the alteration is then transmitted as far as the moving members.[37]

In this alternate theoretical account of nerve action, the substance within the motor nerve is considered to be solid or of such a consistency that, when excited by an impulse, it acts mechanically – the pulling or tugging of the nerve, causing the muscle to act. He gives as an example the phrenic nerve, which he considers to act by pulling up on the diaphragm during expiration.[38] Another example he gave is of the recurrent laryngeal nerves. They turn around the innominate and subclavian branch of the aorta, much like a rope turning around a pulley wheel, ascending to the larynx to pull on the inferior laryngeal muscles to produce sound. This action of nerves, along with the superior laryngeal nerves acting on the superior laryngeal muscles of the larynx, was analogized to the action of the *glossocomion* (Figure 2.2). This apparatus was used by surgeons to reduce a dislocated joint or the fracture of a broken limb.[39] Placed in the box with nooses above and below

---

[37] (De Lacy, 1978), p. 449.
[38] We know expiration to be a passive movement in ordinary respiration.
[39] (Galen, 1968), Volume 1, p. 364 ff.

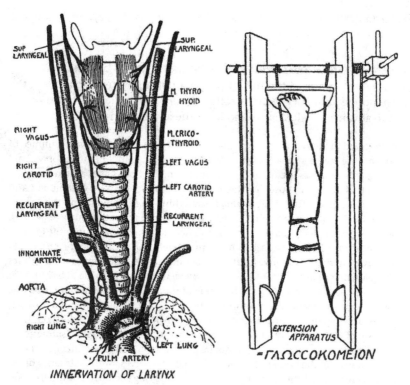

INNERVATION OF LARYNX

2.2. The glossocomion. An apparatus used to reduce a fracture or, as in this figure, to put tension on the upper and lower legs to pull them from a dislocation. The figure on the right shows how turning the screw pulls the leg one way and the thigh the other way from the knee. This is analogized to the innervation of the larynx. The superior laryngeal nerves innervate the upper muscles of the larynx. The recurrent laryngeal nerve of the right side runs under the innominate vessel to innervate the lower part of the muscle; the recurrent nerve of the left side runs under the aorta to innervate the muscle. **PULM**, pulmonary; **SUP**, superior. From (Singer, 1925), Figure 29.

the site of a dislocation or fracture, a pulley arrangement served to pull the bones away from the site to reduce it. Perhaps this alternate mechanical model of Galen represents some lingering identification of nerves with tendons. But still in accord with his more usually expressed view of a flow of animal spirits in hollow nerves, Galen noted at one point that the power of the nerve is not related to its size (thickness), as would be expected if the nerve were acting on the basis of a mechanical action.

Galen's account of sensation other than vision is based on an analogy to wax tablets (commonly used in antiquity for scratching quick notes). The softer sensory nerves were said to receive the impress of sensory objects at their ends and transmit such imprints to the brain, where they are perceived by the *sensorium commune*. Echoing the early discussions regarding sensory channels (Chapter 1), he wrote that some sensory inputs may not be discerned if the sensory passages are too wide for the sensory imprint and it consequently cannot hold its course through the channel, or too narrow so that the sensory imprint is unable to enter the channel in the first place. How the sensory imprints might move in the nerve channels was left unexplained.

Looking back on the achievements of Galen, it would appear that his reasoning and the fruitfulness of his experimental findings in accounting for normal and pathological functions would stimulate others after his death to further such advances. But this was not to be. Experimental studies almost completely came to a halt through nearly all the Middle Ages. What remained in force for a much longer time were Galen's medical teachings that, with some exceptions, were slavishly followed. Some of the social factors responsible for this regression are taken up in the following chapter.

3

NERVE, BRAIN, AND SOUL IN THE MIDDLE AGES

After the death of Galen, little progress was made in studies of anatomy and experimental physiology. No successor approaching his level appeared until the Renaissance. The works he left remained the chief guide for the whole of the Middle Ages, with his teachings petrified into the dogmatism known as *Galenism*.[1] Although he was lauded as representing the highest authority in medicine, on a par with Hippocrates, he was also derided for being opinionated and argumentative. He was even charged with being responsible for holding back progress in the Middle Ages. Modern historians in the twentieth century have given a more evenhanded account of his contributions. Some of his writings have only now been revealed. An important work on the brain, the later book of his anatomical work – *On Anatomical Procedures* – was translated and published as recently as 1962.[2] Galen's concept of the vascular system lasted into the seventeenth century until it was overturned by Harvey's establishment of the circulation (Chapter 4).[3] The question remains as to why anatomical and physiological investigations remained dormant for so long. The answer must lie in the political and social upheavals after the breakdown of the Roman Empire, during which – alongside the regrouping of the secular power of the nobility – the Church emerged as the dominant intellectual presence in the Middle Ages. Whereas the Church Fathers, theologians, and the clergy

---

[1] (Temkin, 1973).   [2] (Galen, 1962), Books (IX–XV).
[3] Galen's clinical and therapeutic teachings still can be found in India and other parts of the Eastern world in medical schools devoted to the teaching of what is called "Hunayn medicine." The designation Hunayn medicine is traced to *Hunayn ibn Ishaq (al-Ibadi)* (808–873), a Nestorian Christian who was appointed chief physician at the court in Baghdad, at the time the center of learning in the Islamic Empire. Hunayn had traveled widely – to Syria, Palestine, and Egypt – gathering Arabic, Syriac, and Greek manuscripts that he and his students translated. Of special importance were 129 Greek manuscripts of Galen works they translated and that would otherwise have been lost.

contemplated the nature of man, what they were chiefly concerned with was the philosophical and theological problems of the relationship of the soul to the body, especially the resurrection of the soul with or without the body after death.

TRANSFORMATION OF POLITICAL AND RELIGIOUS STRUCTURE IN THE MIDDLE AGES

The Roman Empire, long enduring and believed to be imperishable, became weakened by incursions of Germanic tribes from the north until, with the increasingly unwieldy control by Rome over its widespread empire, its Western part declined (England, France, Germany, and Italy) and then totally collapsed in the fifth century. With social cohesion weakened and ordinary life in turmoil, the Church moved in to fill the void. After the official recognition of Christianity in the year 312 by the emperor Constantine and the transfer of the seat of government from Rome to Constantinople in 330, the Empire was split into a poorer Latin West, which crumbled into feudal fiefs only later to slowly become reorganized into kingdoms. In the East, a richer, though shrunken, Byzantium Greek Empire remained in place with Constantinople as its capital. Threatened by Muslim expansion and occupied for a time by the Crusaders, Byzantium lingered on with an ever-shrinking dominion until Constantinople was finally overwhelmed by the Ottoman Turks in 1453.

During the rise of the Church in the West, it was preoccupied with strengthening its rule and developing its teachings of the spiritual life.[4] This was carried out in the major churches, as well as in conventicles that included numerous Pythagorean, Hermetic, and Platonic communities. Intellectual life tended to become diverted by mystery cults with varied sets of formulas, rites, and sacraments that the Church had to contend with. Some of the early Fathers of the Church in the West also showed hostility toward Greek learning, regarding it as pagan and subversive to the main concern of the Church to build and strengthen its teachings. In the East, the ancient Greek heritage remained intact, but in an indolent condition with little inclination to advance the study of physiology and the nervous system to add little more beyond that found in Galen. With time, as the Church grew and had to deal with medical matters, it found Greek science and medicine of use. But, primarily, the *iatrotheology*[5] of the Church in the early

---

[4] (Bréhier, 1965), p. 1–2. See also (Luscombe, 1997).
[5] (Rothschuh, 1978). Rothschuh uses the term *iatrotheology* for the medicine carried out by the church, its most important duty being the care taken for the relation of man to God. Disease was conceived of as punishment for those who break God's commandments, the entry of an unclean spirit, or some unfathomable purpose of God.

Middle Ages was aimed mostly to further the Church's religious teachings by catering to the mundane needs of its flocks. What was of central importance to the Church's ministry was the soul, its immortality, and its resurrection with the body.[6] To the extent that this involved knowing more of the nature of the relation of the soul to the body, attention was given to Greek philosophy and science. The Latins in the West debated, as did the Greeks, whether the seat of the soul, especially its conscious reasoning part, was located in the brain as Plato taught, in the heart according to Aristotle, dispersed throughout the body, or even that it had no locus in space at all. They struggled against the Atomists and materialistic Epicureans, who taught that the soul consists of rarified atoms that were dispersed on death; and those who conceived of the soul as a special organization of the body, a position that would logically lead to its dissolution on death with no possibility of resurrection.

Tertullian (155–220), a native of Carthage, trained in the law, was one of the most important and original of the Fathers of the early church who helped raise the status of Latin to become the language of the Church and then the *lingua franca* of the West. He inveighed against Greek learning as pagan. The philosophers, with Plato in the chief place, he called the "patriarchs of heretics."[7] He even extended his animadversion to Socrates,[8] who in the Platonic dialogue *Phaedo*,[9] put forward the view that the soul is immortal – a position that would obviously be favorable to Christian belief. Nevertheless, Tertullian believed that "each man bears witness for the existence and attributes of God within himself and had no need for philosophical reflection and instruction."[10] Nature, he said, is the teacher of the soul to the effect that she is an image of God. The soul makes man a rational being capable in the highest degree of thought and knowledge. He wrote: "You [the soul] are not, as I well know, Christian; for a man becomes a Christian, he is not born one."[11]

However, he was not consistent in his disapproval of the Greeks and to some extent approved of Stoic philosophy: "I call on the Stoics also to help me, who, while declaring almost in our own terms that the soul is a spiritual essence – in as much as breath and spirit are in their nature very near to each other – will have no difficulty in persuading us that the soul is a corporeal substance."[12] He identified the soul with spirit and considered the mind to be a function of parts or powers of the soul. For Tertullian, the soul consisted of a fine material spread throughout the body resembling it in form. And, it is immortal. It does not exist as an essence or as a necessary

---

[6] (Moore, 1963).   [7] (Tertullian, 1980), p. 35 and (Tertullian, 1947), p. 5.
[8] (Tertullian, 1947), p. 81.   [9] (Plato, 1963).
[10] (Tertullian, 1947), pp. 81–82, 96–98.   [11] Ibid., Chapter 1.
[12] (Tertullian, 1986), p. 184.

consequence of its nature, but as a gratuity of God. He refutes the preex-istence of the soul, teaching that the soul and body come into existence simultaneously, with not a moment's interval occurring from their concep-tion. The act of generation, which produces the entire man, soul and body, entails a "soul-producing seed which arises at once from the out-drip of the soul." Tertullian gives arguments against the view of Plato that the soul is immaterial.[13] The soul is corporeal and with death separates from the body and kept in Hades until its resurrection.

Yet, while Tertullian's fulminations were directed against Greek pagan teachings, the literature and philosophy of classical Greece and Rome were sufficiently appreciated by others in the Church during the darkest days of the Middle Ages, so much so that the Church has been called the "foster mother" of the sciences.[14]

St. Augustine (354–430), the most influential of the early Church Fathers, was classically educated and much influenced by Plato and the Neoplaton-ists. Rebuffing the charge current at the time that the rise of Christianity was responsible for the fall of the Roman Empire, he wrote "The City of God" in which he erected the vision of a Spiritual City as a goal for man in opposi-tion to the merely earthly entity that was Rome. St. Augustine not only did not have an aversion to Greek pagan philosophy and science, but also used it to further his conception of man and his soul. God, in his Neoplatonic view, illuminates the mind of man and without that influence, ideas would not be understood. He considered the soul to be immaterial and to occupy no space. It is in the body where it can effect control of "all the members of the body and serves as a pivot of action, so to speak, for all the motions of the body."

St. Augustine took up the question as to whether the soul is coextensive with the body as Tertullian taught.[15] He observed, as did Aristotle before him (Chapter 1), that when a worm is cut into two parts, each part of the worm moves swiftly away. Carrying the experiment further, he cut the pieces of the worm again, and each of them was able to move about as if it were an independent creature. Asked to explain the phenomenon, St. Augustine's position was that the soul is not contained in any place. It can be divided so that each portion remains with a piece of the body. That was shown by their movement, which was taken as evidence of life. But how this can be

---

[13] Ibid., p. 189.

[14] (Waddell, 1934). Waddell has given an engaging account of the remaining vitality of classical literature in the Middle Ages and its inspiration for some in the church along with the wandering scholars and poets who played a part in spreading its influence.

[15] (St. Augustine, 1964), pp. 89–92, and see (Luscombe, 1997) for a brief view of St. Augustine's platonic-neoplatonism.

was not understandable, and St. Augustine states that further investigation should be sought rather than accepting a false explanation.[16]

The possibility expressed by St. Augustine that the soul need not be in the body at all could be a reference to the teaching of the Neoplatonist, *Plotinus* (205–270),[17] for whom the soul is an illumination, an emanation from the absolute, from the "One," which is radiated through the world soul. The "One" or "Good" is a transcendent source beyond the reach of thought or language. It can be found through purification and the contemplation of its emanation. According to this doctrine, the original creative or expressive act of the One is the *nous*, the soul that operates in the material universe and is the intermediary between the world of the intellect and that of the senses.

For St. Augustine, the soul is by nature intellectual and in communication with realities that are intelligible and immutable.[18] The soul can give true answers in regard to the arts, such as medicine and astrology, that it learns from this life (i.e., through the natural world). Although at first St. Augustine taught that learning was nothing more than remembering, the position held by Plato in the dialogue *Meno* (Chapter 1), he later modified this view as not being in harmony with the teaching of St. Paul, who he noted had written that "no one had done good or evil before birth."

Like Galen, St. Augustine marveled at the small size of the pupil of the eye that yet has the power to see and survey half the sky "whose dimensions are beyond expression."[19] Mere bulk is not to be equated with intelligence. As an example, he points out that, if this were so, an elephant would be more intelligent than a man, and an ass would be more intelligent than a bee (the bee was generally regarded as showing great wisdom). And, as a further argument against mere size, he noted that the eye of the eagle is smaller than ours and yet its vision is more acute.[20]

Galen's influence in the Middle Ages was promoted by *Oribasius* (325–403), who produced an important collection of writings on medicine, with Galen his chief source.[21] But Oribasius did not simply copy Galen in every respect. He criticized Galen's alternate mechanical theory that held that soft nerves subserve sensation and hard nerves motor function (Chapter 2); and he emphasized the view that the animal spirits elaborated in the brain move down within the hollow nerves as being the true position of Galen.

---

[16] Ibid., pp. 89–91.
[17] Plotinus was the pupil of the philosopher *Ammonius*, who also taught the Christian Father *Origen*. Plotinus' Neoplatonic doctrine is found in the work known as the *Enneads*, which was arranged by his disciple Porphyry. St. Augustine was influenced by Plotinus, reading Christian interpretations into his philosophy (Pegis, 1944).
[18] (St. Augustine, 1964). In *Retract*, 1.8.2.   [19] (St. Augustine, 1964) p. 41.
[20] Ibid., p. 42.
[21] (Temkin, 1973), pp. 62–64. His major opus was a large medical encyclopedia.

Galen's teaching was also widely represented in the Middle Ages by the *Isagoge*, a popular simplified introduction by *Iohannicius* to the larger *Ars medica* that itself was an outline of medicine for the use of the medical practitioner.[22] The Isagoge contained a theoretical part that dealt with things natural, what we would describe as basic science, in which the presence of three spirits in the body (the natural, vital, and psychic) were put forward as the Galenic view.[23]

CELL THEORY: A SIMPLIFIED REPRESENTATION OF
THE BRAIN VENTRICLES

Nemesius, Bishop of Emesa (fl. 380/390) – widely known to the Middle Ages through his book *On the Nature of Man*,[24] making extensive use of the writings of Galen – he delineated a simplification of the structure of the brain ventricles pictured as the *cell theory* (Figure 3.1). The 1st and 2nd ventricles, the lateral ventricles, were collapsed into one forward cell that was taken to be the site of imagination, with understanding located in the middle cell and memory in the posterior cell. *Poseidonios of Byzantium* (fl. 370) also localized imagination in the forward cell, understanding in the middle cell, and memory in the posterior cell; this pictorialization was followed by a large number of commentators on the brain throughout the fourth to the sixteenth centuries,[25] many of those illustrations given by Clarke and Dewhurst.[26] The general view held by the cell theorists was that animal spirits flowed from their initial formation in the forward cell, where they acquired the power of imagination; back to the middle cell, where the power of thought (understanding) is located; and then to the posterior cell, which has the power of memorization. Understanding would judge what was valid in the imagination, changing what was necessary before sending the modified spirits to the posterior cell to be retained in memory. Memories, when recalled, could also be sent forward to the middle cell for understanding. In many of the illustrations, control over the movement of the animal spirits between the cells was exerted by the cerebellar vermis acting as a valve, as Galen had proposed, with the vermis, however, pictured by some

---

[22] Ibid., the *Ars medica*, pp. 101–104.   [23] Ibid., pp. 105–107.
[24] In the Middle Ages, Nemesius was confused with *Gregory of Nyssa* (Nemesius, 1955) p. 203 ff. Nemesius' book was later made available in an English translation entitled, *On the Animal Spirits*, by George Wither published in 1604. The author was misnamed Gwither by Isaac Disraeli (the father of Benjamin Disraeli, prime minister of England), who in his book published in 1866 related the belief that a man's face, like soft wax, shows the imprint of objects or ideas moving his affections (Disraeli, 1866). The translation into English of Nemesius' book published by Telfer (Nemesius, 1955) is a much more useful scholarly recension.
[25] (Manzoni, 1998).   [26] (Clarke and Dewhurst, 1972).

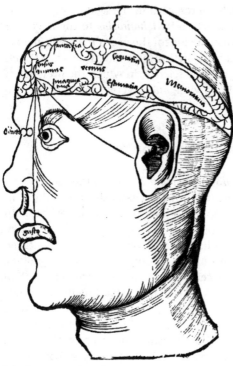

3.1. Schematization of the cell doctrine. This example, given by Reisch in 1504, shows the cell at the front labeled *sensus communis, fantasia, imaginitiva* (common sense, imagination); and the middle cell labeled *cogitava, estimativa* (cognition, judgment), with the flow of animal spirts between them controlled by the vermis. The posterior cell is where *memorativa* (memory) is located. From (Singer, 1925), Figure 41.

between the anterior and middle cells, in a position unable to perform that function.[27]

The Greek Christian Aristotelean and literary scholar *John Philoponus* (fl. sixth century), who taught Greek philosophy and medicine at Alexandria, shows an acute awareness of Galenic texts. He was one of the few known at the time to have carried out animal experimentation.[28] He verified Galen's finding that pressure on the brain of animals can cause immobility, with an evident loss of sensation. He also confirmed that the spinal cord, when injured or "bound," caused a loss of motor control in only the part of the body below that level, whereas the upper part continued to function normally. As did Nemesius, Philoponus assigned the imaginative faculty to the forward brain cell, the rationative (intellectual) to the middle, and memory to the posterior cell. Philoponus held a Neoplatonic view of the soul that

[27] (Manzoni, 1998).   [28] (Todd, 1985).

he thought to preexist and descend from the One to take up a temporary sojourn in the physical body. The rational soul, complemented by the irrational soul with its sensations and desires, inheres in the pneuma as it descends into the body.

Philoponus does not follow Galen entirely in his extramission theory of vision. In Philoponus' view, optic pneuma passes from the optic nerve into the eye, where it reaches its terminus in the "crystalline body" (lens). Discrimination of the visual objects occurred there, rather than in the surrounding air as Galen had theorized.

THE TRANSMISSION OF GREEK LEARNING TO THE WEST

A momentous upheaval of the Roman world was brought about by Mohammed and the Arabic conquests of the seventh century he inspired. After defeating Syria and Persia, Islam spread west, bypassing the Byzantine Empire to overrun North Africa and expand into Spain. Greek medical and scientific texts, and those that had been translated from the Greek into Syriac and Persian representing much of the bulk of medical science known to us from classical times, fell into their hands. With their rise as a world power, Islamic universities were founded, and Syriac, Persian, and Greek texts were translated into Arabic. The heritage of Greek science and medicine consisting of manuscripts, commentaries, manuals, and encyclopedias based on them, containing much of the physiological and medical writings of Galen, was thereby preserved. In the ninth through the eleventh century, Christian, Jewish, and Arabian scholars in the East, in Salerno and elsewhere in Italy, Sicily, and Spain, began the translation of those manuscripts into Latin that were then spread widely throughout France, England, Germany, and other Western European countries. In the translation of Greek into Arabic,[29] philosophical thinking colored the texts, adding a layer of obscurity. Pagan gods became angels; Greek customs, usage by Shari'a (religious law); Greek inspiration, revelation; active intellect by the spirit of holiness; the struggle to convert the world to the rule of philosophy and reason, jihad; the First cause or First mover, Allah. The increased awareness of early Greek philosophy, a mixture of Neoplatonism and Aristotelianism, was to serve as a guide, as was seen in the writings of *Al-Farabi* (870?–950) who rationalized the teachings of Islam with philosophy, even putting philosophy in first place ahead of theology.[30]

A similar approach was later made by *Maimonides* (1135–1204) for Judaism in his classic book *Guide of the Perplexed*.[31] His command of Arabic and entrée through it to Greek sources allowed him to interpret the Bible in rational terms. Angels in the Bible were interpreted as Platonic forms acting as intermediates between the intelligence of the divinity and the governance

---

[29] (Walzer, 1962).   [30] Ibid., pp. 18–19.   [31] (Maimonides, 1956).

of matter. In the main, Maimonides followed Aristotle but for some fundamental differences. While Aristotle viewed the matter of the Cosmos as being eternally present, Maimonides followed Genesis wherein the world was created by God *de novo*, from nothing. Most notable was his difference with Aristotle who taught that man shared some of his faculties in common with other animals while Maimonides followed Biblical teaching that held God to have created man in His image and thus was unique in all respects.[32] His general orientation toward Aristotelianism had some later influence on Christianity through the teachings of St. Thomas.

The Crusaders played a significant role in this intercourse and transmission of ideas between the East and the West. The aim of the first crusade (1095) was to rescue Jerusalem from the "heathen Muslim," with the fourth crusade (1202–1204) ending with the Crusaders plundering the riches of Christian Orthodox Constantinople instead. The plunder consisted not only of jewels and gold, but also treasure of greater worth – precious religious relics for devotional use.[33] But even of still greater worth for learning were the Greek books taken back to the West that helped advance its science and culture. There was not yet, however, the widespread transfer of Greek books to the West that occurred later on with the fall of Constantinople to the Ottomans in 1453. Because of the lack of substantial Greek or Arabic-speaking communities in the West, translations of the earlier importations were done by a relatively few gifted individuals. Beginning in the eleventh century, and increasingly so in the twelfth century, men capable of such translation were found wherever the receding tide of Moslem or Byzantium power (in Spain, Sicily, and Southern Italy) left behind men who knew Arabic and Greek and were able to serve as intermediaries between Christendom and the outside world. Slowly, scholars built libraries of Greek manuscripts that were translated into Latin without an arabic overlay.

Aside from the Carolingian period beginning in the ninth century, where a limited revival of learning in Northern Europe occurred in Frankish lands, the South Italian health resort and medical center at Salerno, which had originated in the ninth century, or perhaps earlier, had its greatest prestige as the first organized medical school in Europe. Flourishing in the period of the high Middle Ages (1050–1300), it was known as the *civitas Hippocratica*, the City of Hippocrates.[34] Here, Latin culture came into contact with the medical thought flowing into it from the Greek schools of lower Italy, where the tradition of ancient Greek medicine was still extant, along with input from Alexandria and the rest of the East. Medical students and teachers from all of Europe intermingled with Arabs, Jews, Syrians, and

---

[32] (Maimonides, 2000).   [33] (Southern, 1953), pp. 63–68.
[34] (Garrison, 1929), pp. 147–150.

Persians in Salerno to create an amalgam of medical lore.[35] There, *Constantinus the African* (1020–1087), famed as a great translator of oriental languages, translated Arabic, Syrian, and Greek manuscripts into Latin. Translations of Galen and Hippocrates, along with a host of other Greek writings, were carried out by *Gerard of Cremona* (1114–1187). He delivered a library of medical knowledge to the West of which no single work would have been known to the Latin western world two centuries earlier.[36] From Salerno, medical learning spread further in the West as a succession of new universities arose in Italy, first at Bologna, then elsewhere in Italy, to Montpellier in the south of France, and soon after widely throughout Europe. Most useful in that growth of the universities was the availability of increasing numbers of translations into Latin that by then had become the *lingua franca* of the West, serving the needs of the growing international community of scholars.

ISLAMIC CONTRIBUTIONS TO MEDICAL THOUGHT AND THE THEORY OF VISION

One of the most famous of those in the East who found a high place in Western thought was *Avicenna (Ibn Sina)* (980–1037), a noted Persian scientist, physician, and philosopher whose extensive writings on medical matters – translated into Latin in the twelfth century as the *Canon of Medicine* – were made available to the West. His philosophical analysis of psychology, the relation of the soul to the body, has been translated into English.[37] His view was essentially Aristotelian blended with Neoplatonism and the thoughts of St. Augustine. The soul for him was "form," acting as an entelechy that perfects the member of a given species. It is like the pilot of a ship, a substance separable from the body that can transcend it.

The Muslim world not only served to transmit Greek learning dealing with the philosophical analysis of the soul, mind, and sensation – along with the bulk of Greek scientific and medical knowledge – it made its own contributions to early chemistry and to medicine, most significantly to ophthalmology. *Al-Kindi* (b. late eighth century–866) wrote a work on optics *De aspectibus* that had a strong influence on Western optics throughout the Middle Ages.[38] Basing his studies on a optics work by Euclid, Al-Kindi asserted that:

Everything that has actual existence in the world of the elements emits rays in every direction, which fill the whole world ... [and] binds the world into a vast network in which everything acts upon everything else to have natural

[35] (Castiglioni, 1947), pp. 229–320.   [36] (Southern, 1953), pp. 66–67.
[37] (Rahman, 1952).   [38] (Lindberg, 1976).

effects. Stars act upon the terrestrial world; magnets, fire, sound, and colors act on objects in their vicinity.[39]

Al-Kindi argued against the intromission theory of vision on mathematical principles, identifying himself with the Galenic-Stoic extramission theory of vision.[40] His contemporary *Hunain ibn Ishaq* (d. 877) also held to the extramission theory, writing that, "when [the visual spirit in the eye] meets the air in the moment in which it goes forth from the pupil, it transforms it immediately it encounters it, and that which arises from the change runs through it [the air] for a very long distance."[41] On the other hand, it was through the support given by *Avicenna* (980–1037), *Alhazen* (965–1039),[42] and *Averroes* (1126–1198),[43] for the intromission theory of vision, that it became the generally accepted view.

THE RELATION OF SPIRIT TO SOUL DURING THE HIGH MIDDLE AGES

Rather than the Middle Ages existing as a period of uniform darkness, there were periods when fundamental changes occurred, one described by Southern in the period from 1000 to 1350 that was known as the High Middle Ages.[44] Major social changes are recorded as kings and nations vied with the Church and then merged with it, the Crusaders fought the Muslims, the black death, and then the Inquisition, put a black mark on the age. Scholastic learning continued to deal with philosophical issues that related to the view taken of the nerve, the nervous system, and the role of the soul.

Primarily, the Galenic doctrine was followed, and the air taken into the body on inspiration was believed to be transformed into animal spirits. The spirits viewed as material in nature, albeit of a rarefied fiery nature, became differentiated from the immortal immaterial soul. This was the position taken by *Costa ben Luca* (c. 900) in his book *de differentia*, written in 912 that was translated from the Arabic into Latin by Constantinus Africanus in 1140. The book was much cited in the twelfth and thirteenth centuries by important figures, such as Albertus Magnus, and incorporated into the standard medical teaching of the time. Putscher gives the Latin text of his book. Freely translated, Costa ben Luca wrote:

> Who would wish to know the difference between two things, of necessity must know the first thing and what it consists of as it is impossible to differentiate between two things unless one knows it and their place. After this one can know their difference. And because we wish to reveal the soul, and its difference from spirit, it is necessary that we first speak of the soul and then of the spirit and

---

[39] Ibid., p. 19.   [40] Ibid., pp. 33–42.   [41] Ibid., p. 31 and (Eastwood, 1982).
[42] (Lindberg, 1976), pp. 58–80.   [43] Ibid., pp. 52–56.   [44] (Southern, 1953).

the difference between the both. But that the work be not too difficult, we will begin first with spirit, afterwards we will speak of the soul.[45]

In Part I of his treatise, Costa ben Luca follows Galenic thought concerning the spirit as being a subtle material substance that arises in the body from the heart. Passing through the veins, it vivifies the body. In the brain as animal spirits, it moves through nerves to subserve sense and motion. In Part II, he writes that the soul is different and points out the great difference between what the philosophers write concerning it, in particular Plato and Aristotle. Plato defines the soul as an immaterial substance that moves the body, whereas Aristotle views the soul as the perfecting agent of the potentiality of the body, and others viewed animal spirits as material, albeit of a finer nature than the body.

The difficulty with Plato's position for Costa ben Luca was that whatever moves a substance is itself a substance, whereas for him the soul is immaterial. The view of Tertullian and others who espoused a view of the soul as a material substance would overcome the problem but it entailed a difficulty for Costa ben Luca.[46] He viewed spirits not as just a rarified form of matter like air, but also as a kind of substance essentially different from the ordinary matter out of which the body is made. This modification of the conception of spirit allowed it to serve as an intervention between the body and soul, where body "rises" through the spirit to attain the soul and the soul "descends" through the spirit to act on the body. This intermediation appeared to answer the problem of how an immaterial soul can interact with a material body, although it did not remain the final thought on the subject. With the Renaissance and the expansion of science, the problem was sharpened by Descartes' model of man in the seventeenth century (Chapter 4) and continues in our day as the mind-body problem (Chapters 14–16).

THE LATER MIDDLE AGES

The universities of Europe, which developed in the later Middle Ages, served the needs of the increasing institutionalization of the Church – the growth of a large number of cloisters, retreats, and churches. Their need for a supply of documents and teaching tracts, as did the developing civil bureaucracies, required a great increase in the numbers of scribes. In their *scriptoria*,[47] they wrote out the large number of manuscripts needed by the clerics, political officers serving the church, legal functionaries, and along with them, professors and students of the medical profession. In the darkest days of the Middle Ages, only the elementary subjects of grammar, rhetoric, and logic, with a smattering of arithmetic, astronomy, and music, had sufficed. C. F.

---

[45] (Putscher, 1973), pp. 145–150.   [46] (Bono, 1984).   [47] (Diringer, 1982).

Lewis pointed out that, in the early Middle Ages, only one book of Plato was at hand: the dialogue *Timaeus*. With the resurgence of learning in the later Middle Ages, there was an increased interest in the works of Plato, Aristotle, Hippocrates, and Galen. As their works became available, they were copied to serve the needs of the universities.[48] The ever-burgeoning need for labor to copy manuscripts by hand was later resolved by the invention of the printing press,[49] one of the most important factors in bringing about the Renaissance (Chapter 4).

The writings of *St. Thomas of Aquinas* (1224–1274), a mendicant religious in the Dominican order, are of major significance in that they have been accepted in our time as the basic position of the Catholic Church. When Thomas Aquinas arrived at the University of Paris, the influx of Arabian-Aristotelian science from the East, chiefly the Latin translation of Aristotle's writings and the comments on them by *Averroes (Mohammed ibn Roshd)* (1126–1198), had been taken up enthusiastically by schools in Italy, the faculty of arts in Paris, and elsewhere in Europe. The Averroes-Aristotelian philosophy appeared to be more attuned to the new experiences then developing in the world. However, the emphasis on reason and experience as the way of furthering the truth appeared to confront Scholastic philosophy, which dominated the earlier Middle Ages and aroused a sharp reaction by Church theologians. Thomas was attracted to the then newly recovered scientific ideas and the philosophy of Aristotle that his master Albertus Magnus and Roger Bacon lectured on in Paris. But, in this philosophy, there was a fundamental theological conflict: the view that man possesses only a disposition for receiving the intellect from without. That a unity of active intellect exists present in all men was recognized by Thomas as being incompatible with the notion of personal immortality. Thomas had William of Moerbeke make a translation of Aristotle's works to which he added his extensive comments outlining his philosophy.[50] Thomas took over Aristotle's thesis of the existence of a primary unmoved mover, which for Thomas was the Judeo-Christian god. He taught that the soul is man's form, the ultimate value of the individual and immortal. And, as in Platonic thought, that the soul in its highest part is responsible for rational thought (Chapter 1). The soul is believed to know the world of ideas (this proved in the Platonic

---

[48] (Haskins, 1965). Haskins points out that the early university were a collection of students with no fixed buildings who hired the professors to teach them. Later, university buildings arose with the professors then in control through their granting of licenses to teach. The Cathedral School of Paris evolved into a university that served as the model for the rise of other universities in northern Europe. The earliest establishment of university in Italy, that of Bologna, revived the law, basing it on Roman law.

[49] (Eisenstein, 1979).   [50] (Aristotle, 1965).

dialogue *Meno* where ideas are recollected from a previous life), the basis of the concept of its immortality.[51]

With the Renaissance, thought turned ever more toward the new findings in the world, the new knowledge of the structure and function of the body – in particular the brain and its relationship to the body through the nerves – and how it might interact with the soul (Chapter 4).

---

[51] (Durling, 1993), pp. 237–238.

4

RENAISSANCE AND THE NEW PHYSIOLOGY

The recovery of long-lost Greek literature and science manuscripts and their translation into Latin led not only to an admiration of the achievements of past masters, but also before long a desire to emulate and surpass them. Starting first in Florence, where the "Florentine Renaissance" began in the mid-fourteenth century with the enthusiasm of Petrarch in his search for ancient manuscripts, the restoration of learning spread elsewhere in Italy and then to all of Europe in the fifteenth and sixteenth centuries. The Renaissance gave rise to the remarkable literary and artistic achievements of such men as Botticelli, Donatello, Michelangelo, Leonardo da Vinci, Erasmus, among others. The enthusiasm for learning and the expression of individualism characterized those of the Renaissance, differentiating them from the conformity to church doctrine of the preceding centuries that came to be called the "Middle Ages" or the "Dark Ages."

A contrast developed between the universities, with their need to provide training for clerics and where church matters took precedence, and the new humanistic circles in which laymen and clerics mingled freely under the protection of princes, popes, and such powerful families as the Medici, who gave support to the arts and sciences. Through them, "the modern age began with the acceptance of the autonomy of intellectual methods and problems" to fulfill the desire of knowledge for its own sake, as well as for worldly gain.[1] New findings were investigated, and new things brought to light. Ships rounded Africa and penetrated into the exotic Far East; and sailing west the Americas were discovered. The explorers meeting with new people and different cultures stirred the imagination, and brought about a greatly enlarged world view with new opportunities for the advancement of knowledge.

A major contribution to the changed outlook of the period was the introduction of printing by Gutenberg in the mid-fifteenth century. Instead

[1] (Bréhier, 1965).

of the limited number of handwritten manuscripts laboriously copied by scribes, the larger numbers of identical copies produced by the press gave rise to an easier access to the learning of the past and, significantly, the means to widely broadcast new discoveries and concepts. The increased flood of ancient Greek, Arabic, and Syriac manuscripts carried into Italy and the other western countries after the fall of Constantinople to the Ottomans in 1452 were soon translated into Latin and printed. Latin by then was the *linga franca* of western Europe. Greek, the language of science and philosophy in the ancient world, still held its special place in the Renaissance. Greek manuscripts were printed and circulated along with the Latin texts as part of the new humanistic outlook. It was held that by going back to the original sources, a truer insight into the thought of the ancient masters could be had, one not overlaid by interpolations and mistranslations from the Arabic and other Eastern intervening languages.

The incorporation of the classic scientific literature into Western thought acted as a spur to emulate the ancients, to make direct observations, and to experiment. Public dissections of executed criminals were undertaken to explicate the anatomy of Galen, and autopsies were made to determine the causes of disease. An anatomy book by *Mondino de Luzzi* (1270–1320) that appeared in 1360 became an early textbook and was credited with bringing human dissection into the medical curriculum. The common opinion that the Church was entirely opposed to dissection was not the case.[2] As noted by Roger French, the fourth Lateran council of the Church in 1215 had decreed that, when the soul is resurrected, the body is made whole again. This allayed fears that a body mutilated in the course of dissection would make it unfit for resurrection. Though reluctance to perform dissections still remained, the practice was continued. Discrepancies were noted between Galen's descriptions and the structures directly observed in the body. Some of the differences were the result of Galen's use of animals instead of humans, a point that he himself had mentioned. Human autopsies showed findings that conflicted with his earlier medical teaching. For example, Galen teaching that epilepsy is caused by the accumulation of a thick humor in the brain ventricles was not confirmed by autopsy.[3]

Leonardo da Vinci (1452–1519) personified the Renaissance man, trusting to his own vision and rejecting the dogmatism of medieval scholasticism. However, this was not a complete revolution. His thinking shows an evolution as he freed himself from the past. He carried out remarkable anatomical

---

[2] (French, 1999).

[3] (Temkin, 1973), pp. 136–138. Galen's thought on this point could be saved by his adherents, for he also taught that an excess of black bile in the ventricles (which would escape detection) could account for convulsions.

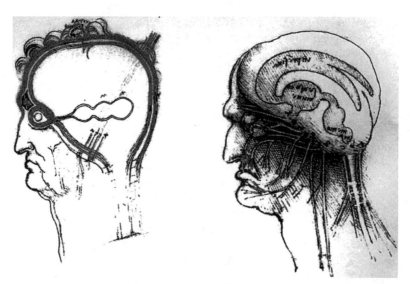

4.1. Leonardo da Vinci at first represented the brain ventricles on the basis of the cell doctrine as shown in the figure on the left. On the right is his later, more realistic representation of the ventricles made on the basis of wax injections into the brain. Modified from (Esche, 1954), Figure 54.

studies of the human body.[4] At first, Leonardo depicted the brain ventricles in accordance with medieval cell theory and then later showed them to be closer to their true form by making a wax cast of them (Figure 4.1). Yet, he still retained the functions that had long been assigned to the cells: The term *imprensiva* (the perceptual center) was given to the lateral ventricles (taken together as the first cell), *sensus communis* was given to the 3rd ventricle (the second cell), and *memoria* to the 4th ventricle (the third cell).[5]

Scientific inquisitiveness informed his anatomical studies. This can be seen in the directions he gave for the dissection of the *reversive nerve* (the vagus nerve innervating the heart). To demonstrate the role it might play in controlling the beat of the heart, he directed the dissector to:

> Follow up the reversive nerves as far as the heart, and observe whether these nerves give movement to the heart, or whether the heart moves of itself. And if its movement comes from the reversive nerves which have their origin in the brain then you will make it clear how the soul has its seat in the ventricles of the brain, and the vital powers [spirits] derive their origin from the left ventricle of the heart. And if this movement of the heart originates in itself then you will say that the seat of the soul is in the heart and likewise that of the vital

[4] (Singer, 1925), pp. 90–94.
[5] (Keele, 1963), p. 24, and (Clarke and Dewhurst, 1972).

powers; so you should attend well to these reversive nerves and similarly to the other nerves because the movement of all the muscles springs from these nerves which with their ramifications pour themselves into these muscles.[6]

Although Leonardo's skill as an artist and sculptor was widely known and won great admiration in his day, as well as in ours, his anatomical studies had little influence on his contemporaries. These studies had been mislaid and unknown until they were recovered long afterward.[7]

New discoveries of things and properties not found in Galen's writings led some to radically reject all of his teachings *in toto*. After the introduction of metals into medical practice, substances unknown to Galen, *Paracelsus* (1493–1541) concluded that Galenic teachings were worthless; and he symbolized his rejection of them by throwing Avicenna's book into a bonfire.[8] Although Galen's scientific position was weakened by the new discoveries, Galenic doctrine still remained in force as the foundation of medical practice. This is exemplified by *Fernel* (1497–1558) in his book *On the Natural Part of Medicine*, a subject he later named "Physiology," (Physiologia) published in 1452.[9] It was a popular work, widely accepted as the basis of medical practice, with numerous editions published even into the following century. The anatomical descriptions were basically those of Galen, with no figures to show the body structures. It was Galenic physiology with additions of some newer mechanical concepts, such as the action of pumps applied to the brain:

> the soul sends out spirits like servants...into all the external senses, into all the bodily parts that lie outside [the skull]...this [animal] spirit completely fills the anterior ventricles...distributed from them into the senses and muscles, on this general plan: the body of the brain is in unending movement, like the heart, of itself and its own volition, now expanding and now contracting in the course of this movement, and narrows its internal cavities by squeezing their sides, spirit pours out from the anterior ventricles, both into the posterior ones and into the instruments of the senses. But when it enlarges and opens itself out, from the arteries of the structures it entices and pulls vital spirit, and from the nasal cavities, air. At that very moment the pathway that leads from the third ventricle to the posterior one closes down completely, and is blocked by the descent of the process that is named after a worm [the vermis lobe in the middle of the cerebellum], settling and slipping down between the "buttocks" [corpora quadrigemina], so that no spirit can retrace its path from the posterior ventricle to the middle [third] one. The reverse occurs when the brain contracts: the process is raised, and the "buttocks" themselves draw further apart, so that the way

[6] (Keele, 1952), p. 49. Keele gives an extensive treatment of Leonardo's anatomical and physiological studies of the heart and blood vessels.

[7] (Singer, 1925), and (Keele, 1979).

[8] (Pagel, 1982), p. 20, and see also (Paracelsus, 1951). Paracelsus' writings show the entanglement of alchemy and older medieval concepts.

[9] (Fernel, 2003), and (Sherrington, 1946).

is more open for spirit to run from the third ventricle into the fourth. The "worm" and the "buttocks" contribute as much to the brain as the valves do to the heart...the posterior ventricle...is always filled with spirit, which...flows into the cavity [of the spinal cord] that runs the length of the spine. All this spinal cord ["medulla"] is hollowed out by a central innermost canal; and the larger nerves that stem from it are also strikingly hollow at their origins. Consequently animal spirit, running down from the spinal cord ["medulla"], glides into them too, and ends by diffusing into the final slenderest nerve fibers.[10]

This passage is interesting in that Fernel, basing his account on actual brain ventricles rather than on medieval cell theory, points to the pineal gland as a valve that prevents a backward movement of animal spirits from the 4th ventricle into the 3rd ventricle, when the brain expands and opens when the brain contracts to force animal spirits down into nerves and muscles. As will be seen later (Chapter 5), Descartes points to the pineal gland as the site where the soul is located and, by its action, selectively moves animal spirits into one or the other of the various nerves controlling the muscles.

In Fernel's physiological thought, different "faculties"were present in the liver, heart, and brain – these representing the activities of a unified indivisible "life-soul" – the *anima* that was present in every act of the body. The anima descends from God directly down into each individual's brain. A figure later given by Fludd shows its descent from the *Mundus intellectualis*, in which Deus (God) is pictured, the anima moving down into the middle cell of the brain into the *Mens intellectualis*.[11] There, reason has its seat with the vermis, contra Fernel, shown between it and the first cell, the *Mundus imaginabilis*, in which the senses are located. Sherrington noted that:

> Fernel is careful to maintain that the soul, despite its [having] different faculties, is one. The soul is utterly simple. It cannot be split. It cannot be subdivided. It is an individuum, a spiritual unity...it cannot perish. It is immortal...[while] the manifestations of the soul are multiple. Here Fernel's faith is struggling against Aristotle....The different functions of different parts of the body come from the differently prepared nature of those material parts (organs) in which the one and single soul is diffused. It is as when one and the same craftsman uses different tools.[12]

Sherrington notes that Fernel seems not to have apprehended any difficulty about how the spiritual and corporeal substances could interact, the mind-body problem that was brought into sharp relief by Descartes (Chapter 5).

---

[10] (Fernel, 2003), pp. 489–491. See also (Rothschuh, 1958) for an alternate German translation of this passage. Rothschuh interpolates 'pineal' for 'vermis' though the latin text clearly gives the term as 'vermis'.

[11] (Clarke and Dewhurst 1972), Figure 57, p. 38.

[12] (Sherrington, 1946), pp. 79– 80.

In the very same year of 1542 that Fernel's book was first published, the book *De corporis humani fabrica* by *Andreas Vesalius* (1514–1564) appeared which greatly altered the view of human anatomy and further questioned the authority of Galen.[13] The book was based on Vesalius' extensive series of human dissections, a number of them carried out as public demonstrations attended by large and enthusiastic crowds. The lectures he gave at Bologna in 1540 were written down by a medical scholar from Germany, who gives us an eyewitness report (the only one we have) of the procedure that was followed.[14] The dissections were done over a number of days after lectures on a given part of the body by the well-known anatomist Mattaeus Curtius. The lectures were based on the anatomy of Mondinus and Galen. These were then followed by Vesalius' anatomical demonstration on the cadaver given either later that day or the next morning. Twenty-five lectures and twenty-six demonstrations in all were given on this occasion, with the last demonstration a vivisection made by Vesalius on a dog to show the function of the *nervi reversi* (recurrent laryngeal nerves) in control of the voice. He said, "you will hear how the dog will bark as long as these nerves are not injured. Then I shall cut off one nerve, and half the voice will disappear, then I shall cut the other nerve, and voice will no longer be heard... and [when] he did so... the bark of the dog disappeared."[15]

Vesalius' *Fabrica* not only represented a great escalation in the detailed knowledge of human anatomy, it also showed the body as an artistic representation. Although it breathed the spirit of the Renaissance, some remnants of Galenic teaching remained. In his earlier work, *Tabulae Anatomical Sex*,[16] Vesalius pictured some anatomical features in man that were in fact animal structures that had been described by Galen. For example, the liver was drawn as five-lobed.[17] A truer picture of the organ corresponding to that in the human was later figured by Vesalius in his *De corporis humani fabrica*. Vesalius' struggle to free himself from Galenic doctrine as given by Church teaching is indicated by the description he gave of his education at the University of Louvain. There, in what he referred to as "the leading

---

[13] (Vesalius, 1543), and (O'Malley, 1964).

[14] (Eriksson, 1959). The study of medicine at Bologna could be traced as early as the twelfth century and was independent of the school at Salerno (Chapter 3). When Vesalius was called to give the anatomical demonstrations at Bologna, it was because of the competition with Padua for medical students; Vesalius was a star attraction. Soon after, Padua – with the support of the richer city of Venice – housed the preeminent medical school.

[15] Ibid., pp. 291–293. As noted in Chapter 2, Galen – in his demonstration of the action of the laryngeal nerves – used ligatures to compress the nerves rather than cut them. This allowed the return of the voice when the ligatures were relaxed.

[16] (Singer and Rabin, 1946), and (Singer et al., 1946).

[17] (Temkin, 1973), Figure 5 facing p. 108.

and most distinguished college," he was taught by "a theologian by profession and therefore, like the other instructors at that Academy, ready to
mingle his own pious views with those of the philosophers – the brain was
said to be equipped with three ventricles. The first was in front, the second in the middle, the third behind, with names given them accorded with
their position and other names derived from their functions." Vesalius later
clearly pictured the true anatomical relations of the ventricles in the human
brain.[18]

Vesalius entered on the question as to what difference between human
and animal brains might account for man's power of reasoning. In examining the brains of quadrupeds, he found them to "closely resemble those
of men in all their parts." He asks, "Should we on that account ascribe to
these [beasts] every power of reason, and even a rational soul?" Vesalius
concluded that it was the relative size of the different brains that accounted
for the greater intellectual powers of humans:

> Certainly in the brain of sheep, goat, ox, cat, ape, dog, and of such birds as I have
> dissected, there is a shaping of the parts corresponding to that of the human brain,
> and specifically is this so of the ventricles. There is hardly any difference that we
> have detected except in bulk, the brains do vary according to the intelligence with
> which the animals are endowed. For to man has been given the largest brain; next
> after him the ape, the dog, and so on, according to the order that we have learned
> of the power of reason in animals. And to man's lot falls a brain not only bigger
> in proportion to his bulk of body, but actually bigger than the brain of any other
> animal.[19]

In the transformation of vital spirit into animal spirits in the ventricles, he
included the substance of the brain as playing a role:

> From the air which has entered the brain, and from that vital spirit which, by its
> devious course, becomes progressively more assimilated in the ventricles to the
> action of the brain, the animal spirit is elaborated by the cerebral power (*virtus*).
> We believe that this power depends on the opportune balancing of the elements
> of the brain substance.[20]

How the animal spirits are distributed in the nerves he is "not over-anxious
to decide – whether through passages in the nerve as is vital spirit in arteries –
or along the surface of the nerves, like [rays of] light along a column, or
simply through the community of the nerves."[21]

Vesalius carried out physiological experiments on animals using ligatures,
as Galen did on the laryngeal nerves to show their role in producing the
voice, and he similarly used ligatures on other nerves to show the dependence of a given muscle's movements on its innervation. He demonstrated

---

[18] (Singer, 1952). Shown in plates taken from the *Fabrica*.
[19] Ibid., pp. 6–7.   [20] Ibid., p. 3.   [21] Ibid., p. 4.

the falsity of the supposed function of the sheath around nerves as the means by which animal spirits move down the nerve by removing the sheath and finding no change in the nerve's function.

With regard to the nerves, Vesalius relates that he has shown "that the nerves originate in the brain. [That] nerves are to the brain as aorta [and its branches] to the heart, and vena cava (and its branches) to the liver. They deliver the [*animal*] *spirits*, which the brain has prepared, to those instruments [muscles] to which it is to be conveyed. They are diligent servants and messengers of the brain," but as to how the intellectual functions act in relation to the brain, he confesses ignorance:

> I can in some degree follow the brain's functions in dissections of living animals, with sufficient probability and truth, but I am unable to understand how the brain can perform its office of imagining, meditating, thinking, and remembering, or, following various doctrines, however, you may wish to divide or enumerate the powers of the Reigning Soul.[22]

While Vesalius still remained within the framework of Galenic physiology, some of his observations threatened its foundations. In the first edition of *De corporis humani fabrica*, Vesalius expressed some doubt that blood with vital spirit was supposed to pass from the right ventricle of the heart through the interventricular septum to the left ventricle. He examined the septum and found no visible pits and supposed that invisible channels could account for such passage. In the second edition of *De fabrica*, he expressed much more skepticism as to the reality of such channels. Another change in his concepts was the picture he gave showing the *rete mirabilis* in his *Tabulae sex* in 1538, whereas in the *De fabrica* in 1543, he commented that it did not exist in the human; nevertheless, he still pictured it.[23]

THE NEW EXPERIMENTALIST VIEW OF THE WORLD

The landscape of science was undergoing rapid changes as new discoveries in the physical and chemical sciences continued to increase ever more in the sixteenth and seventeenth centuries. An influential spokesman for the new age of scientific discovery was *Sir Francis Bacon* (1560–1626). His enthusiasm for the usefulness of experimentation struck a responsive chord in many, freeing them from the confining feeling that Aristotle and Galen, and their medieval partisans, were the fount of all knowledge. That the wisdom of the ancients could not be bettered was replaced with the feeling that science

---

[22] Ibid., p. 4.
[23] It was suggested that he did this to pay homage to Galen (Clarke and Dewhurst 1972), legend to Figure 75 p. 58. However, given Vesalius' independence of mind, this could have been an error or an indication that he was not yet sure of the generality of his findings and was hedging his bets.

was now developing rapidly, offering great material benefits for mankind. It was the moderns against the ancients.[24]

A major figure in the development of the new sciences was the great physicist *Galileo Galilei* (1564–1642). In his modernization of physics, he replaced Aristotelian concepts, such as "bodies seeking their natural place," with simple mechanical concepts of force, weight, and acceleration to account in a lawful way for the behavior of bodies. The shift to an experimentally verifiable view of nature, with an emphasis on the use of mathematics, was accompanied by an analytic attitude – one that aimed to provide an explanation for phenomena with emphasis on clarity, simplicity, and elegance. This view gained increasing favor not only with physicists, but also with chemists, biologists, and medical scientists.

CIRCULATION OF THE BLOOD: THE OVERTHROW OF (MOST OF) GALENIC PHYSIOLOGY

Among those in the medical sciences imbued with the new experimental philosophy was *William Harvey* (1578–1657). His book on circulation, published in 1628 (Harvey, 1970), has been hailed as "the most momentous event in medical history since Galen's time."[25] Whereas Galen postulated that different spirits were carried by the blood vessels (in the veins to nourish the body's tissues and in the arteries to confer vitality on the tissues), Harvey convincingly demonstrated that the same blood circulates in both the arteries and veins, thereby cutting the ground from under Galenic physiology and calling for a new physiological paradigm to replace it.

Others before Harvey have been put forward as the discoverers of circulation. *Michael Servetus* (1511–1553), an anti-Arabist, had hypothesized a pulmonary circulation as early as 1553.[26] Galen had allowed that a small amount of blood is carried via the pulmonary vessels into the left heart, but that the largest portion passes to the left ventricle via pores in the interventricular septum. This was reversed by Servetus. He had the bulk of blood passing by the pulmonary vessels through the lungs into the left heart with only a small amount trickling through pores in the interventricular septum into the left ventricle in which:

> the vital spirit . . . composed of a very subtle blood nourished by the inspired air. The vital spirit has its origin in the left ventricle of the heart, and the lungs assist greatly in its generation. It is a rarefied spirit, elaborated by the force of heat, reddish-yellow (*flavo*) and of fiery potency, so that it is a kind of clear vapor from very pure blood, containing in itself the substance of water, air and

[24] (Jones, 1965).   [25] (Garrison, 1929), p. 248. See also (Bates, 1992).
[26] (Castiglioni, 1947). A draft was supposed to have appeared in 1546. See the translation by O'Malley of Servetius' book *Christianismi restitutio*, which contains his views on circulation (O'Malley, 1953).

fire. It is generated in the lungs from a mixture of inspired air with elaborated, subtle blood which the right ventricle of the heart communicates to the left. However, this communication is made not through the middle wall of the heart [interventricular septum], as is commonly believed, but by a very ingenious arrangement the subtle blood is urged forward by a long course through the lungs; it is elaborated by the lungs, becomes reddish-yellow and is poured from the pulmonary artery into the pulmonary vein. Then in the pulmonary vein it is mixed with inspired air and through expiration it is cleansed of it sooty vapors. Thus finally the whole mixture, suitably prepared for the production of the vital spirit, is drawn onward from the left ventricle of the heart by diastole.[27]

The emphasis Servetus placed on the vital spirit in blood was guided by his reading of scripture: "the divine spirit is itself the blood, or the sanguineous spirit. It is not ... principally in the walls of the heart, or in the body of the brain or of the liver, but in the blood, as is taught by God himself."[28] He related this passage of blood to the belief in three kinds of spirits:

that ... which is communicated through anastomoses from the arteries to the veins ... [via the lungs] which it is called the natural [spirit], its seat is in the liver and in the veins of the body. The second is the vital spirit of which the seat is in the heart and in the arteries of the body. The third is the animal spirit, a ray of light, as it were, of which the seat is in the brain and nerves of the body. In all of these there resides the energy of the one spirit and of the light of God.[29]

O'Malley characterizes Servetus as being more a theologian than a medical scientist. However, although it is true that he was fundamentally driven by his religious beliefs, nevertheless the position he advanced was firmly based on the physiological thinking current at the time. Servetus, ever the zealot, antagonized Calvin to the point where he was declared a heretic, and Calvin had both him and his books burned at the stake. Fortunately, a few copies escaped the conflagration, and it was later republished to come down to us.

Servetus' concept was that of the "lesser circulation." The "major circulation" carrying blood from the left heart through the body to return to the heart via the veins was conceptualized by *Andrea Cesalpino* (1519–1603). In 1593, he portrayed blood as continuously flowing from the heart via the arteries throughout the body, with its return to the heart by the vena cava. This led to his being taken in some quarters as the true discoverer of circulation.[30] Historians have concluded, however, that whereas Cesalpino and Servetus had advanced the idea of parts of the circulation, it was Harvey's experimental studies demonstrating the truth of it for which he is

---

[27] (O'Malley, 1953), p. 204. The change in color from sooty vapors to reddish-yellow we recognize as due to the oxidation of venous blood.
[28] Genesis 9, Leviticus 7, and Deuteronomy 12.
[29] (O'Malley, 1953), p. 204.   [30] (Arcieri, 1945).

given credit. Cohn put it succinctly: "Harvey's achievement was not in the discovery of circulation but in its demonstration beyond reasonable doubt: the result is science."[31]

It was while a student of medicine at Padua, where Galileo had taught mechanics and hydraulics, that Harvey imbibed the new physical and mathematical thinking of the time. His teacher, *Frabricius ab Aquapendente* (1537–1619), had demonstrated that the valves in the veins of the arms were so oriented as to direct the blood in the veins *toward* the heart and not *away* from it as was required by the Galenic paradigm. Harvey used this information and the idea that too much blood was flowing out into the arteries than could be accounted for by its rate of production in the liver, as was held to be the case in Galen's physiology. Harvey set out to prove that the blood pumped from the heart through the arteries was returned to the heart via the veins, thus constituting a continuous circulation containing both the greater circulation through the body and the lesser circulation through the lungs. He did this by measuring the blood flow on hydraulic and mechanical principles à la Galileo, as well as taking into account all the anatomical evidence bearing on the mechanical properties of the heart and vessels.

One gap still remained in Harvey's account, the connections between the arteries and veins at their terminations needed to provide for the passage of blood between them, but were too small to be seen with the naked eye. Nevertheless, overwhelming evidence for circulation advanced by Harvey demanded their existence. Using the then newly invented microscope, *Marcello Malpighi* (1628–1694) revealed the presence of small *capillaries* in the webbed foot of the frog. He saw blood passing from the arteries into their small arteriolar endings, then into the capillaries, and from them into venules and back into the veins to be carried to the heart.[32]

The view of the heart as a muscular pump led Harvey to address the problem of the action of muscles in general and the relation of nerves to them in a study that he planned to publish, but didn't. His manuscript notes for that book were recently found and translated by Whitteridge.[33] In them, Harvey is seen trying to rationalize Aristotle's description of the heart as the source of nerves with the bands in the heart; the chordae tendineae connecting the

[31] (Cohen, 1957). In his history of medicine, Castiglioni gives a spirited defense of his fellow countryman Cesalpino as being the true discoverer of circulation (Castiglioni, 1947), pp. 436–440. However, the translator and editor of his book, E. B. Krumbhaar, pointed out that despite the anticipation of Cesalpino and others of some aspect or other of the concept of a circulation, it was Harvey's experimental results that were instrumental in firmly establishing its reality.

[32] (Harvey, 1970), Leake refers to Malpighi's demonstration of the presence of capillaries in the foot of the frog foot and in the lung (Young, 1929), i.e., at the junctures of arterial and venous vessels in the major and minor circulations.

[33] (Harvey, 1959).

cusps of the valves to the papillary muscles in the ventricles were designated as nerves. This suggests the ancient confusion of nerves and tendons (Chapter 1). For Aristotle, the animal spirits located in the heart were the prime source of movement, the pivot around which the body revolved, much as the sun represents the center of the universe. Harvey wrote that, according to Aristotle, "the principal organ of movement is the motive spirit for the reason that it is capable of contracting and relaxing. Because of the spirit the power of pushing and pulling is conveyed to all parts of the body, it is the medium between the soul and the body. By the spirit is occasioned the beating of the heart."[34] Harvey referred sympathetically to a Dr. Flud, who most certainly was his friend *Robert Fludd* (1574–1637), a fellow physician and mystic, who drew a parallel between the mind of man with the light of the sun, which for him had supernatural properties. Witteridge could not find Harvey's precise reference, but she quotes a passage from Fludd's book that she thinks expresses the substance of the view with which Harvey sympathized,

> the purest portion of elementary matter is aire, the purest sublimity of the aire is the vitall form, in which is the mentall beam, and in it is the Word which is God. The soul in the aire is conducted into the body, where it operateth the effects of vivification, and internal multiplication of the species. We must imagine, that the aire which is drawn into our heart by inspiration, is full of the divine treasure of life, the which residing in the heart of man, sucketh and draweth his life into it, by a magnetick force and virtue. . . . The heart is the precious storehouse of the active treasure of life.[35]

Yet Harvey does not follow Aristotle completely, accepting the Galenic view that the brain is the ruling organ of the mind and the source of the nerves and not the heart. Quoting Fabricius, Harvey indicated that there is a place in his physiology for the special nature of the brain and its command of the body through the nerves of motion and sensation,

> that as the magnet attracts iron while it itself remains motionless, so nerve draws muscle towards itself and toward the brain and imparts motive power to every part of the muscle . . . and in the manner as light sheds radiance and the sun illumens the hemisphere, so nerve transmits the command from the brain. It is proved that if a nerve be injured, blocked or cut, movement is destroyed.[36]

With regard to sensation, Harvey noted that, "when one eye (of a cat) is shut, the pupil of the other eye is seen to be dilated [and he asks himself] as to

---

[34] (Harvey, 1970), p. 95.
[35] (Harvey, 1959). See note 2 on p. 94, where Witteridge refers to Fludd's Mosaical Philosophy, London, 1659. A figure attributed to Fludd shows a radiation from the sun (Deity), the intelligence of the world, onto the middle cell in the brain where intellect resides. (Clarke and Dewhurst 1972), Figure 57.
[36] (Harvey, 1959), p. 109.

whether spirit is conveyed from the brain through the nerves. Here we must consider whether the sensitive faculty is derived from the brain."[37] Animal spirits are their agency: "Nerves are only like pipes and not like reins or the sources of movement acting on the motive organs. This is proved from their weakness and lack of power, their smallness and fineness, their oblique insertion and because there are many in one." This passage is an argument against the alternate position of Galen, where motor action is proposed to take place by a mechanical pull. Harvey clearly supports the concept of a flow of animal spirits in the hollow nerves.

The persistence in Harvey's mind of some Aristotelian concepts somewhat blurs the view that Galenic physiology was completely overturned with the establishment of the circulation. Thomas Kuhn had proposed that one paradigm accounting for a science is overturned at a stroke by another when sufficient contrary evidence has been gathered.[38] The example Kuhn gave of such a revolution was the replacement of the Ptolemaic with the Copernican theory of the solar system. Although the discovery of the circulation is often termed a revolution, that part of Galenic physiology relating to a production of animal spirits in the brain that flows out in the nerves to activate the muscles and account for sensation remained a viable concept for Harvey and others for a long time afterward. How this view of animal spirits was later replaced as further understanding of the nature of nerve developed will be described in the following chapters.

[37] Ibid. This derives from the observation given by Galen (Chapter 2).
[38] (Kuhn, 1996).

5

NEW PHYSICAL AND CHEMICAL MODELS OF NERVE IN THE ENLIGHTENMENT

Great advances were made in physics after the studies of Galileo, as the findings of other scientists in the sixteenth century led to ever greater accretions of knowledge of physics and chemistry and awareness of the complexities of the nervous system. The latter part of the seventeenth and eighteenth centuries, the period known as the "Enlightenment"or the "Age of Reason," was characterized by a new critical examination of received teachings. This was the age in which superstition and dogmatic religion were rejected by the *Philosophes*, the popular philosophers of France.[1] The views on religion held by them and other educated intellectuals included those who believed in a providential God that continually acted on behalf of the affairs of man, the *theists*, and the *deists* – those who held that a God may have created the universe, but then left man and the world to run their course without Him. For the *atheist*, there was no reality at all to God and the soul of man was a material entity that operated by the inherent motion of its substance. This mechano-materialist view was forcibly presented by the English philosopher *Thomas Hobbes* (1588–1679). For him: "Thought is a form of motion of matter; my 'ideas' are vibrations in the matter of my brain and nerves."[2] In France, the atomism of the ancients was revived by the French philosopher

---

[1] But these were a small group, compared with the great mass of humanity for whom adherence to religion was strong and included a deep belief in the supernatural. Even the educated classes were not immune from belief in witches who were feared as real presences, even in the late seventeenth century. This is attested to by the Salem witch trials of 1692. The occult and sorcery had deep roots in medieval Europe, flowering in the sixteenth and seventeenth centuries, abating in the eighteenth century. It has not disappeared; the belief in the supernatural still has its adepts today.

[2] (Willey, 1952), p. 102. Willey pointed out that, for the Deists, God was replaced by Nature which played the role of a divinity.

Pierre Gassendi (1592–1655)[3] and used by leading figures in science to advance their theories. The English philosopher *John Locke* (1632–1704), in his seminal work *Essay Concerning Human Understanding*, viewed the mind as a *tabula rasa*, a blank sheet, on which the senses act, with all ideas coming from the senses and with no innate ideas, that of God or otherwise present at birth.[4] Such a naturalistic approach was anathema to the mainstream who believed that only a God-given soul could be the source of human judgment, consciousness, and ideation ruling over the body. A theory as to how that could come about was given by *Rene Descartes* (1596–1650). In his dualistic theory, there are two different substances – that of the soul and that of the body (brain and nerves). Nerves carry sensory information to the soul residing in the brain – in the pineal gland – by which the will of the soul could be exerted. This dualistic view enjoyed wide acceptance at the time and is still held by a considerable portion of those in religious orders and by some thinkers today.

DESCARTES' CONCEPTION OF THE MOVEMENT OF ANIMAL SPIRITS IN THE NERVES AND BRAIN

Descartes was famous for his philosophical and scientific writings and for the advances he made in mathematics. He took a mathematico-deductive approach to construct a theory in which the body acts as a machine controlled by the mind. It was an elaboration of the Galenic concept of animal spirits flowing in the nerve fibers that were directed by the soul in the pineal gland of the brain. His was a dualistic view in which "men who like us are composed of a soul and a body."[5] The soul, the rational cognitive thinking soul (*res cogitans*), expresses its actions through the pineal gland. Sensory inputs passing by means of animal spirits in nerve tubules project them onto the pineal gland. There, sensations are perceived and animal spirits are directed by the soul into the proper nerve tubules to effect muscle movements. In his dualistic concept two radically different substances – the immaterial thinking soul (the *res cogitans*) and the material of the body (brain) (the *res extensa*) – somehow interact: "The two natures of soul and body brain, would

---

[3] While Gassendi was an ordained priest, as an atomist he viewed the soul to be constituted of matter, albeit of finer atoms. He avoided the charge of being an Epicurean and materialist by avowing his Christian belief in the existence of God, who he said, was demonstrated by the harmony of nature and man's awareness of moral values and his power of reason. Tracts promoting the concept of a material soul that were considered to be sacriligious, was circulated clandestinely (Niderst, 1969).

[4] (Locke, 1959).

[5] (Descartes, 1972), p. 1. Hall suggested that Descartes' construction of "a man like us" as a machine was offered as a theoretical model in order to shield himself from Ecclesiastical opposition (Descartes, 1955).

have to be joined and united to constitute men resembling us." How two such different substances can interact is the archetypical difficulty of the dualistic position, the philosophical *mind-body problem* remaining with us today.[6] To show how the body can act through its nerves, Descartes drew on the mechanical amusements installed in the grottos of the King's gardens, where figures of humans and animals were made to move by water-powered contrivances:

> even to make them [the machines] play certain instruments or pronounce certain words according to the various arrangements of the tubes through which the water is conducted. And truly one can well compare the nerves of the machine that I am describing to the tubes of the mechanisms of these fountains. Its muscles and tendons to diverse other engines and springs which serve to move these mechanisms its animal spirits to the water which drives them of which the heart is the source [and] of the brain cavity the water main. Moreover, breathing and other such actions which are ordinary and natural to us and which depend on the flow of the spirits, are like the movements of a clock or mill which the ordinary flow of water can render continuous. External objects which really by their presence act on the organs of sense and by this means force them to move in several different ways depending on how the parts of the brain are arranged, are like strangers who entering some of the grottos of those fountains, unwittingly cause the movements that then occur, since they cannot enter without stepping on certain tiles so arranged that for example, if they approach a Diana bathing will cause her to hide in the reeds.[7]

Descartes held that all change and movement in the physical world are purely mechanical, with God giving the initial impetus to bring the physical system of the world as a whole into existence. Once set, it proceeded on its own according to mathematical laws. The movements of animals are those of inanimate automatons, pushed and pulled about by forces in accordance with natural laws. The essence of physical bodies is to occupy space. However, for man, the soul (the *res cogitans*, the thinking substance that occupies no space at all) can be affected by and can control the physical substance extended throughout the body, the *res extensa*.

NERVES COMPOSED OF SMALL HOLLOW TUBULES
Descartes, in his schematization, considered the control of the body to occur through animal spirits flowing in a large number of small hollow tubules of the nerves. Whether such tubules had been seen at the time he wrote is in question. Microscopy at the time was in a primitive state. Although

---

[6] Hall pointed out that Descartes, educated in the Jesuit tradition, was familiar with the problem of how it was possible that an immaterial soul could interact with the material body, a problem that was much debated in Patristic and Scholastic schools.

[7] Ibid., p. 22.

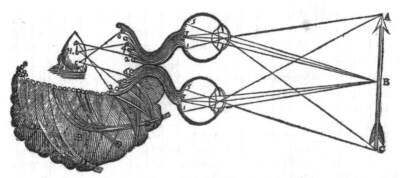

5.1. Light rays from an arrow (**A,B,C**) in the visual field are shown projected onto the retina of the eyes (**1,2,3**). There, animal spirits in nerve tubules carry the sensory impress to the ventricle with a representation of the image and project it onto the surface of the pineal gland (**H.a,b,c**). Blood vessels under the pineal gland give it extraordinarily great mobility, accounting for the quickness with which the soul can move the gland and direct animal spirits into other tubules. From (Descartes, 1972), Figure 29.

single lenses were used in spectacles since at least 1300, they were of low power.[8] The development of microscopes of increased power was promoted by the Academy of the Lynx, founded in Rome in 1601.[9] Galileo became a member in 1609 and is credited with devising the compound microscope in 1612.[10] This offered more magnification, but also increased the spherical and chromatic aberrations inherent in the lenses then used. It is of no little interest that Descartes is said to have ground lenses and to have recognized that a parabolic curvature was needed to overcome spherical aberration. In 1637, he published a diagram of a simple microscope.[11] It is possible that he could have had one constructed of sufficient magnification to enable him to observe tubules in the nerve. On the other hand, it is likely that his concept of tubules was advanced to suit the needs of his theoretical construct.[12]

Descartes gives a picture of how rays from an object in the visual field are focused onto the retina at the back of the eye, where open ends of nerve tubules are spread over its surface, to initiate a movement of animal spirits (Figure 5.1). Animal spirits from the tubules pass out into the inner surface in the ventricle of the brain (pictured as a single chamber) to impinge on the pineal gland. The pattern of animal spirits projected onto the surface of the pineal gland is there perceived as an object in the visual field.

[8] (Singer, 1914).
[9] The Society was named after the lynx, an animal that was thought to have excep-
tionally keen eyesight.
[10] (Singer et al., 1954), p. 232.   [11] (Singer, 1914), p. 250, Figure 1.
[12] (Descartes, 1972), note 9, p. 5.

THE PIVOTAL POSITION OF THE PINEAL GLAND IN THE CONTROL OF ANIMAL SPIRITS

An object is not only perceived by the pattern projected onto the pineal gland by the soul present in the gland, but the soul is also capable of moving the gland to effect an appropriate motor response in accordance with its will. The pineal gland is:

> composed of matter which is very soft and that it is not completely joined and united to the substance of the brain but only attached to certain little arteries whose membranes are rather lax and pliant, and that it is sustained as if in balance by the force of the blood which the heat of the heart pushes thither . . . [and] therefore very little [force] is required to cause it to incline and to lean, now more now less, now to this side now to that, and so to dispose the spirits that leave and make their way toward certain regions of the brain rather than toward others. . . . Just as a body attached only by threads and sustained in the air by the force of fumes leaving a furnace would incessantly float here and there as the different particles of the fumes acted differently against it, so the particles of the spirits that hold up and sustain this gland.[13]

The vessels supporting the pineal come from "those [blood] vessels which divide into a myriad of very small vessels . . . [that] carpet the cavities of the brain."[14] Thus, by its rapid movements, the pineal gland can readily respond to the overall sensory pattern received and through it the will (of the thinking substance), can direct the flow of animal spirits into nerves controlling various muscles so as to give rise to appropriate responses of the body.[15]

The animal spirits reflected from the pineal gland flow into open nerve tubules in the ventricles that lead down to the muscles of the arm. In the example given (Figure 5.2), it causes the arm to be brought up and the finger directed to a point in the visual field. The animal spirits that enter the muscles cause them to expand and shorten, thus pulling on the limbs to which they are attached and moving them in an appropriate manner.

---

[13] Ibid., p. 91 and Figures 31 and 32.

[14] Ibid., p. 80. These vessels are those of the choroid plexus lining the ventricles and presumed to act like the réte mirabile that Galen had described.

[15] Galen had previously mentioned the possibility that the pineal gland could be controlling the movement of animal spirits between the ventricles to modify higher functions, but discounted that possibility in favor of the cerebellar vermis playing that role, because he found the pineal gland to lay outside the ventricle. (Galen, 1968), Book I, Chapter 8, pp. 418–419. Fernel had also pointed to the pineal gland as playing this role. Rediscovery of the knowledge of the true location of the pineal gland outside the ventricle led to the rejection of the role it played in Descartes' model.

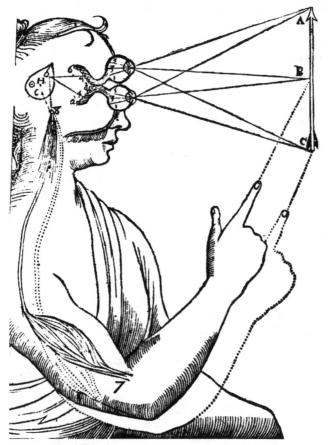

5.2. The flow of animal spirits from the optic nerve fibers onto the pineal is directed from it into the openings of the tubules leading to the appropriate muscles to bring the finger to a desired point in the visual field. Letters as in Figure 5.1 with, in addition 8 indicating tubules leading down to the arm muscle with its attachment 7. Finger is moved from **C** (dotted outline) to **B** under command of the will (see text). From (Clarke and Dewhurst, 1972), Figure 93.

Descartes likened animal spirits to a wind or a subtle flame:

> No power propels them but the inclination they possess to continue on their movement according to the laws of nature. They are mobile and subtle but they do not lack the strength to inflate and tighten the muscles in which they are enclosed just as the air in a ball hardens and stretches its skin.[16]

---

[16] Ibid., p. 28.

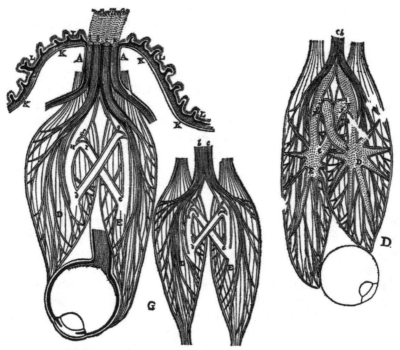

5.3. Reciprocal action of the rectus muscles of the eye moving it from side to side. Animal spirits flowing into one muscle of a pair are added by valve arrangements between the opposite muscle so that its animal spirits add to those contracting entering from the nerve to contract the muscle. Eye movements in the right (D droight) and left (G gauche) directions. Other letters indicate valve action for animal spirits (see text). From (Descartes, 1972), Figure 3.

RECIPROCAL ACTION OF MUSCLES

Not all the swelling of the muscles causing contractions is due directly to the entrance of animal spirits into them from the nerve. Descartes pictured valves or gates between a reciprocal pair of muscles (i.e., those having antagonistic actions on a limb) so that animal spirits can move from one of the muscles to the other. Reciprocal innervation is exemplified by the rectus muscles responsible for the side-to-side movement of the eye (Figure 5.3). One muscle of the pair (e.g., the lateral rectus), when contracted, pulls the eye outward in the lateral direction. As it does so, there is an accompanying relaxation of its reciprocal muscle, the medial rectus, that acts to pull the eye inward. A similar reciprocal action occurs when the medial rectus contracts, pulling the eye inward as the lateral rectus relaxes.[17] Descartes

---

[17] Ibid., pp. 24–28.

pictured valves in these reciprocally innervated muscles arranged so that, as animal spirits flow from the nerve into one muscle of the pair to inflate it, the valves in the nerve leading to the other muscle close off. Animal spirits are drained from it and added to further inflate the other muscle of the pair. In this way, a relatively small amount of animal spirit entering from the nerve can have a large effect. If the animal spirits flow only a little or not at all into the muscles, the two sets of valves remain ajar with, as a result, both muscles becoming lax. Only when animal spirits enter forcibly into one or the other of the set of reciprocally innervated muscle making for an imbalance does a reciprocal movement ensue.[18]

THE REFLEX

Descartes elaborated the theory of nerve tubules to account for reflexes in which motor control is directed back down to the region of the body from which sensory input arises to initiate the reflex. Very fine fibrils are pictured as present in the marrow of each tubule (Figure 5.4).[19] These were considered to extend from its peripheral end of the nerve tubules in the muscles or skin to their termination in the brain ventricle. Upon sensory activation of the fibrils, they pull open trap doors covering the ends of the nerve tubules in the ventricle. This allows animal spirits to flow out and project onto the pineal gland and, after reflection from the pineal, to move down within those same tubules to contract the muscles controlling the reflex action. The fine fibrils do not completely fill the tubules leaving enough room for animal spirits to easily flow around them. Descartes describes how this mechanism can:

> move its members in a thousand different ways . . . a) the filaments . . . [which] come from the innermost part of the brain and compose the marrow of the nerve are so arranged in every organ of sense that they can very easily be moved by the objects of that sense and that b) when they are moved, with however lit-tle force, they simultaneously pull the parts of the brain from which they come, and by this means open the entrances to certain pores in the internal surface of

[18] Our present understanding of reciprocal action of muscles is conceptually similar to Descartes', except that it is not effected at the peripheral level in the muscles, but in the central nervous system where interneurons with excitatory and inhibitory effects synapse on the motoneuron cell bodies that innervate the reciprocally innervated muscles (Eccles, 1964), and (Sherrington, 1947).

[19] The figure is one supplied by Schuyl, Figures vi and xiii in his Latin version of Descartes' manuscript that appeared in 1662 (Descartes and Schuyl, 1662). In the French version later published by Clerselier, in 1664 [(Descartes, 1664) and (Descartes, 1972)], this figure does not appear. Schuyl's figures were replaced by those of Gutschoven and La Forge who were commissioned by Clerselier to redraw the figures used earlier, (Descartes, 1972), p. xxiv. Descartes had supplied only two of the figures in his manuscript.

5.4. Nerve showing tubules labeled **b** to **l**, each containing a fine fibril running up within them. These serve to signal sensations (see Figure 5.5 legend). A sheath (**c**) encloses the tubules. From (Descartes and Schuyl, 1662), Figure vi.

the brain [allowing] c) the animal spirits in the cavity [to] begin immediately to make their way through these pores into the nerve, and so into muscles that give rise to movements in this machine quite similar [in their movements] to which we men are naturally incited when our senses are similarly impinged upon.[20]

Here, Descartes is using the mechanism of fibrils within the nerve tubules to account for reflex actions, – the quick, invariant, and often appropriate machine-like movements directed toward the source of the initiating stimulus made even before conscious awareness of it.[21] Descartes gives an example of a reflex (Figure 5.5) in which heat-producing pain is the stimulus that mechanically pulls on the threads to initiate the reflex:

If fire (A) is near foot (B), the particles of this fire (which move very quickly, as you know) have force enough to displace the area of skin that they touch; and thus pulling the little thread (cc) which you see to be attached there, they simultaneously open the entrance to the pore [or conduit] (de) where this thread

[20] (Descartes, 1972), pp. 33–34.
[21] Although some elements of the concept have appeared before Descartes, he is the originator of the reflex concept as we know it (Canguilhem, 1955).

5.5. Reflex action based on fibrils and flow of animal spirits within the same nerve tubules. The sensory stimulation of pain in the foot (**B**) by fire (**A**) is signaled via the fibrils (cf. Figure 5.4) within the tubules (**C**) leading to the brain. The fibrils open lids over the tubules allowing animal spirts to flow out and impinge on the pineal (**F**), where the perception of pain is realized by the soul. Animal spirits are directed down to those same tubules causing the muscles to withdraw the limb from the fire. From (Descartes, 1972), Figure 7.

terminates [in the brain]: just as, pulling on one end of a cord, one simultaneously rings a bell which hangs at the opposite end . . . now the entrance of the pore
or small conduit (de), being thus opened, the animal spirits from cavity (F) enter
them and are carried through it – part into the muscles that serve to withdraw the
foot from the fire, part into those that serve to turn the eyes and head to look at
it, and part into those that serve to advance the hands and bend the whole body
to protect it . . . the threads I speak of are very thin, yet they extend safely all the
way from the brain to parts [of the body without any breaks] . . . even though the
parts are bent in myriad ways: because [a] these threads are enclosed in the same
tubules that carry the animal spirits to the muscles and [b] these spirits, always
somewhat inflating the tubes, protect the fibers against crowding and keep them
always maximally taut all the way from the brain whence they arise to the places
where they terminate.[22]

Descartes elaborates on how his schematization can account for pain. A
sensory stimulus causing a movement or a pull with sufficient force to break
the fibrils "will cause the soul to experience a feeling of pain."[23] If the force is
not quite so great as to break the filaments, its movement will cause the soul
in the pineal gland to feel "a certain corporeal sensual pleasure referred to as

---

[22] Ibid., pp. 34–36, Figure 7.   [23] Ibid., pp. 37–38.

tingling which though very close to pain in its cause, is quite the opposite in its effect." The basis of this part of Descartes' theory is the idea that pain is evoked by an excessively strong stimulation of a sensory modality. Only in recent times has it been shown that there are specific pain fibers that exist as part of a nociceptive system with stimulation just below the pain threshold sensed by other fibers as itching and tingling.[24]

Although the concept of fibrils within the nerve tubules could account for how a reflex response could be directed to the appropriate muscles, his solution of a tug or pull on the fibrils within the nerve tubules comes at the price of physical inconsistency. For the fibrils to act this way, the nerve tubules cannot be lax. They would need to be taut or at least straight, which he recognizes not to be the case for nerves. They are seen as bent and have a wavy course. Nevertheless, the needs of his theoretical concept overrode the physical evidence.

THE BREAKDOWN OF DESCARTES' MODEL

The importance of theory over physical reality in Descartes' schematization is seen by his concept of a freely mobile pineal gland controlling the flow of animal spirits able to account for the rapidity of reflexes and willed movement. He had gained some knowledge of the anatomy of the brain, even by bringing brains from the butcher's stalls home for study. However, his view of the position of the pineal with respect to the brain ventricles was simply wrong. Possibly, he may have been led astray by the difficulties inherent in examining fresh brains, where the soft structures are readily disrupted. But, more likely, his anatomical views conformed to the demands of his theoretical concepts. Although there were distinguished anatomists at the time who were aware of the true position of the pineal gland outside the ventricle and in a fixed position in the brain, the Cartesians, particularly those in France, were a dominant philosophical and scientific group able to resist challenges to the views of the master. This situation changed in France when, at an important meeting held at the house of Thevenot[25] in Paris, *Nicolaus Steno* (1638–1686) gave a lecture on the brain to the leading scientists and philosophers of the day.[26] Steno clearly demonstrated that the pineal gland is firmly joined to the surrounding tissue outside the ventricles and that it is not delicately suspended on blood vessels as Descartes

[24] (Keele, 1957). For the modern concept of pain differentiated from pain-related sensations, see (De Reuck and Knight, 1966).

[25] At the time, France was under Louis the XIVth, the Sun King supreme in Europe who fostered the arts and sciences. It was at gatherings of scholarly circles in the houses of important figures, such as Thevenot's, that ideas were exchanged and the latest scientific results examined. Such groups were the nuclei of the Royal Society in France, and from similar groups meeting in England, the Royal Society.

[26] (Steno, 1965).

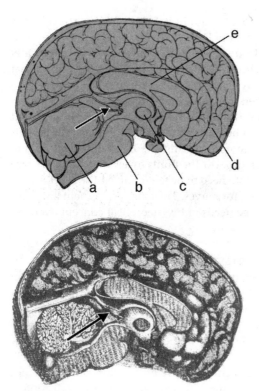

5.6. Figure of the brain in sagittal section given by Steno. (TOP) Arrow points to the pineal gland arising from adjoining brain structures that is clearly outside the ventricle. The ventricle figured is that of the 3rd, which has as its landmark the massa intermedia (**c**) joining the thalami of the two sides. (Compare with the massa intermediate shown in the sagittal section of the monkey brain in Figure 15.10.) The 3rd ventricle is further indicated by its extension down toward the pituitary below the brain. Other orienting landmarks are: (**a**) the cerebellum, (**b**) the brain stem, (**d**) the frontal pole of the cerebrum, and (**e**) the corpus callosum joining the two cerebral hemispheres. (Bottom) This figure, anatomically more detailed, also has an arrow pointing to the pineal gland clearly outside the ventricle. Modified from (Steno, 1965) and (Faller, 1968), Figure 11.

had envisioned (Figure 5.6).[27] Steno's lecture aroused great admiration, and his findings were quickly accepted. A letter written by an attendee at the lecture wrote:

> M. Stenson the Dane...has even forced the obstinate and dogmatic Cartesians to admit the error of their leader with regard to the gland of the brain

---

[27] Galen had already written that the pineal gland lay outside the ventricle and that it could have an action like the pylosis of the stomach, (Galen, 1968), Book I, Chapter 8, pp. 418–419.

and its function and this in the presence of the most highly respected people in the city whom he bombarded with the deductions of a calm, reasonable intelligence.[28]

WILLIS AND HIS CIRCLE

In England, *Thomas Willis* (1621–1675) and his colleagues[29] were making remarkable advances in studies of the structure of the brain of the human and lower animals.[30] They had also shown the pineal gland to lie outside the ventricle and firmly fixed to the brain substance.

Rather than the *rete mirabile*, which Galen had extrapolated from animal studies to the human, Willis depicted the interconnections of branches of the carotid vessels with other blood vessels at the base of the brain in the human, which is now referred to as the *circle of Willis*. This pattern of vascularization had been described earlier by *Johann Jakob Wepfer* (1620–1695). In a dispute with the Galenist *Riolan*, who believed in the presence of the *rete mirabile*, Wepfer instead viewed the internal carotid arteries entering the skull to divide into many communicating branches, but not to form a rete in man. As noted previously (Chapter 4), Vesalius had already observed that the rete is found in some animals, but not in man. With respect to its function, Wepfer declared that, from the communicating vessels at the base of the brain, the vital spirit was carried by other vessels into the brain ventricles, where the animal spirits are distilled. He stated that "each carotid artery proceeds intact and not divided along the side of the sphenoid bone and pituitary gland... [and that it] perforates the dura intact... to provide the different parts of the brain with nourishment."[31]

Willis also regarded these vessels as playing a role in a process analogous to distillation. The animal spirits extracted from them are led into the

---

[28] (Steno, 1965), p. 70. Part of the intransigence of the Descartians had to do with national politics, the opposition of the French to the English, and vice versa. Descartes had become the exemplar of French brilliance in science and philosophy, and an attack on him was felt to be an attack on France.

[29] Among those who assisted Willis in his studies of brain were *Richard Lower* (1631–1691), who performed many of the dissections and *Christopher Wren* (1632–1723), of later fame as the chief architect in the rebuilding of St. Paul's cathedral. It was Wren who drew the illustrations of Willis' book on the brain (Willis, 1965). Another colleague of future great fame for his discoveries of the properties of gases was *Robert Boyle* (1627–1691). He immersed brains in wine spirit (alcohol) as a fixing agent to overcome the difficulties and erroneous interpretations inherent in working with fragile fresh tissue.

[30] (Willis, 1965a) The tercentennial publication of the English edition of Willis' book was edited by Feindel in a handsome production worthy of the importance of Willis' contribution to neurology.

[31] (Tieleman, 1996).

cerebral cortex, where the spirits are further distilled in the gray matter and
stored:

> There is a notable provision made in all the carotids about the base of the skull:
> because their crooked inbowing and branching into infoldings, hinders the too
> great and rapid approach of the blood. Then, lest the passage of it should be shut
> up, the mutual ingrafting of all the vessels on either side, do help or provide
> for . . . the business of extracting the animal spirits is performed even as a chemical
> elixir . . . exposed to distillation . . . The fourfold chariot of the arteries . . . brings the
> matter to be distilled . . . to the cortical substance out of which the spirits are dis-
> tilled . . . through the serpentine channels of an alembic [a still] is made extremely
> subtle.[32]

Man with his erect head was thought to receive the purer volatile blood
after it passes through the crooked course of the vessels at the base of the
brain in the circle of Willis, where extraction of the vital spirits begins, much
as in a chemical *elixir* (retort). From these vessels, others arise to carry the
spirits through the whole circumference of the brain. Out of these vessels,
the animal spirits are distilled in the gray matter that was described as the
universal cortical or "shelly substance" of the brain. Although the animal
spirits are generated and stored in the gray substance of the brain, the white
matter (the corpus callosum) and the streaked regions (the striate bodies)
constitute tracts in which the animal spirits are distributed.

Willis reassigned the functions that had classically been associated with
the ventricles in the old cell theory to certain brain regions: perception
to the *corpora striatum*, imagination the *corpus callosum*, memory the *gyri*
of the cerebral hemispheres, and instinct the *midbrain*. The ventricles were
relegated to the role of draining off the effete matter from the brain. The
excretion of ventricle contents to appear as nasal mucous fluid is an old concept that can be traced to Hippocrates. Vesalius had described mucous fluids
passing down to the base of the middle ventricle guided by the downward
taper of the cleft-like form of that part of the ventricle, the *infundibulum*,
the term itself meaning "funnel."[33] The mucous fluid was then supposed to
pass into the pituitary gland (the term for which is derived from "mucus")
and from the pituitary through small foraminae at the base of the skull into
the nasal passages. The similarity in consistency of mucoid nasal secretions
to the substance of the brain may very well have encouraged this concept.
However, Willis' associate Lower made a special investigation of this theory
and showed that nasal mucous discharges had no relation to the brain, that
it was *sui generis*, a secretion of the pituitary.[34]

[32] (Willis, 1971b), p. 54.   [33] (Singer, 1952a) p. 52.
[34] (Hunter and MacAlpine, 1963a).

WILLIS' CONTRAST OF THE SOUL OF BRUTES WITH THE RATIONAL SOUL OF MAN

In opposition to the localization of the soul in the pineal gland, as Descartes had imagined, Willis stated that, "there is no median [intermediation] between the body and the soul but that the members and parts of the body are the organs of the soul and therefore are the distinct portions of the same extended soul [that] actuate the several members and parts of the body."[35] In support of his view that the soul is extended throughout the body rather than being located in the pineal body, Willis referred to his observation that – when worms, eels, and vipers are cut into pieces – each piece will show movement. This indicated to him that the soul, which is responsible for the faculties of motion and sense, is present in each of the cut portions. They were seen to move for a time, and each piece when pricked "will wrinkle up themselves together." Aristotle had earlier observed that certain insects go on living when divided into segments, writing that "plants and many animals continue to live even when divided, and seem to retain in these fragments a soul specifically the same as before."[36] St. Augustine also had observed the same phenomenon (Chapter 3), but he felt that he could not come to a conclusion as to what this meant other than that the soul is not contained in any one place in space.[37]

Willis viewed the soul as a material substance, most thin, as it were, the ghost or the shadow of the body, and of a subtle active fiery nature.[38] But this fire was not taken to be that seen when burning fuel, such as wood, but something like it – "an heap of most subtle contiguous particle, and exiting in a swift motion." Engendered in a furnace or fireplace (the body) is being supplied constantly by food, "a sulfur or some other nitrogenous thing in the air . . . out of the food . . . the particles being most minutely resolved, and agitated with a most rapid motion, [results] in fire and flame."[39] The corporeal soul shut up in the body agitates its thick bulk, actuating all its members and arteries, in some animals "with wonderful agility." For Willis, as for Descartes, the soul of brutes is corporeal and passive, and does not move unless moved by other bodies – actions, as it were, of an artificial machine. Cognition is caused by something of another nature, one not granted to brutes; otherwise, Willis notes, one must yield conscience, deliberation, and a knowledge of universal and spiritual things to them. Man has, in addition, a *rational soul* that is not composed of material particles and is located in the brain. By its means, man has the powers of intellect, judgment, discourse, and other acts of reason, far excelling any faculty or science of the brute and the powers of the corporeal soul.[40] The rational

---

[35] (Willis, 1971b), p. 5, second paragraph. [36] (Aristotle, 1964).
[37] (St. Augustine, 1964), pp. 90–91. [38] (Willis, 1971b), p. 6. [39] Ibid., pp. 5–6.
[40] Ibid., p. 38.

soul is immaterial, allowing for its speculation on the nature of humanity, rationality, temperance, fortitude, spirituality, whiteness, and the like –
speculation carried higher to God, angels, infinity, eternity, and other notions far remote from sense and imagination. This "argues that the substance or nature of the rational soul is immaterial and immortal because of
this aptness or disposition [for such higher conceptual thought]."

This view of an immaterial and immortal rational soul bears on the vex-
ing mind-body problem; where and how can the immaterial substance of
the mind interact with the material machinery of the brain? Willis places
the interaction in the middle part of the brain, where fantasy and imagination are placed. There, sensory input, which is governed by the corporeal
soul that we share with animals, is received where the rational soul can
exert its control over the animal soul. "The sensible species are received
impressed on it by the organs of sense apprehended according to exterior
appearances." However, these impressions "can be fallacious and deception
is possible – the sun appears no bigger than a bushel. The horizon of the
heaven and sea appear to meet. There does not appear to be any antipode,
the images in a glass, [or] an echo is felt to be a . . . voice coming from some
other place," and so on. The superior capability of the spiritual soul allows
it to correct such fallacious impressions by using notions of congruity or
incongruity and the disposition of a series of notions and speculations that
offer a restraint on fantasy, enabling the spiritual soul to correct false impressions of sense. Science is a product or creature of the human mind,
and Willis argues that the work of men, if not divine, at least has a particle
of the divine breath – a spiritual substance that is wonderfully intelligent,
immaterial and, therefore, for the future, immortal.[41]

Invoking the ancient view of a macrocosm and microcosm (Chapter 1),
the immortal soul is said to come from the soul of the world and be infused
into the body, where it is attached to and presides over the fantasy part
of the animal soul. The will, which is attributed to the rational soul, can
thus gain control over the emotions "lust and wrath," which are relegated
to the irrational animal soul of brutes. The concept advanced is that the
rational soul enjoys serenity, in contrast to the perturbations belonging to
the irrational soul below it. All, however, may not always be serene. The
struggle between the rational soul and the corporeal lower soul can develop
to the point in which "the lower soul growing weary of the yoke of the
other, frees itself from its bonds . . . This kind of intestine strife, does not
truly cease, till this or that champion becoming superior leads the other
away clearly captive."[42]

---

[41] Ibid., p. 40.   [42] Ibid., p. 43.

THE MOVEMENT OF ANIMAL SPIRITS IN NERVE

In respect of the means by which animal spirits carry out their functions in the body, Willis conceives of a dual mechanism of their movement in nerve fibers:

> We have already shewed that the animal spirits are procreated only in the brain and cerebral, from which they continually springing forth, inspire and fill full the medullary trunk: (like the chest of a musical organ, which receives the wind to be blown into all the pipes) but those spirits being carried from thence into the nerves, as many pipes hanging to the same, blow them up and actuate them with a full influence; then what flow over or abound from the nerves, enter the fibres dispersed every where in the membranes, muscles, and other parts, and so impart to those bodies, in which the nervous fibres are interwoven, a motive and sensitive or feeling force...
>
> Indeed the animal spirits flowing within the nerves with a living spring, like rivers from a perpetual fountain, do not stagnate or stand still: but sliding forth with a continual course, are ever supplied and kept full with a new influence from the fountain. In the meantime, the spirits in the rest of the nervous system, especially those abounding in the membranes and musculous stock, are like ponds and lakes of waters lately diffused from the chanels of rivers whose waters standing still are not much moved of their own accord; but being agitated by things cast into them, or by the blast of winds, conceive divers sorts of fluctuations.[43]

This passage relates to the concept of spirits moving in hollow nerves, but then Willis invokes another, mechanical principle of nerve action. The nerves are said to contain "little bodies" by which the nerve impulse is transmitted. The animal spirits flowing from the medullary substance into the nerves are activated much as beams of light rays agitate the ethereal little bodies existing in the pores of the air. "In the fibers of the nerves, are as it were rays diffused from the light...stirred up by them into motion, [to] perform the acts both of the sensitive and locomotive faculty."[44] He elsewhere speaks of when "the nervous parts are struck, or receive a motion from this or that kind of matter, the impulse is transferred by continued undulation or '*wavering*' into their respective sensory or motor parts."

How the structure of the nerve can support this kind of movement is discussed:

> The passages of the nerves are not bored through as the veins and arteries; for the substance of those [nerves] are not only impervious to any bodkin [stiletto or sharp needle], but no cavity can be seen in them, no not by the help of spectacles or a microscope. As to what belongs to the smelling little pipes, they seem to be so made, not for the passage of the animal spirits, but that some serosities [thin watery constituents] might slide down that way: but the spirits themselves are

---

[43] (Willis, 1971a), p. 105.   [44] Ibid., p. 106.

carried in the sides, not in the cavity of either pipe; but the substance of the other nerves appears plainly firm and compacted, that the subtil humor, which is the vehicle of the spirits, may pass through their frames or substances, even as the spirits of wine, the extended strings of a lute, only by creeping leisurely through. Hence it may be argued, that because the animal spirits require no manifest cavity within the nerves for their expansion; neither is there need of the like for them within the substance of the brain.[45]

Thus, the fibers being solid and "without cavities," the spirits are carried along their sides with another fluid subjoined to it, a "watry latex" in the pores in the nerves. But these pores are the spaces between the solid fiber capillaries:

The nerves themselves (as may be discovered by the help of a microcosm [microscope] or perspective-glass) are furnished throughout with pores and passages, as it were so many little holes in a honey-comb, thickly set, made hollow, and contiguous one by another; so the tube-like substance of them, like an Indian cane, is every where porous and pervious. Within these little spaces the animal spirits or very subtle little bodies, and of their own nature ever in a readiness for motion, do gently flow; to which is joyned, both for a vehicle, as also for a bridle or stay, a watry latex, and that is itself of very subtle parts. This humor diffuses with its fluidity the spirits through the whole nervous system; also by its viscosity retains them, that they be not wholly dissipated, but as it were in a certain *systasis* [harmony of the two] ... otherwise the spirits would vanish away into the air. ... This nervous juyce being derived from the brain and cerebrel into the medullar appendix, is carried from thence by a gentle sliding down through the nerves ... and waters its whole system. ... As occasion is offered, the same spirits, as a breath moving upon the waters ... for as in a river, from winds or any thing cast in, divers undulations or wavings are stirred up; so the animal spirits being raised up by objects for the performing the offices of sense and motion, do tend this way or that way to and fro within the nervous stock, and are agitated hither and thither by other means.[46]

Willis thus invokes two different mechanisms for nerve action. One is the propagation in the solid capillaries of "undulations" or "waverings" to transmit rapid sensory and motor impulses. (This concept will be carried out in more detail by Newton, as will be described later.) The other is of a fluid movement on the sides of the "little pipes" of nerves, actually the spaces between the solid capillaries that are slower and described as a sliding down – a leisurely creeping movement. This seems to be an attempt by Willis to account for the longer lasting influence of the nerve on muscle, indicated by the atrophy of muscle seen to follow its denervation (Chapter 9).

[45] Ibid., p. 107.   [46] Ibid., pp. 107–108.

THE PRODUCTION OF MUSCLE CONTRACTIONS

Whereas Descartes viewed muscle contraction as being brought about by its inflation with animal spirits, a more complex process was hypothesized by *William Croone* (1633–1684). He viewed a "subtle, active and highly volatile liquor of the nerves" flowing via hollow nerve fibers into muscle, interacting there with blood to produce a kind of fermentation causing the muscle to swell.[47] Croone rejected the idea that the animal spirits of nerve are an ethereal "wind." He viewed animal spirits as a "subtle, active and highly volatile liquor of the nerves, in the same way as we speak of spirit of wine or salt or others of this kind."[48] The central thought is that there is an interaction of the spirits – the "liquor" or "juice" supplied by the nerve to the muscle – which meeting with blood from the circulation in the cavity of the muscle produces an effervescence that swells the muscle, thus causing it to shorten and pull on its tendon to move the bones of a limb. Croone added that "the impulse was transmitted along the nerve in the same way as vibrations along the tightened string of a musical instrument"[49] – a complicated theory that included vibrations, as did Willis' theory.

That muscle contraction is the result of a chemical process, a fermentation or an ebullition, was a concept earlier advanced by *John Mayow* (1640–1679). He theorized that a *nitro-aërial* spirit, or matter consisting of fine particles, is present in air.[50] On inspiration, it is carried by the blood to the brain where, "The structure of the brain seems to be such as to render it specially fit for separating the nitro-aërial spirit from the blood."[51] The nitro-aërial spirits are identified as the animal spirits that are "elaborated in the brain … disseminated from that source to the spinal marrow and to the nerves originating in it." In the muscle, it combines with a *sulphureous* material (the fuel) that supports combustion and muscle contraction. If an animal is decapitated, "the influx of animal spirits into the spinal marrow is altogether shut off, so that the parts of the decapitated body at once collapse and are deprived of animal motion."[52] That life depends on the nitro-aërial particles in the air was shown experimentally:

> If the trachea is obstructed, and respiration is suppressed, or if the motion of the heart and of the blood stops, or even if the brain is disordered, the nitro-aërial particles will not be transmitted to the brain for the preparation of animal

[47] (Kardel, 1994), p. 8. (Borelli, 1989).
[48] (Wilson, 1961), p. 160. The analogy of spirit of wine (alcohol) to animal spirits had been made by van Helmont and Sylvius and others (for references, see note 11 in Wilson's paper).
[49] Ibid., p. 161.
[50] (Mayow, 1907). Fourth Treatise, pp. 230–240. Mayow's nitro-aërial particles or spirits in the air are functionally equivalent to oxygen, anticipating the identification of oxygen by *John Priestley* (1733–1804) and *Antoine Lavoisier* (1775–1780).
[51] Ibid., p. 250.   [52] Ibid., p. 254. See Chapter 14.

spirits, and therefore the animal will speedily die. From these things I conclude that ... the nitro-aërial particles, transmitted by means of respiration to the mass of blood and thence to the brain, are the animal spirits themselves. And this is in accordance with the fact that animals placed in a glass vessel from which the air exhausted by means of Boyle's air pump, will perish miserably in convulsions. For the animal spirits being deprived, by the removal of air, enter, as is their wont, upon disorderly movements, and rushing tumultuously into the nervous system, excite convulsive movements, and at last the animal dies for want of air, and of spirits.[53]

Mayow held that whereas the nitro-aërial particles are the animal spirits, they are not the sensitive soul that is "something much different from the animal spirits ... the sensitive soul consists of a special subtle and ethereal material while the nitro-aërial particles, i.e., the animal spirits, are its chief instrument ... as to the sensitive soul, I can form no other notion about it than that it is some sort of divine *aura*, endowed with sense from the first creation and coextensive with the whole world."[54]

The mechanism by which the nitro-aërial spirits are extracted from the blood by the brain was considered by Willis to be caused by an action of the thicker meninges (dura mater) surrounding the brain "which undergoes a sort of pulsation, and that by its contraction the blood driven to the brain is compressed: and that thus the nitro-aërial particles are pressed out of the mass of the blood, and driven into the brain."[55] The dura mater is so designed that "by its diverse motion various effects are produced in the bodies of animals ... for according as the membrane contracts itself more strongly or more weakly, the nitro-aërial particles, i.e., the animal spirits, are driven in greater or in less abundance into the brain and thence into the nervous system."

PLETHYSMOGRAPHIC EVIDENCE AGAINST AN INFLATION OF MUSCLE CAUSING CONTRACTION

To determine if a volume change by an influx of animal spirits occurs in a frog muscle on stimulating its nerve, *Jan Swammerdam* (1637–1680) developed a plethysmograph in which a frog's muscle with its nerve attached was placed in the closed chamber (Figure 5.7).[56] The movement of the drop of fluid placed in a thin tube at the top of the chamber would show small volume changes of the muscle. When the nerve was excited to cause muscle contractions, no increase in muscle volume could be seen. This was taken

---

[53] Ibid., pp. 254–255.   [54] Ibid., p. 259.   [55] Ibid., p. 260.

[56] Swammerdam is given credit for introducing the nerve-muscle preparation of the frog, typically the gastrocnemius muscle with a length of the sciatic nerve that innervates it remaining. This hardy preparation, able to survive on atmospheric oxygen, has served generations of physiologists well in their investigations of the properties of muscle and nerve (Bastholm, 1950).

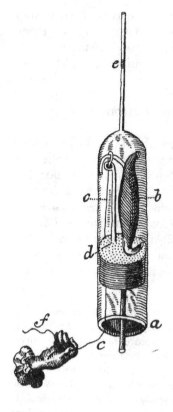

5.7. Swammerdam's plethysmographic method to show lack of swelling of muscle on activation of its nerve. A drop of water in the thin tube (e) registers volume changes in the vessel (a) containing the muscle (b). On pulling the thread (c), it compresses the nerve through the loop of (d) to excite it and cause the muscle (b) to contract. No movement of the water drop indicative of a volume increase was seen in the thin tube. From Swammerdam (Swammerdam, 1758) as shown by Wilson (Wilson, 1961), Plate 13.

as evidence against a volume of nerve fluid passing from the nerve into the muscle causing it to swell and shorten, nor for that matter to an increase in muscle volume by an ebullition produced by gas formation. Swammerdam's work was printed in Dutch and was not widely known until a century later when, through the efforts of Boerhaave, it was translated into Latin and published in 1758.[57] Steno, a close friend of Swammerdam, was aware of his experiments at the time and used plethysmography to measure volume changes in the human arm. After a willed contraction of the arm muscles, no increase of muscle volume was seen. The results of these experiments, and similar ones by others, argued against the entry of a nerve fluid into the muscle or an evolution of gas to cause an increase in the volume of the muscle.

It should be noted that Descartes had envisioned the means whereby only a small amount of nerve fluid or spirit need enter a muscle to cause its contraction, with the bulk of the swelling coming about by a valvular

[57] (Swammerdam, 1758).

relation between pairs of muscle antagonists so that nerve fluid could shift from the inactive to the active muscle of the pair (Figure 5.3). This hypothetical valvular mechanism, however, would not take place when a single muscle was caused to contract, as in the case of Swammerdam's plethysmographic experiments on the isolated frog muscle.

BORELLI'S MECHANICAL ACCOUNT OF NERVE ACTION
AND MUSCLE CONTRACTION

Giovanni Alphonso Borelli (1608–1679), eminent iatrophysicist and member of the Italian Academia del Cimento (experimental academy), applied the then new laws of statistics and mechanics to show how muscles could move the limbs of the body by a system of levers.[58] He assumed the structure of muscle to be composed of a series of rhomboids that change in shape to a rectangle on activation, accounting thereby for the shortening of muscle without an overall increase in volume. He called these small muscle components *machinulae*.[59] He further analogized muscle to a sponge. When its empty pores are filled with fluid, it becomes hardened without an increase in volume. The muscle on activation shows

> a pure contraction of its length. The extremities are brought closer together by moving in opposite directions with swelling and turgescence in width. Such action cannot be conceived without a wedging action since these spirits are driven into the pores of the muscles like iron wedges [along their length] and they contract the fibres with so much force that the muscles can raise heavy weights. Such an operation, however, requires firstly wedges, secondly hardness of these wedges, thirdly a motive force to drive them in very violently . . . I do not see whence so much "spirituous gas" would come and in what cavity these spirits would be hidden. If, however, they are assumed to be present, they will have no hardness since they are looser than sponge, wool or cotton.[60]

Borelli considers muscle contraction to come about by a process

> like a fermentation or an ebullition, [which] produces this instantaneous contraction of the muscles. Such [is observed in] chemistry . . . acid spirits mixed with fixed salts suddenly boil by fermentation. Consequently, in muscles some similar mixing may occur. This entails sudden fermentation and ebullition which fill and expand the pores of the muscles, resulting in turgescence and contraction.[61]

This does not cause an overall increase of volume, but acts to fill in the spaces to make muscle turgid. Borelli set out to "investigate what is sent through the nerves, with what force and how it is pushed, and through which canals." While "the nerve appears as fasciculate, or capillary, composed of several thin fibres bound by some membranous envelope," he asks,

[58] (Borelli, 1989), and see (Bastholm, 1950).
[59] Ibid., p. 218.   [60] Ibid., p. 222.   [61] Ibid., p. 232.

rhetorically, "who can see if they are hollow or full and solid?" But he notes that the ducts in fleas or small worms are exceedingly small and yet are able to transmit blood within them so that the nerve tubules can be considered to pass fluid in them:

> We can more readily admit that they are tubules full of some spongy and wet substance similar to the marrow of elder, giant fennel or sugar cane. This is confirmed by the fact that nervous fibres are soft, flexible, slippery and always wet. They accept and absorb wet nutrient and they secrete some juice.... The spongy cavities of these nervous fibres are conceived as being always soaked and filled up to turgescence by some juice or spirit transmitted from the brain. As in a bowel [a length of intestine] full of water and closed at both ends, [an] impulse at an extremity [when] compressed and slightly percussed is instantly transmitted to the other extremity of the bowel. The adjacent elements of the liquid are aligned in a row. By pushing and shaking each other, they transmit the movement to the extremity of the bowel. Similarly, as a result of some slight compression, jolt or irritation at the origins of the canals of the nervous fibres which are in the brain itself, these fibres thus shaken and activated must secrete some drops of this juice which swells their internal spongy substance, into the fleshy mass of the muscles.[62]

Borelli surmises that the motive and sensitive spirits carried in the nerve fibers differ in nature from the juice that is used for its nutrition. The difference in their rapidity of action shows that they cannot be carried through the same channels. The spongy and porous nerve fibers are filled with *succus nerveus*, the rapidly acting agent, whereas the spaces between the fibers contain another fluid, *succus nerveus nutritivus*, which provides for their nutrition.[63] This nutritive nervous juice "seems to be different from the motive and sensitive spirits." The latter, which are concerned with action, "are most noble, bitter, sulphurous, saline and very active, like spirit of wine. The former is very sweet, inducing quiet sleep rather than dissolution and exhaustion of strength." The passage of nutritive juice in the spaces between the nerve fibers is comparable with the movement of liquids in the interstices of bundles of thin glass threads that can "absorb liquids and raise them like a sponge."[64]

He describes how the same mechanism that underlies muscle activation can also serve to carry sensory information to the brain:

> When the extremities of the sensitive nerves which end in the skin, tongue, nostrils, ears or eyes, are slightly compressed, stricken or tickled, the shaking, waving and tickling of the spirituous juice contained inside the tubules must be

---

[62] Ibid., p. 233.
[63] (Bastholm, 1950). The conception of two different movements in nerve – a rapid one for conduction of sensation and motion, and a slower one for nutrition – is similar to that given by Willis.
[64] (Borelli, 1989), p. 369.

immediately transmitted over the length of the nerve to the well-determined area of the brain to which the nervous fibres are attached. There the sensitive faculty of the mind can judge the cause of the movement from the place attained in the brain, from the violence of the blow and the characteristics of the motion.[65]

This "mechanical concussion" causes the movement of fluid in the nerve tubules and he asks:

How can two opposite shaking movements occur together at the same time in the same nerve? The tongue can move and feel at the same time....I think that this difficulty can be explained in two ways. Firstly, the two opposite shakings do not occur in the same fibrous canals but through distinct canals so that the fibres which transmit the movement of the order of the will do not accept the movement of pain [or other sensation] which is transmitted through other canals. Secondly and more likely, a wave of the juice contained in the tubes cannot be conceived without reciprocal movement, pushing to and fro towards the two opposite extremities of the nerve alternately as happens in shivering. Actually, these opposite shakings do not occur simultaneously but at different times which cannot be distinguished and are thus obscured by their brevity and frequency.[66]

Borelli considers how the fluids in the nerve fibers relate to the animal spirits. The juices in the fibers terminating in the brain act there on the animal spirits, such interaction occurring because they are both physical bodies. The nerve juices

however spirituous and active, are always corporeal and cannot act at a distance, and cannot, without physical contact, increase, intensify, or depress the animal spirits; it is by means of their corporeal presence that they either increase the animal spirits which are also corporeal, mixing themselves with them, or expel them, or transform them.[67]

MALPIGHI AND HIS GLANDS IN THE BRAIN

Marcello Malpighi (1628–1694), the great pioneer microscopist who had revealed the capillaries of the vascular system and demonstrated the histological structure of glandular tissue in the spleen, liver and kidney, concluded that the brain is a glandular organ as well. He described cells in the cerebral cortex and considered them to elaborate a fluid, *succus nervosa*, which then moves down within the tubular fibers of the white matter of the brain (Figure 5.8).[68] Could Malpighi have been the earliest forerunner of the neuron doctrine? To answer this question, it is necessary to examine his findings

---

[65] Ibid., pp. 366–367.   [66] Ibid., p. 367.   [67] Ibid., p. 370.
[68] This figure did not accompany the publication of his communication to the Royal Society; it was later provided by Biddlo.

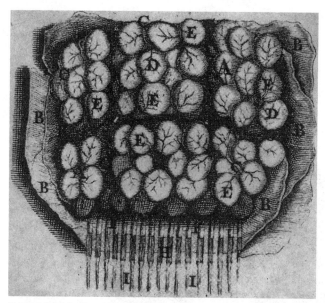

5.8. Malpighi's view of brain glands (**A**, **D**, and **E**) in the cortex secreting nerve fluid into the fibers (**I**) of the white matter. **B** shows dura mater reflected from the cortex and **C** its upper margin. From (Belloni, 1968), Figure 16.

critically. First, he found the fibers of the white matter of the brain to be similar to the fibers he had seen in the peripheral nerve:

As regards the cerebral white matter or corpus callosum, it has been described as firmer than the cortex and irrigated from all sides by arteries and veins which occupy the intermediate spaces. This, however, is wrong, as I have been able to observe by the microscope, certainly in the brains of fishes, though less clearly in other higher animals. It is quite evident that this whole white portion of the brain is divided into small, almost round fibres, which are not dissimilar to those white bodies or intestinulae which make up the tissue of the testicle. These are so conspicuous in the fishes' cerebral ventricles, that, if inspected in frontal illumination, they present themselves like an ivory comb or like the pipes of which church organs usually consist ... [and] the fibrous bodies ... finally end in the much reduced threads or prolongations in a gyrus and are immersed or implanted, like large roots of plants into the cortex which seems to serve as their bed or soil. . . . They do not easily catch the eye, because of tortuous cavities of the ventricles, and of the involved substance of the cortex in red-blooded higher animals; and because of their very thinness, frequency and friability.[69]

Like peripheral nerve, the fibers in the brain when cut show a similar emission of a fluid resembling egg albumin and like egg albumin is coagulated by

[69] (Meyer, 1967). See also (Swedenborg, 1918).

heat. This fluid, he inferred, is what is supplied by the cells in the cerebrum to the fibers connected to them.

Malpighi found the "glandular cells" in the cerebral cortex (Figure 5.8) by boiling the brains of mammals and fish to fix them and then brushing their surface with India ink. Viewed under the microscope, he saw what looked like glandular cells in the outer gray (*cinereus*) brain substance. However, the cells that Malpighi so identified were far too large to be identified with our nerve cell bodies. In what they called an exercise in "practical history," Clarke and Bearn repeated Malgiphi's procedure of fixation by boiling and using a microscope similar to one that would have been available to him, saw that the glandular cells were clearly artifacts related to the pattern of vascularization of the cortical surface.[70]

Yet, in other observations in which Malpighi examined sections of the cortex, his description suggests that he could indeed have seen nerve cell bodies:

> Through repeated *sections* [my italics] and long toil I...found in the brain of the higher red-blooded animals that the cortex is swamped with the most minute glands of which a multitude can be observed. They are found in the cerebral gyri which are spread out [on the surface] like intestines [gyri], and in which end white roots of fibres or, rather from which the latter take their origin. They are arranged in relation to each other in such a way that the external surface of the brain [cortex] is formed by their mass. They have an oval shape which is exposed to the mutual pressure of adjacent structures: hence the somewhat blunt angles and the almost equal intermediate spaces [between them].[71]

The deviation in the form of the cells from that of a round shape that Malpighi described, their "oval shape and blunt angles" with almost equal spaces between them, is significant. This is just what we would see in a section passing vertically down through the cortex containing the large pyramidal cells of the cortex. This impression is reinforced by his description of the relation of the cells to their fibers:

> Their [cortical] external portion is in contact with the pia mater and its blood vessels, which penetrate from above into their substance; the interior part gives off a white nervous fibre, like a special vessel.... Out of the joining and bundling together of these multiple fibres, emerges the cerebral white matter.... The cerebral glands of this kind are seen with difficulty in the fresh state, because, in the large brain of higher animals, they are destroyed by the removal of the pia mater; and their contours...because of their softness – are, thus, not easily

---

[70] (Clarke and Bearn, 1968). I have also repeated these experiments with similar findings. The regions of swellings taken by Malpighi to be cells are little regions isolated by surface vessels, the swellings arising most likely by differences in heat transfer between the vessels and the cortical substance when boiling the brain.

[71] (Meyer, 1967), p. 188.

distinguished. They are seen with greater ease after boiling the brain; this, *while it increases their size* [my italics], renders the interstitial spaces more patent....An intimate connection and continuation between these cortical glands and nerve fibres is observed after boiling the brain.[72]

This passage indicates the likely sequence of his thinking. He could have seen cortical cells, but with difficulty when taking sections of fragile fresh brain. On resorting to boiling to fix the tissue, the "glands" so revealed, though larger, were similar to other gland cells he had observed in sections of brain and other organs. The inference then made by Malpighi was that boiling the brain causes the cortical gland cells to simply increase in size rather than that the cells seen in sections and the "cells" he saw on the surface of boiled cortices were different entities. Meyer concluded from Malpighi's descriptions that he had seen not only the large pyramidal Betz cells in the cortex, but also cells elsewhere in the nervous system that he visualized; in the cerebellum, basal ganglia, throughout the brain stem, in the periventricular gray and probably also the spinal cord, at sites where the cells are seen intermingled with fibers. The definitive identification of cell bodies in the nervous system was made later in the first half of the nineteenth century by Purkinje and Valentin on the basis of the nucleus they found in the cells. This opened the struggle carried throughout much of that century as to whether cells and fibers were independent entities, or if fibers emerge from the cells as integral appendages, the view later established as the neuron doctrine (Chapter 8).

NERVE CONDUCTION BY ETHEREAL VIBRATIONS

Isaac Newton (1642–1727), the preeminent physicist of his day, whose opinions were given great weight by his peers, had proposed an *ether* to be present everywhere throughout space. By the ether, he accounted for optical phenomena and a wide range of mechanical properties. Among the phenomena to which he applied his theory was that of nerve action, viewing it as caused by the vibrations of the ether in nerves that are composed of solid filaments. He asks rhetorically if in nerve there is not

a certain most subtle spirit [the ether] which pervades and lies hid in all gross bodies by the force and action of which...all sensations [are] excited and the members of animal bodies move at the command of the will, namely, by the vibrations of this spirit mutually propagated along the solid filament of the nerve...from the outward organs of sense to the brain and from the brain into the muscles.[73]

---

[72] Ibid., p. 189.
[73] (Newton, 1952). In the Scholium at the end of Book 3, last paragraph.

He gives as an example light rays exciting vision asking:

> do not the rays of light falling upon the bottom of the eye excite vibrations in the
> tunica retina? Which vibrations, being propagated along solid fibers of the optic
> nerves into the brain cause the sense of seeing ... [and further in query 14] may
> not the harmony and discord of color arise from the proportions of the vibrations
> propagated through the fibers of the optic nerve into the brain as the harmony
> and discord of sounds arise from the proportions of the vibrations of the air? For
> some colors as they are viewed together, are agreeable to one another, as those of
> gold and indigo, and others disagree.[74]

That visual sensation can follow a mechanical stimulus applied to the eye
as is shown:

> When a man in the dark presses either corner of his eyes and his finger and turns
> his eye away from his finger he will see a circle of colors like those in the feather of
> a peacock's tail. If his eye and finger remain quiet these colors vanish in a second
> minute of time, but if the finger be moved with a quavering motion they appear
> again. Do not these colors arise from such motions excited in the bottom of the eye
> by the pressure and motion of the finger, as, at other times they are excited there
> by light for causing vision? And do not the motions once excited continue about
> a second of time before they cease? And when a man by a stroke upon his eye sees
> a flash of light, are not the light motions excited in the retina by the stroke?[75]

He makes an analogy of the outward propagated circles seen in a pool of
water after a stone is dropped into it to the vibrations he postulates as taking
place in nerve fibers and asks:

> Is not animal motion performed by the vibrations of this media excited in the brain
> by the power of the will and propagated from thence through the solid, pellucid
> and uniform capillamenta [fibers] of the nerves [and from them] into the muscles,
> for contracting and dilating them? I suppose that the capillamenta of the nerves are
> each of them solid and uniform, that the vibrating motion of the ethereal medium
> may be propagated along them from one end to the other uniformly, and without
> interruption [except] for obstructions in the nerves [which] create palsies.[76]

His concept of a vibration of the ethereal medium in nerve as the basis
of vision was generalized to account for the other senses. Like vision from
vibrations, hearing is evoked by vibrations, either of the ether or of some
other medium in the auditory nerves as: "by the tremors of the air are
propagated through the solid pellucid and uniform capillamenta of those

---

[74] Ibid., queries 12 and 13, pp. 345–346.   [75] Ibid., query 16, p. 347.
[76] Ibid., query 24, pp. 353–354.

nerves into the place [in the brain] of sensation and [this is] so of the other cases [of the various sensory inputs]."[77]

NERVE VIBRATIONS; A MUSICAL INTERLUDE

Giorgio Baglivi (1668–1707) was impressed by the curious affliction observed in the natives of Apulia (Puglia), at the "heel," of the Italian peninsula. Balglivi saw patients there who had fallen into a state of deep melancholy and had become lethargic as a result, they said, of the bite of the *tarantula* spider. Death was believed to be the likely outcome unless a certain kind of lively music was played – that which we know as the *tarantella*. This was performed by groups of itinerant musicians called on to provide the musical therapy. As they struck up the music, the fingers, arms, and legs of the afflicted individual began to twitch and then they perform a frenetic dance that could last for hours, or in some cases, with a midday break, all day until, sweating profusely, the dancer would fall exhausted. This could be repeated for days until a cure was pronounced. Once contracted, the disease could recur year after year, presumably because the bite produces a venom that lies dormant only to reappear again at that particular time of the year when the disease is rampant. Baglivi advanced a theory to account for the affliction and the therapeutic value of the music.[78] He proposed that the poison of the tarantula spider causes a coagulation of humors and spirits within the nerves with, as a result, the symptoms of lethargy and melancholy. The efficacy of the tarantella music to effect a cure is due to the ability of the particular vibrations of that lively music to "dissolve the coagulum in the nerve fibers" and restore the nerve juices to their normal fluidity.

To account for the restriction of the disease to Apulia, when the tarantula spider is also found in other parts of Italy, the peculiar hot and swampy regions of Apulia were invoked as a necessary adjunct in causing the disease. This region was in ancient times part of Greater Greece colonized by settlers from Greece, including the brotherhood of the Pythagoreans, a mystical society in which music and ecstatic practices figured.[79] Orgiastic practices, mostly carried out by women, were endemic in the area. When in medieval times the church tried to suppress the practice, it was later resumed under the guise of a disease caused by the bite of the tarantula spider, with the frenetic dancing and the particular music necessarily accompanying the dance claimed to be therapeutic. Hecker considered tarantism to be an hysterical phenomenon related to the mysticism current at the time in medieval Europe – a dancing mania seen along with others, such as St. Vitus's dance.[80]

---

[77] Ibid., queries 23 and 24, p. 522. See queries 11–15, pp. 518–519 and the quote from the mathematical principles, p. 372.

[78] (Baglivi, 1710), and (Marchand and Hoff, 1993).   [79] (Sigerist, 1948).

[80] (Hecker, 1884), pp. 107–133.

Although the bite of the tarantula was focused on, its occurrence without bites had been observed since the Middle Ages with its annual reappearances; "wherever the merry notes of the Tarantella resounded . . . inquisitive females joined the throng and caught the disease, not indeed from the bite of the spider, but from the mental poison which they eagerly received through the eye. The musical cure of the *Tarantati* gradually became established as a regular festival of the populace, one which was anticipated with impatient delight."[81] This became a season of dancing and music that was called "the woman's little carnival," for it was the women who conducted the arrangements, including the paying of the musicians.[82] Hecker, in an appendix to his chapter on Tarantism, gives the lively music for the *Dance of the Tarantati* that had been written down by Kircher.[83] The social and psychological motivations for the syndrome have been more recently described.[84] A much watered down version of the dance has turned into a tourist attraction, one characterized as a "lively flirtatious folk dance . . . full of grace and originality . . . teasing flirtation between partners . . . with measured steps," a far cry from the frenzy of the earlier form of the dance with its orgiastic affiliations.[85]

[81] Ibid., pp. 117–118.   [82] Ibid., p. 122.   [83] Ibid., pp. 167–174.
[84] (Russell, 1979) and see (Carmichael, 1997) for a more medically oriented view.
[85] In a tourist brochure of 1980, a water painting by Gaetano Dura made in 1840 portrays the dance with a man and woman curtseying and bowing. Part of his picture is given in the *Encyclopedia Brittanica*, (Brittanica staff, 1974), Volume 9, p. 821.

6

NEW SYSTEMATIZATIONS OF NERVE FUNCTION IN THE ENLIGHTENMENT

With the overturn of the old physiology underpinning Galenic medicine, new physiological foundations for medicine were searched for. Systems based on new physical principles proposed by Galileo and other physicists, and those on chemistry following its transformation from alchemy, were advanced: on physical principles by the *iatrophysicists* and on chemical principles by the *iatrochemists*. Both groups proposed agents to replace the ancient concept of animal spirits, but they essentially represented only a change of name. Alongside the nervous system by which sensations were perceived and motor nerves innervating muscles expressed the will, an involuntary nervous system was recognized – one by which the various bodily organs and its movements were carried out independently of the will (autonomously). Since Galen, the intercostal nerve chains were known, but were thought to be an offshoot of the vagus nerve originating from the brain, and it was so figured by Vesalius. These chains were then recognized as not connected directly to the brain, having its origin in neural connectives from the spinal cord. How the ganglia associated with this system, the intercostal chains and the ganglia found elsewhere with the involuntary nerves in the abdomen, their relation to the voluntary nervous system, and mode of action, became a matter of inquiry.

BOERHAAVE'S SYSTEM OF NERVE FIBERS
One of the most eminent clinicians of his day who incorporated the developments in physics and chemistry into a new system of physiology and medicine was *Hermann Boerhaave* (1668–1738), whose influential six-volume compendium, *Institutiones medicae*,[1] was aimed at eliminating metaphysics from medicine and medical science. He considered the body to be

---

[1] (Boerhaave, 1743).

made up of indivisible and invisible elements in the form of fibers.[2] These joined together form membranes, and membranes in turn join to form blood vessels. As the blood is pumped through the larger vessels and then through successively smaller vessels, the blood becomes separated into cells and fluid, with only the serous and lymphatic fluids left in the smallest vessels. It was these he designated "nerves."[3] These nerves were not those we recognize as such today, but a theoretical construct advanced to account for the specific properties of the various body tissues at a time before their cellular nature was understood and selective uptake mechanisms into the cells became known.

The cerebral cortex was considered to consist of a network of such vessels from which an extremely subtle, highly mobile juice is separated and, passing out into the nerve fibers, to be distributed throughout the body by the nervous system. This view of the cortex as the site of extraction of nerve fluid from blood vessels was based on the studies of *Fredrick Ruysch* (1638–1731). He had developed a technique of injecting colored waxes mixed with other materials into the bloodstream to delineate the finest vasculature of organs in animals and recently deceased humans.[4] The cerebral cortex was shown to be particularly rich in its vasculature, and Ruysch was led to the view that the cortex was not a glandular organ secreting nerve fluid into the tubules of white matter as Malpighi had proposed, but that the cortex consists of fine blood vessels that debouch directly into the nerve tubules of the white matter.[5] Boerhaave was greatly influenced by Ruysch's view, and incorporated it into his system.[6]

In Boerhaave's system, the vasculature acts as a succession of filters, the smallest being the finest and simplest vessels that he called "nerve," the vessel that is the "ultimate and most delicate," which can only pass the smallest particles. Those less than a fraction of the blood cell separated the yellowish serous component of the blood from the cells, which then flowed in smaller vessels as a colorless lymph.[7] It is this colorless lymph carried in the fibers of the white matter of the brain and peripheral nerves that mediates

---

[2] The history of the concept of fibers as the elementary constituent of the body structures presented by Berg (1942) describes other concepts than the one given by Boerhaave.

[3] (von Haller, 1966), pp. xxviii–xxxi.

[4] Ruysch's injection preparations were unsurpassed and evoke admiration even today. The material he used was a secret, the *arcanum Ruyscii*, which only after his death was found to consist of wax, talcum, and cinnabar (Schulte and Endtze, 1977), p. 21.

[5] Ibid., p. 23.

[6] Ruysch and Boerhaave were close friends. Ruysch would spend his vacations with Boerhaave (Bayle and Thillaye, 1855), p. 149.

[7] (von Haller, 1966), p. xxvi. It is of interest that Galen had used the term "exanguiis," without blood, to characterize and differentiate nerves from arteries and veins.

sensation and muscular activity.[8] By the time the small nerve tubules are reached, the force has decreased and the nerve fluid moving in the nerves slows to become a relatively gentle stream.[9] This slow flow has superimposed on it a faster flow – one with a very high velocity; a "moving force" derived from the brain when the mind wills a voluntary muscle action.[10] The slow movement of nerve fluid accounts for the late development of muscle atrophy after transection of its nerve supply (Chapter 9).

At the time, pulsations of the dura mater covering the brain were known to occur. They were viewed by Pacchioni and others as a *vis a tergo*, a force from behind, which propels the nerve fluid into the fibers. The driving force proposed was due to a rhythmic contraction of muscles present in the dura. This view was one generally ascribed to at the time, but Ruysch and Boerhaave were convinced that there was no musculature in the dura, that the pulsations seen were hemodynamic in origin.[11] Yet, dural movements as a driving force of nerve fluid still had its adherents, as will be recounted later.

Although Boerhaave, as a representative of enlightenment thought, presented a naturalistic view of man and animals as acting in accordance with natural laws (with an emphasis on physics and chemistry), his philosophical conception of how the mind acts on the body was not that of a materialist, nor was it that of the dualism of Descartes. He espoused the philosophical position known as *occasionalism*, the view developed by the priest and philosopher *Malebranche*. What one called causes were "occasions" in which God acts to produce effects. The human condition for Boerhaave is one that ultimately is dependent on God.[12] Boerhaave's close associates and followers did not follow him in this. They had widely divergent views of his philosophy, as well as proposing different views of nerve and the action of the nervous system.

VON HALLER'S PHYSIOLOGY OF NERVE

Albrecht von Haller (1708–1777), a pupil and later a colleague of Boerhaave's, was the most prominent of his group.[13] He annotated the six-volume opus of Boerhaave's *Institutiones medicae* – a massive compendium of the medical science of the time – and supplied critical notes where he thought corrections or additions were needed. He also oversaw its publication. Von Haller wrote what has been called the first modern textbook of physiology, *Primae Lineae Physiologiae (First Lines of Physiology)*.[14] Later, he published his own massive and authoritative summarization in eight volumes of what was

---

[8] (Jackson, 1970), pp. 307–310.   [9] Ibid., pp. 315–316.
[10] Ibid., Volume II, p. 317.   [11] (Schulte and Endtze, 1977), p. 20.
[12] (Boerhaave, 1959), p. 410.   [13] (von Haller, 1966), See King's Introduction.
[14] (von Haller, 1966).

known of physiology up to his time, *Elementa physiologiae corporis humani*: The fourth volume deals entirely with nerve, the central nervous system, and muscle.[15] Von Haller was not just an expositor of physiology, he also carried out an extensive series of experiments on the nervous system that he incorporated into his books. In the section on nerve in his *First Lines of Physiology*, he pointed to the individuality of the senses when he wrote:

> Every nerve, therefore, that is irritated by any cause, produces a sharp sense of pain. But we must reckon the mind to be changed, when any change happens to the body. It is the medullary part of the nerve which feels the pain. If the nerve was endued with any peculiar sense, that sense perishes when the nerve is compressed or dissected: the senses of the whole body are lost by a compression of the brain; and of those parts whose nerves originate below the seat of pressure, if you compress the spinal marrow. If certain parts of the brain are compressed from which particular nerves arise, then those senses only are lost which depend on the nerves, [such] as the sight or hearing. Those parts of the body which receive many nerves, as the eyes and penis, have the most acute sensation; those have least sensibility which receives few nerves, as the viscera; and those which have no nerves, as the dura mater, tendons, ligaments, secundines [the afterbirth], broad bones, and cartilages, have no sensation.[16]

Von Haller drew a distinction between the *sensible* and *irritable* parts of the body based on a large series of experiments performed by himself and his pupils. Those parts supplied with nerves possessed "sensibility," whereas "irritability" is a property inherent in muscle fibers – one that depends on the different properties of the two tissues.[17] The term irritability was later replaced by the modern term *contractility* with the same meaning. As pointed out by Temkin, the separation of sensitivity from contractility was one of von Haller's enduring contributions to neural science.

With respect to the hypothesis that impulses are transmitted in nerve fibers by vibrations when struck, von Haller gave strong arguments against that view:

> But here a controversy begins concerning the nature of this fibril, which with others of the like kind composes the substance of the medulla and of the nerves. That this is a mere solid thread... asserted by many of the moderns... when it is struck by a sensible body, a vibration is excited, which is then conveyed to the brain.[18] ... [However], the phenomena of wounded nerves will not allow us to imagine the nervous fibers to be solid. For if an irritated nerve is shaken (and that happens after the manner of an elastic cord, which trembles when it is taken hold of), the nerve ought to be made of hard fibers and tied by their extremities to hard bodies: they ought also to be tense for neither soft cords, nor such as

---

[15] (von Haller, 1762).   [16] (von Haller, 1966), Section 365.
[17] (von Haller, 1936). This thesis was delivered to the Royal Society of Sciences of Goettingen and published in two papers.
[18] (von Haller, 1966), Section 375. A reference to Newton's concept.

are not tense, or such as are not well fastened, are ever observed to tremulate [vibrate]. But all the nerves, at their origin, are medullary, and very soft, and exceedingly far from any kind of tension.... Finally, that the nerves are destitute of all elasticity, is demonstrated by experiments, in which the nerves cut in two neither shorten nor draw back their divided ends to the solid parts; but are rather more elongated by their laxity, and expel their contained medulla in form of a protuberance.[19] Again, the extreme softness of the medulla in the brain, with all the phenomena of pain and convulsion, leave not room to suspect any sort of tension concerned in the effects or operations produced by the nerves.[20]

If a vibratory movement cannot serve as the agent of nerve action, von Haller felt that one is left with the flow of nerve fluid in hollow nerve fibers:

The only probable supposition that remains is that there is a liquor sent from the brain, which, descending from thence through the nerves, flows out to all the extreme parts of the body; the motion of which liquor, quickened by irritation, operates only according to the direction to which it flows through the nerve; so that convulsions cannot thereby ascend upwards, because of the resistance made by the fresh afflux [efflux] of the fluid from the brain. But the same liquid being put in motion in an organ of sense, can carry that sensation upward to the brain; seeing it is resisted by no sensitive torrent coming from the brain in a contrary direction.[21]

It is therefore probable, that the nervous fibers and the medullary ones to the brain, which have the same nature are hollow. Nor is the objection which arises from the smallness of these tubes, not visible by any microscope,[22] of any force against the proposed arguments; to which add the absence of a swelling in a tied nerve.... If they are tubes, it is very probable that they have their humours from the arteries of the brain.[23]

Von Haller discussed the nature of the nerve fluid, conceiving it to be

a watery and albuminous nature [that] is common to most of the juices in the human body, and may be therefore readily granted to the juice of the nerves; like the water which exhales into the ventricles of the brain from the same vessels; also, from the example of a gelatinous or lymphatic juice, which flows out in cutting through the brain in fish, and the nerves of larger animals; to which add, the tumour [the swelling just above] ... tied nerves. But are these properties sufficient to explain the wonderful force of convulsed nerves, observable in the dissections of living animals, and even in the lesser insects, with the great strength shown by mad and hysterical people?

The nervous liquor then, which is the instrument of sense and motion, must be exceedingly moveable, so as to carry the impressions of sense, or commands

---

[19] This could refer to the axoplasm seen to be extruded from the cut ends of nerve fibers (Chapter 8).

[20] Ibid., Section 376.   [21] Ibid., Section 377.

[22] It is curious that von Haller apparently was unaware that Leeuwenhoek had described hollow fibers in nerve (Chapter 8), a finding that would have added force to his view.

[23] Ibid., (von Haller, 1966), Section 378.

of the will, to the places of their destination, without any remarkable delay: nor can it receive its motions only from the heart. Moreover, it is very thin and invisible, and destitute of all taste and smell; yet reparable from the aliments. It is carefully to be distinguished from the visible, viscid liquor exhaling from the vessels in the intervals between the nervous cords. That this liquor moves through a spongy solid, we are persuaded from its celerity, and the analogy of the whole body; of which all the liquids, the fat excepted, run through their proper vessels.

Therefore, upon the whole, it seems to be certain, that, from the vessels of the cortex, a liquor is separated into the hollow pipes of the medulla, which are continued with the small tubes of the nerves, even to their soft pulpy extremities, so as to be the cause both of sense and motion. But there will be a twofold motion in that humour; the one slow and constant, from the heart; the other not continual, but exceedingly swift, which is excited either by sense or any other cause of motion arising in the brain.[24]

In von Haller's model, the same nerve fibers preside over both sense and motion. To counter the objection that sensation can sometimes remain after motion is destroyed – this indicated by cases of "dying people who can hear and see but are incapable of motion" – he proposes that this is caused by a quantitative difference, that more strength is required for motion so that it can be lost while sensation remains. With regard to the fate of nerve fluid, he says: "If it be asked, What becomes of this nervous juice, it may be answered, It exhales probably through the cutaneous nerves; the lassitude both with respect to sense and motion, which may be overcome by spirituous medicines, shows that this liquid may be both lost and repaired." This statement suggests a similarity in the properties of spiritous medicine (alcohol) and that of nerve fluid.

LA METTRIE'S MATERIALISTIC ACCOUNT OF THE BRAIN AND MIND

While in his philosophical position, von Haller, unlike Boerhaave, was a dualist; another student of Boerhaave, who translated the *Institutions* and other writings of the master into French, *Julien Offray de La Mettrie* (1709–1751) was a materialist. While serving as an army physician, suffering from a fever, he had an epiphany. It appeared to him that the mind is an expression of the organization and function of the brain, the theme he presented in a book published in 1748, *Man a Machine*.[25] The attack on the theological metaphysics of the time caused a sensation and led to his ostracism as a

---

[24] Ibid., (von Haller, 1966), Section 378.

[25] (La Mettrie, 1912). This book followed an earlier publication with the title *The Natural History of the Soul*, which expressed this concept arousing the ire of theologians. What is known of La Mettrie's life is given in the eulogy to him delivered by Frederick the Great, published along with this book on pp. 3–9.

materialist and atheist.[26] In opposition to the dualist position taken by Descartes that there exists two substances, mind and matter (Chapter 5), for La Mettrie there was only matter. But matter for him is not inert, it has the capacity of movement and sensation, with thought, the higher powers of the mind, being a further, higher organization of matter:

> The soul is therefore but an empty word, of which no one has any idea, and which an enlightened man should use only to signify the part in us that thinks. Given the least principle of motion, animated bodies will have all that is necessary for moving, feeling, thinking, repenting, or in a word for conducting themselves in the physical realm, and in the moral realm which depends on it.[27]

La Mettrie looked to the brain as the source of all our feelings, pleasures, passions, and thoughts: "for the brain has its muscles for thinking, as the legs have muscles for walking."[28] He lists a number of instances that show actions not dependent on mind: reflexes such as the pupil contracting when exposed to daylight, the regularity of the heart beat remaining in pieces cut from it, and the ability of cut pieces of polyps to each regenerate into an individual whole animal.[29]

La Mettrie's thought was based on Atomism, a restoration of the Atomism of classical times, the position that Lucretius expressed in his book *De rerum natura* (*The Nature of Things*).[30] As did Lucretius, La Mettrie desired to free men from the burdens of religious fear.[31] In La Mettrie's time, religious thought was highly conservative and mostly dualistic: the soul having a real existence with its substance fundamentally different from that of the body. The monism and Epicurianism, if not the atheism of La Mettrie, was not only rejected and denounced by the larger society, but also by many of the *philosophes* for being too radical.[32] La Mettrie's position does not appear too exceptional in our time for neurobiologists to express the view

---

[26] (Wellman, 1992). A large share of the animosity of physicians came from those whom he had satirized. He nevertheless did find favor by a number of prominent personages, among them Frederick the Great, for his humanitarianism and philosophical writings. Wellman has pointed out that his philosophical writings were motivated by the desire to alleviate human suffering (p. 5).

[27] (La Mettrie, 1912), p. 128.   [28] Ibid., p. 132.

[29] The regeneration of cut pieces of polyps into whole animals was the surprising discovery of Trembly (Lenhoff and Lenhoff, 1986). It had implications for how the soul is related to the body, opening up such questions as: Does each piece of the regenerated polyp have the same or a different soul? Is the soul divisible? (Dawson, 1987). This was a problem touched on by Aristotle (Chapter 1) and St. Augustine (Chapter 2) following upon their observations of cut worms and insects.

[30] (Lucretius, 1937), and (Chapter 1).   [31] (Campbell, 1967).

[32] (Wellman, 1992). The Epicurianism that La Mettrie espoused was not the Hedonism with which it is commonly confused. Its moral injunctions were as strong as those of the Stoics.

that it is the brain's neuronal organization that is responsible for the higher functions of mind.[33] Some have even taken the logical leap in supposing that a sufficiently complex computer could have a mind of its own.

La Mettrie was not alone as an avowed atheist in the eighteenth century. *Paul-Henry Thiry Baron d'Holbach* (1723–1789) was a strong antagonist to organized religion in general and the Catholic church in particular.[34] His was the foremost example in the Enlightenment of one who advanced a direct and unequivocal defense of atheistic materialism. The soul for him was an illusion. His views were widely spread through the influential meetings held at his house with his wide circle of eminent friends in the sciences, government, literature, the arts,[35] and his many writings, which included his contributions to the *Encyclopedia* of *Diderot*, helped prepare the ground for the French revolution.

FORCES ACTING ON THE BRAIN CAUSING MOVEMENT OF NERVE FLUID

But not all thinkers during the Enlightenment were atheists or Deists. One who was very much a devoted religious was *Emanuel Swedenborg* (1688–1772), a polymath who wrote on the whole range of the science of his day.[36] His profession was in the field of mining, but he studied a wide variety of sciences, in particular the nervous system in which he undertook some limited anatomical study of the brain on his own. For the most part, he collected and systematized the work of others. He was deeply interested in the nervous system, his motivation being the hope that his studies would lead him to a better understanding of the soul.[37]

---

[33] (Churchland and Sejnowski, 1992). Some of the different positions taken on the mind-body problem will be referred to in Chapter 16.

[34] (Vartanian, 1967), and see (Willey, 1940).

[35] Among them were Diderot, Helvetius, d'Alembert, Rousseau, Condillac, and Condorcet, and contacts with those outside of France; David Hume, Edward Gibbon, Adam Smith, Joseph Priestly, David Garrick, Lawrence Sterne, Beccaria, and Benjamin Franklin. These men held enlightened views that were strong in educated circles in America. It should be noted that the first three American presidents were Deists.

[36] He published voluminously, his books dealing with the animal, vegetable, and mineral kingdoms. In the third volume of his work on the *Economy of the Animal Kingdom* (Swedenborg, 1918), he dealt at length with the nerve fiber. He treated the central nervous system in a two-volume compilation: *On the Brain* (Swedenborg, 1883). See also (Swedenborg, 1899).

[37] (Toksvig, 1948) and (Sigstedt, 1952). Swedenborg was raised in a highly religious family environment. His father was a prominent member of the Swedish clergy, a professor of theology who became a bishop. Later in his life, Swedenborg had a spiritualistic revelation in which he saw visions of departed souls with whom he had conversed. He put aside his scientific studies to enter into an intensive study of biblical exegesis and other religious works that he published on extensively.

In his early book, *On Tremulation*,[38] Swedenborg dealt in a general way with the physics of vibrations (tremulations) and its presence in the body. In his later writings he used the concept to consider that the pulsations of the heart transmitted to the brain can act there as the *vis à tergo*, the motive power from behind, propelling the animal spirits out of the cortex into the nerve fibers supplying the body. He later decided that the respiratory movements transmitted to the brain act as the motive force:

> In the cortical substance, this [force] is affected [effected] by means of the expansion and constriction of the glands [cortical cells of Malgighi], that is, by means of the animation or alternate respirations of the cortical brain. That it is the cortical and cineritious [gray] substance that is expanded and contracted, or that is the origin of the brain's animation, [is evident]; for each of its spherules [cells of the cortex] is like a little heart prefixed to its fiber, not unlike the great heart of the body prefixed to its arteries. And since there are as many origins of motion as there are spherules of this substance, therefore, when these are expanded, the whole mass of the conglobated viscus, namely, its surface and blood vessels and its whole inter medulla, is contracted, and vice versa....In the medullary and nerve substance likewise; for each fiber, like an artery, drives its fluid into the parts that follow, and like the cortex it has its alternation of expansion and constriction.[39]

This passage refers to the work of Malpighi, who had described what he saw as glandular cell bodies in the cortex supplying nerve fluid to the fibers (Chapter 5). With respect to the nerve fluid or humor, Swedenborg wrote that it "courses through them . . . that it may excite the motor forces and convey to the mind the modes, forces, and forms of sensation."[40] Swedenborg drew a strong association between respiration and animation; the lungs breathe synchronously with the brain so that the spirit may be more expeditiously promoted through the fibers. The cortex of the cerebrum is for Swedenborg the chief organ and the site of action of the soul to which

> sensations of the body according to the [sensory] fibers, like rays, flash into the cerebrum, and by the cerebrum into the cortical substances, which are the last and first termini of the fibers; and from these they flash into the soul, which

His followers organized a church in his name, The Swedenborg Church, which is extant today, and under whose auspices many of his works are still being reprinted and commented on in its publications.

[38] Vibrations (Swedenborg, 1899), written originally in 1719.

[39] (Swedenborg, 1918), Section 206, pp. 127–128. Pulsations of the fiber as a motive force has been considered in recent times as a possible account of axonal flow but then discounted (Chapter 12).

[40] Ibid., pp. 128–129.

resides in these substances as in its principles, endowed with the faculty of perceiving, thinking, and judging.[41]

Swedenborg viewed the sensory nerves as being soft because perception depends on their passive yielding to external impressions, whereas motor nerves are hard enabling them to produce active mechanical movements, following Galen's alternate view of nerve action (Chapter 5). Swedenborg related nerve differences to the different roles played by the cerebellum and cerebrum. Those fibers that end in the cerebellum are firmer and therefore the function of the cerebellum is to control motion. Those fibers that end in the cerebrum are softer and relate to sensation. In the nineteenth century, the anatomist Bell made much of the difference between the cerebellum and the cerebrum, relating the dorsal roots of the spinal cord, which he took to have a motor function, to the cerebellum and the ventral roots, which he believed to subserve sensory functions, to the cerebrum (Chapter 9).

THE SYMPATHETIC (AUTONOMIC, INVOLUNTARY) NERVOUS SYSTEM
Along with the somatic nervous system with its voluntary control of skeletal muscles, another system of nerves was related to the muscle-like responses of the visceral organs in the thorax and abdomen. Their actions were regulated without conscious control and in emotional states could show exaggerated activity; palpitations and increased heart rate, labored respiration, flushing, sweating, and urinary bladder and bowel discharges. These effects were ascribed to the province of the large nerve [our vagus nerve], with many branches distributed to the viscera that Galen designated as the 6th pair (of seven pairs of cranial nerves he enumerated). He included the glossopharyngeal and spinal accessory nerves along with the vagus nerve. Galen did not count the olfactory and trochlear nerves and the abducens nerves were included as part of the optic nerves – this accounting for his numbering it the 6th pair rather than the the the 10th nerve in our modern tally of twelve cranial nerves. For long after Galen, the 6th nerve [vagus] was not clearly discerned and was thought to be the origin of the intercostal chain of nerves and ganglia that we recognize as forming our sympathetic nervous system. The Galenic concept was figured by Vesalius[42] in his *De humani corporis fabrica* of 1543, who was said to have "adopted to the letter" Galen's description of it.[43] Galen's description is given in the book *The Use of the Parts*:

> This sixth pair of nerves . . . is very large, no small portion of it traverses the whole [length of the] neck and descends in the thorax. Here it immediately gives off to the thorax itself the first pair of nerves [sympathetic trunk?] extending along the

[41] (Swedenborg, 1883), Volume 1, Section p. 77.
[42] (Vesalius, 1950), Plate 49, p. 147.   [43] (Sheehan, 1936).

roots of the ribs and then other branches, some to the heart, others to the lung, and still others to the esophagus ... offshoots which these nerves distribute as they pass downward to the stomach, liver, and spleen and which they bestow [their influence] on all other parts in their path.[44]

Vesalius's picturing of the intercostal nerve chains (the sympathetic nerve trunks) issuing from the vagus nerves may have been caused by too great a reliance on the authority of Galen and perhaps Vesalius' lack of expertise at this stage of his career. But, another factor that could account for his mistake was the human material he had to use for his dissections. These were commonly executed criminals hanged or with their head cut off, making the dissection in the neck region quite difficult.[45] The differing origins of the vagus and intercostal nerve chains were shown not long after the publication of Vesalius' book by *Estienne* (1545) and *Eustachio* (1552), though the latter's work was published much later by *Lancisi* in 1714.[46] With Willis's work (1664) came the modern description of the sympathetic nervous chain that was termed the "intercostal" nerves and their function distinguished as "involuntary" in distinction to the operations of the voluntary nervous system.[47] It was distinguished from the "wandering pair," our vagus that he designated in his system as the 8th nerve (our 10th nerve).[48]

Although Willis, as did Galen, thought that the intercostal chains rose from the cranium, *François-Pourfour du Petit* (1664–1771) clearly showed them to originate from below the cranium.[49] He also showed that the intercostals carried "spirits" to the eye. Sectioning them caused the eye changes that were later rediscovered as the Claude Bernard-Horner syndrome while sectioning the vagus nerve, on the other hand, caused an "embarrassment" of respiration.[50]

The basis for the term sympathetic ascribed to the vagus and intercostal nerve chains was that, along with the wide distribution of their fibers over the viscera, the fibers apparently coalesced at places. This was held to account for the wide spread of sensations when these organs were affected by disease, giving rise to pains that could be severe but not well localized. These sensations were differentiated from the precise localization of stimulation felt in the skin and body openings that were held to be the office of the volitional somatic nervous system. The other obvious difference in function between the somatic system and the sympathetic nervous system was that the latter carried out its functions – the beat of the heart,

---

[44] (Galen, 1968), 7th Book, Volume 1, p. 367.
[45] (French, 1971), p. 48. See also (French, 1999).
[46] (Vesalius, 1950). See also (Sheehan, 1936), (French, 1971), and (Pick, 1970).
[47] (Willis, 1965b).
[48] Ibid., Chapter 23, p. 146 ff. The intercostal nerves were described in Chapters 25–27, 157 ff. Shown in detail in Table X.
[49] (Pick, 1970).   [50] Ibid., p. 6

respiration, digestion, bladder evacuation, and so forth – without conscious attendance. Our modern picture of this autonomic nervous system, comprised of the sympathetic (thoracolumbar outflow) and the parasympathetic (craniosacral outflow) with their differing properties, was shown later in the nineteenth century by *Gaskell*,[51] *Langley*,[52] and others.

THE ROLE OF THE GANGLIA

Lancisi (1718) viewed the ganglia of the sympathetic nervous system acting to accelerate the flow of the nerve spirits through its nerves while *Winslow* (1732) described the ganglia as "little brains" that carry their "sympathies" far and wide throughout the body.[53] The thought that they act by muscular contractions was rejected by von Haller, who did not see muscle present in the ganglia and could give no account for their action. A hypothesis for the function of the sympathetic ganglia was advanced by *James Johnstone* (1730?–1802).[54] Pointing to the complex irregular entanglement of fibers in the ganglia through which tortuous path the flow of the neural agent was hindered, he hypothesized that the ganglia acted to "limit the power of volition" by exerting control over the passage of nerve fluid through it – an action similar to the effect seen by a compression of the whole brain or the spinal marrow. In this way, the control by the will of the nerves and their actions on the heart and body organs distal to the ganglia was throttled or checked. Johnstone argued that this was a providential arrangement by the Deity so that one could not simply will one's death by causing the heart to cease beating. Sensations were also diminished, at least when exposed organs were directly stimulated as in the experiments carried out by von Haller and his associates.[55] Yet, the pain of gallstones and kidney stones can be most severe. The pain, when felt, is not well localized by the sympathetic nervous system because of the supposed anastomosis of its nerve branches spread out widely over the viscera.

In reference to the set of ganglia seen associated with the spinal nerves, which from "the spinal marrow have ganglions [and] send off filaments (rami communicans) which communicates with the intercostals," Johnstone proposed that these ganglia act as the "first checks to the usual powers of volition," affecting only the filaments sent to the sympathetic nerves. Experimental evidence was offered in support of his position. In decapitated kittens, when the spinal cord was probed, the limbs were immediately

---

[51] (Gaskell, 1916).   [52] (Langley, 1921).   [53] Quoted in (Pick, 1970), p. 10.
[54] (Johnstone, 1771). First published as a paper (Johnstone, 1764). The 1771 book was reprinted in a book of medical essays (Johnstone, 1795).
[55] (von Haller, 1936). Movements produced without apparent pain were considered due to the power of irritability in those body structures.

convulsed "but the heart was not in the least affected."[56] But one might ask if the ganglia found in the somatic nerves connecting to the spinal cord would similarly prevent the control of the muscles distal to them, the structure of the spinal cord ganglia being no less complex than the ganglia associated with the intercostal nerve chains. This posed a challenge to Johnstone's concept. It was overcome by the careful examination of the spinal cord root by Monro, who found that only part of the fibers passed directly through the ganglion (the ventral roots) and part entered the complexity of the ganglion (situated on the dorsal root).[57] The difference between the spinal ganglia and the sympathetic nerve chain in this respect was taken by Johnstone to save his theory.

However, the agency of the action of nerve and muscle left Johnstone at a loss:

[T]he supposition of subtile fluids called *spirits*, flowing in nerves as canals, or of an electric aura *conducted* by them, or of vibrations like elastic strings, are not only assumed without proofs, but are all equally inapplicable, in all points to appearances, and insufficient to account for the communication of motion from the brain to the muscles.[58]

That the agent was not caused by a vibration as occurs in an elastic cord was pointed out by von Haller: Nerves were soft and not firm nor tied at their ends as needed to propagate vibrations.[59] And, as von Haller found, on cutting nerves, they do not contract as would be expected of an elastic cord. An electrical aura, animal electricity, he felt was also not able to account for nerve action.[60] That electricity was the agent of nerve action required the evidence brought to bear later in the nineteenth century for its full acceptance (Chapter 7). The possibility of an ethereal vibration, such as had been proposed by Newton (Chapter 5), although an attractive theory to some at the time as will be noted in a following section, was not mentioned by Johnstone.

A MECHANICAL THEORY OF GANGLION ACTION

A more mechanical hypothesis was advanced to account for ganglion action by Swedenborg. He thought the dorsal root ganglia along the spinal cord had the nerve fibers in them so compacted within their tunics that unless they were "loosened it would be impossible for them to pour forth their liquid in that abundance which the viscera by reason of their exceeding

---

[56] (Johnstone, 1795), p. 22. So also was the intestine unaffected.
[57] Ibid., pp. 31–32.   [58] Ibid., pp. 52–53.   [59] (von Haller, 1966), Section 376.
[60] This point was inserted into Johnstone's 1771 book when it was reprinted in the 1795 version (Johnstone, 1795), where he referred to the "new experiments and discoveries in animal electricity" – those apparently by Galvani and others, pp. 30–31.

activity continually and urgently demand. . . . Their movements propel their fluid onward . . . for the ganglions are muscular . . . so that when they contract they draw up the continuations of the nerves and when they unfold they relax the nerves." The respiratory movements contributed by the ribs on the ganglia are the means by which the nervous juices are propelled through the ganglia:

> which may be called little succentureate brains[61] [which] . . . aids . . . in the promotion and transfusion of the nervous juice. . . . For the ganglia are attached to the vertebra as to bony fulcrums which are vibrated and moved at every turn of the pulmonary respiration; consequently the appended ganglia are likewise excited to their contractions and expansions, and by these movements the nerves also are drawn up or stretched out, and the nervous juice forwarded through them in a wonderful way.[62]

OTHER EXPONENTS OF VIBRATORY THEORY AS THE AGENT OF NERVE ACTION

Richard Mead (1673–1754) wrote that the ether, "Newton's subtle and elastic fluid," was carried to the brain in the blood and there separated from the blood as a "thin volatile liquor, of great force and elasticity . . . [and was then] lodged in the fibers of the nerve . . . [in which it] became the instrument of muscular motion and sensation." *David Hartley* (1705–1757) also used Newton's concept to account for his associationist psychology: "Newton supposes that . . . the very subtle and elastic fluid, which he calls aether . . . is diffused through the pores of gross [solid] bodies, as well as through the open spaces that are void of gross matter . . . that it is rarer in the pores of bodies than in open space . . . [and serves for] the performance of animal sensation and motion."[63] The vibratory motion is carried out in the fibers. "The white matter and its fibers appears to be everywhere uniform and similar throughout the whole brain, spinal marrow, and nerves . . . so that we suppose vibrations run freely along and pervade the whole, in whatever part they are first excited."[64] In the cortex, only the finest vessels are present and nerve tubules that are rather solid capillamenta according to Isaac Newton, rather than small tubuli according to Boerhaave,[65] and so could support a vibratory movement.

Hartley's account of the medullary substance of the nerve fibers able to support a vibratory transmission and yet act like a nerve fluid lies in

---

[61] *Succentio* is the latin term used in Imperial times for an under-centurion in the Roman army, the centurion being the company commander. The idea is that the brain is higher, with the ganglia acting similarly but in a junior capacity.

[62] (Swedenborg, 1918), Section 239, p. 149.   [63] (Hartley, 1967), p. 13.

[64] Ibid., p. 16.   [65] Ibid., p. 17.

the smallness of the particles supposedly present within the fibers. "These [particles] are subject to the powers of attraction and repulsion [which according to Newton], be the cause what it will, impulse, pressure, an unknown one, or no physical causes at all but the immediate agency of the Deity."[66] The very smallest of particles are the most active and could be considered to act like a fluid that would then approach the concept of animal spirits so that "all the arguments which Boerhaave has brought forward for his hypothesis of a very subtle active fluid in the brain may be accommodated to the Newtonian hypothesis of vibrations."[67] The small particles of the medullary substance vibrate in synchrony with the vibrations of the ether because of the exciting agency, be it sound, rays of light, odors, and so forth. Among the agents that could excite sensation, Hartley includes the "effluvia of electric bodies [which] seem to have vibrating motions," their properties analogized to that of other physical phenomena, such as "motions along hempen string, [which] resemble the motions along the nerves in sensation and muscular contraction."[68]

In conclusion, Hartley considered that the brain may "be reckoned as the seat of the sensitive soul, or the sensorium, in men . . . [and] whatever motions be excited in the nerves, no sensations can arise, unless this motion can penetrate to, and prevail in the brain."[69] With respect to the question as to how the mind is related to the body, how these two incommensurate substances could interact, Hartley supposed "an infinitesimal elementary body to be intermediate between the soul and gross body, which appears to be no improbable supposition, then the changes in our sensations, ideas and motions, may correspond to the changes made in the medullary [nerve] substance."[70] This concept of an intermediate body pushes the mind-body problem back to the difficulty of how they can interact with an intermediate substance, a "solution" that only leads to an infinite regress. Some recent discussions on that subject are dealt with in Chapters 15 and 16.

---

[66] Ibid., p. 20. This passage could have been written to forestall a possible charge of atheism, a threat that was still not to be taken lightly in his day. Or, it could be a reference to the occasionalism espoused by Boerhaave as noted earlier in this chapter.

[67] Ibid., pp. 20–21.   [68] Ibid., p. 28.   [69] Ibid., p. 31.   [70] Ibid., p. 34.

7

ELECTRICITY AS THE AGENT OF NERVE ACTION

In the preceding chapters, various agents of nerve action were put forward to account for the rapidity with which nerves conduct sensations and produce motor responses as occurs in a reflex such as that in the example given by Descartes (Chapter 5), where a foot is burned and rapidly withdrawn even before the pain is sensed. The various new physical principles and chemical entities discovered in the Renaissance were advanced to serve this function, but failed to fit all the properties of nerve action. Electricity had properties that suggested that it might be the long sought-for agent of nerve action. It was invisible and imponderable; acting with lightning speed and having profound excitatory actions on the nerves and muscles. How electricity came to be accepted as the agent of nerve conduction is the theme of this chapter. Its history can be divided into three periods. The first period extended from ancient times to that of Galvani at the turn of the eighteenth century when electricity was generated as a static discharge and its potent effects on the body experienced. The second period extends from the introduction of the battery by Volta after the turn of the nineteenth century, when the flow of current in body tissues was investigated, though not differentiated from electrical conduction in metals. The third period extends from the latter part of the nineteenth century to the present, when potentials were realized as due to ionic movements in electrolyte solutions and the special properties of the nerve and muscle membranes.

STATIC ELECTRICITY
The ancients noticed that when amber was rubbed, it could attract small bits of wool or leaves. The term "electricity," the Greek name for amber, was applied to it by Thales, who is credited as being the first to have mentioned the phenomenon.[1] *Diogenes Laertius* (fl. third century) commented

---

[1] (White, 1886) and (Caley and Richards, 1956), p. 117.

that "Aristotle and Hippias affirm that arguing from the movements produced by magnets and amber, he [Thales] attributed a soul or life to even to inanimate objects."[2] It was not clear to the ancients that amber was a resin exuded by trees. Theophrastus described the material as having been dug from the ground.[3] It could have been confused with a number of other substances that he mentioned in the same context, some of them apparently different terms for amber.

In more recent times, *William Gilbert* (1544–1603) undertook an examination of its properties, comparing it with magnetism.[4] Magnetic bodies, such as a lodestone attracting iron, were seen to have two different poles; one pole attracting the opposite pole of another magnet, or repelling it if the same poles were opposed. The magnetic force remained intact if paper or other materials were interposed, or if the magnets were placed in water. Electric bodies, which included not only amber, but glass, sulfur, sealing wax, precious stones, and so on, needed to be rubbed before they were able to show their ability to attract small bits of paper, lint, and the like. Unlike magnetic attraction, the attractive property of the electrified bodies was destroyed by interposition of paper, a linen cloth, or if the electrified body was immersed in water. Gilbert thought that rubbing released the electrical property that consisted of some kind of a humor held within the electric bodies. Released as an "effluvium" into the atmosphere by rubbing, it could then attract other bodies to it.

The existence of two kinds of electricity was inferred when different materials, such as glass or resin, were rubbed and then seen to have different effects on a gold leaf. The leaf, which at first was attracted and then repelled by rubbed glass, could still be attracted by rubbed resin. Gilbert applied the term "electrics" to these substances able to store charges and exhibit attraction and "nonelectrics" to those substances that did not. Instead of two electrical fluids, each "very subtle and inflammable," Benjamin Franklin supported a one-fluid theory wherein, on rubbing some of these substances, electricity is taken from them; and whereas on rubbing others, electricity is added to them. He called the excess electricity produced by rubbing

---

[2] (Diogenes Laertius, 1966), Volume 1, p. 25.

[3] Ibid., p. 51. Theophrastus wrote that it comes from the urine of wild male lynx, giving it the name of *lygourion*. Whether this is the same material as amber, along with a number of other substances named as having a similar property, is difficult to determine. Theophrastus was apparently unaware that it had to be rubbed to induce its attractive power.

[4] (Gilbert, 1958). Much of his book deals with magnetism and his discovery that the earth itself acts as a magnet. He likened magnetism to the soul of the earth, one purer than the soul of man (Heilbron, 1999).

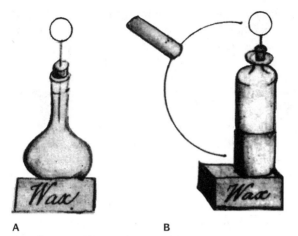

A   B

7.1. Leyden jars used to store electricity. The one on the left (A) shows a flask with a conducting metal rod passing down into it through an insulating stopper contacting a fluid filling the flask. The rod outside the flask is topped with a ring. The flask is held in hand, and after it is charged by contact with a friction machine generating electricity, is set on a nonconducting wax block that holds its charge. On the right (B), a cylindrical form of the jar with foil on the outside is shown being discharged by a conducting arc of metal held by an insulated clamp touching the ring to the outer foil. Modified from Galvani in (Cohen, 1956), Figures 13 and 14.

"positive" electricity, its deficiency "negative" electricity[5] – the electricity acting as a fluid that he considered to consist of "extremely subtle particles."

The small amount of electricity that could be produced by rubbing materials by hand was greatly increased by the use of electric machines. These were made of glass, amber, or some other electrical material in the form of a disk, sphere, or cylinder that was rotated while a cushion was set to rub against. The charge generated could be "drawn off" the cushion of the machine with a metal comb producing visible sparks when discharged to earth. Another device that greatly aided the experimental studies of the "electricians" was the *Leyden jar*, discovered in the 1740s that could store an electrical charge.[6] The jar, usually in the shape of a tall glass bottle or flask with a narrow mouth (Figure 7.1), has a wire passing through a nonconducting stopper in the mouth of the bottle to make contact with water or metal shot on the inside, with the wire emerging from the stopper ending

---

[5] (Franklin, 1941). It should be emphasized out that Franklin's identification of the sign of electricity as positive was purely arbitrary. This causes difficulty for some students today when the direction of current flow in electrolyte solutions is given by convention as moving from the positive to the negative direction when they have learned that current flow in metals is due to the movement of negatively charged electrons.

[6] (Rowbottom and Susskind, '84).

in a hook, ring or knob. The Leyden jar is in effect a capacitor. To charge the jar, the operator grasps it and touches the knob to an electrostatic machine. By holding the jar, the charge that appears on the outside is drained off to the ground leaving the charge accumulated inside. The jar could then be conveniently carried about and discharged at will through test objects by touching the knob to them if they are in contact with the ground.

Discharges from Leyden jars were used to test the ability of the electric charge to pass through various materials. By this means, materials were categorized as either "conductors" or "nonconductors." A distinction was made between "artificial electricity" (produced by rubbing) and "natural electricity" (present in the air and seen as fiery emanations from the top of ship masts and from church steeples and tall buildings during thunderstorms). With his famous kite experiment, Benjamin Franklin showed that atmospheric electricity drawn from the clouds with a kite had the same properties as "artificial electricity" generated by friction. The spark from the kite string could be stored in a Leyden jar.

A person touched with a charged Leyden jar could receive a strong shock. Its dramatic effects captured the popular imagination, and it was used as entertainment. One type was the formation of a chain of people holding hands and on completion of the circuit to a charged jar, all receive a shock; if it was sufficiently strong, it causing all of them to jump in unison. These demonstrations excited investigations into how electricity acted on the body, especially on nerves and muscles. That quickly led to thinking that electricity was in fact the long looked-for agent of nerve, animal spirits. This was first broached by Hales in 1736 and Sauvages in 1753 who named it the "true nerve fluid."[7] Priestley described electricity as having "an occult power and agreeable to a fluid which was analogized to animal spirits."

In a review of electricity in the last quarter of the eighteenth century, Bonnefoy lists a number of authorities who equated electrical phenomena with animal spirits in nerve: Hales, Jallabert, Dufay, among others were mentioned by the eminent von Haller and Priestley in their histories. But Bonnefoy writes that, "one must not judge by the words of a master, it must be on the basis of proof" that the matter is to be ascertained. He therefore sets out arguments for the thesis:

1. The speed of nerve action estimated from muscle contractions in running or in pronouncing letters in rapid speech gave, in one such calculation made for running dogs a value of 19 ft/sec, a value lower but not too far from modern determinations, considering the inaccuracies involved in making such an estimation. The important finding was that the speed of nerve was indicated as being much less than that

---

[7] (Priestley, 1966), p. 473.

of an electrical discharge, but faster than other agencies advanced to account for nerve action.

2. The impossibility of blood furnishing energy to supply muscle contractions for long periods of time as shown by runners able to run for a day without eating or drinking. This leads to the conclusion that only respiration could be the inexhaustible source of the nerve fluid. The hypothesis advanced is that electricity present in atmospheric air is taken up on respiration. That such electricity is constantly present everywhere is shown by the electrical sparks seen in the dark seen when putting on clothes, combing the hair, stroking fur, and so forth.

3. The separation of electricity from the inspired air is carried out in the brain. The brain receives one-fifth part of all the blood circulating in the body, arguing by this large proportion of blood supply to the brain compared with the body the importance of the brain for the process. The separation occurs in the ramifying fine blood vessels of the cerebrum, in particular, the cortex with its rich provision of vessels, as was shown by the perfusion studies of Ruysch (Chapter 6). There, from the fine blood vessels of the cortex, electrical fluid is emptied into the nerve tubules of the white matter to be distributed throughout the body.

4. The movement of the electrical fluid in the nerve tubules is shown by the ligature of nerve preventing motor control beyond the ligation, or the passage of sensation from below it – a result in accord with the movement of a fluid and against a vibration of solid nerve fibers.[8]

Whereas an electrical fluid flowing in nerve tubules was an attractive hypothesis as the agent of nerve action, there were considerations advanced by von Haller that cast doubt on the concept. Von Haller pointed out that electrical fluids could not be contained within the nerve fibers when they were ligated. Electric currents would spread out into the tissues (known as a good electrical conductor) around a nerve ligation to excite the nerve below the ligation and continue to conduct the nerve impulse, whereas experience had clearly shown sensation and motor control to be readily blocked by a ligation. Von Haller wrote:

> Concerning the nature of this nervous fluid, there are many doubts. Many of the moderns will have it to be extremely elastic, of an ethereal or of an electric matter, but the more reasonable part make it to be incompressible and watery, but of a lymphatic or albuminous nature. Indeed it is not to be denied, that we have many arguments against submitting [to] either of these opinions. An electrical matter is, indeed, very powerful and fit for motion; but then it is not confinable within the nerves, since it penetrates throughout the whole animal to which it is communicated, exerting its force upon the flesh and fat, as well

[8] (Bonnefoy, 1782).

as upon the nerves. But, in a living animal, the nerves only, or such parts that have nerves running through them, are affected by irritation; and, therefore, this liquid must be of a nature that will make it flow through, and be contained within, the small pipes of the nerve. And a ligature on the nerve takes away sense and motion, but cannot stop the motion of a torrent of electrical matter.[9]

These considerations led Wrisberg[10] to put the agent of nerve into the category of a "third class of fluids," those of an unknown nature, writing:

There is certainly no physiological dispute, in which, without regard to the parts, a more difficult decision can be offered, than that important question, how do the nerves act in the bodies of animals? The doctrine of the nervous fluid, or animal spirits, is confirmed by many arguments drawn from anatomy but is fully refuted by as many, if not more, of a similar nature. But ocular demonstration of the nerves proves, that such a fluid under the name, and sense, which determines fluids in our bodies, moving in canals and hollow fistulae, does not exist. However, it is therefore not improbable, that [the] substance, which produces the wonderful phenomena in the nerve, is one of the class of the more subtle fluids of nature. I pretend not to give it a name. Since 1766, I have been inclined to think, that it perhaps resembles the electric and magnetical fluids; since which time, long before Mismer [Mesmer], I publicly taught the power of magnetism upon our body. For this reason the nervous principle has been numbered in the third class of fluids.[11]

A TYPE OF ANIMAL ELECTRICITY SHOWN BY ELECTRIC FISH

The third class of substances included the "electrical matter of animals," that of certain fish which were known from ancient times to have effects strikingly like those of an electric shock.[12] The *torpedo* or *electric ray, Gymnotus*, and the *electric eel* could give a shock producing a stupefying and numbing effect.[13] Electric fish are realistically portrayed in ancient mosaics and wall paintings sufficiently so as to identify their species; for example, the Nile catfish *Malopterurus electricus* is depicted on the wall of Egyptian tombs and on ancient Greek vases.[14] In the Platonic dialogue *Meno*, when Socrates' irony proves too strong for him, Meno complains to Socrates, "you are exactly like the flat stingray fish that one meets in the sea. Whenever anyone comes into contact with it, it numbs him, and that is the sort of thing that you seem to be doing to me now. My mind and my lips are literally numb, and I have nothing to reply to you."[15]

---

[9] (von Haller, 1966), Section 379.

[10] von Haller's book (von Haller, 1966) had been edited by the noted anatomist Heinrich A. Wrisberg, who added his own observations and extensive notes to it.

[11] Ibid., p. 221, note 105. This passage indicates Wrisberg's acceptance of von Haller's argument against the electrical nature of the nerve fluid.

[12] (Kellaway, 1946).   [13] (Walker, 1937), and (Kellaway, 1946).

[14] (Kellaway, 1946), Plate V, p. 126.

[15] (Plato, 1920), Volume 1 in the dialog *Meno*, p. 359.

The ancient Greek fishermen knew that it was not necessary to touch the fish to receive a shock. Plutarch wrote,[16] "You know yourself the property of the torpedo or cramp fish, which not only benumbs all that touch it but strikes a numbness through the very net into the hands of them that go about to take it. [The torpedo] shoots forth the effluviums of his nature like so many darts, and first infects the water, then fish [that they prey on] through the water [so that the fish] is held (as it were) in chains and frozen up." The mysterious influence sent out through the water by the torpedo was later analogized to the electrical effluvia generated in the air by electrically charged bodies. "Around an electrified body there is formed a vortex of exceedingly fine matter in a state of agitation, which urge towards the body such light substances as lie within its sphere of activity. The existence of this vortex is more than a mere conjecture; for when an electrified body is brought close to the face it causes a sensation like that of encountering a cobweb."[17]

In 1666, Redi and his pupil Lorenzini theorized on the nature of the organ in the electric fish causing the numbness:

I should be apt to say, that this Numness of Pain . . . proceedeth from on abundance of certain Corpuscles or *Effuviums*, which coming from the Cramp-Fish, enter into his Hand who toucheth it. These Corpuscles or *Effluviums* do not come thence of themselves, but they are driven, and as it were darted forth by the Constriction of those Fibres of which those two hooked Bodies or Muscles, are composed.[18]

The designation "hooked bodies" refers to the muscle-like electric organs on either side of the body. This passage pointing to the source of the electric discharge was neglected until the eighteenth century when the properties of electricity were better known and the discharge produced by the fish shown to be identical to that produced by artificial means: Both charges could be stored in a Leyden jar. Walsh noted in 1772 that, when the torpedo fish attempted to give a shock, its eyes were depressed – a sign that the discharge was dependent on the animal's will.[19] He requested the eminent anatomist *John Hunter* (1728–1793) to undertake an examination of the electric organs. Hunter found them to consist of perpendicular columns reaching from the upper to the under side of the body. Each column measured a fifth of an inch in diameter and in length from a quarter of an inch to one and a half inches. The columns, divided by horizontal partitions placed over each other, appeared to contain a fluid. (In modern times, the columns were shown to be composed of modified muscle cells arranged in a series to give

[16] In Plutarch 978b, quoted from (Kellaway, 1946).
[17] Desaguliers quoted in (Whittaker, 1951), Volume 1, p. 43.
[18] (Walker, 1937), p. 88.   [19] Ibid., p. 91.

rise to the high voltages and with those alongside in parallel, high currents). Hunter found the electric organs to be liberally provided with nerves that:

> are subservient to the formation, collection or management of the electric fluid: especially as appears evident, from Mr. Walsh's experiments, that the will of the animal does absolutely controul the electric powers of its body; which must depend on the energy of the nerves. How far this may be connected with the power of the nerves in general, or how far it may lead to an explanation of their operations, time and future discoveries alone can fully determine.[20]

Priestley related electricity to his theory of phlogiston, which he considered to be a principle in matter responsible for fire and on respiration in animals to heat on which life depends:

> Animals have a power of converting phlogiston, from the state in which they receive it in their nutriment, into that state in which it is called the electric fluid; the brain, besides its other proper uses, is the great laboratory and repository for this purpose; that by means of the nerves this great principle, thus exhalted, is directed into the muscles, and forces them to act, in the same manner as they are forced into action when the electrical fluid is thrown into them *ab extra* [from the outside].[21]

In this passage, Priestley makes the proposition that the electricity produced by electric fish is a general principle present in all animals:

> We shall apparently be driven to have recourse to some other principle; that principle, if it be not common electricity may be something however very analogous to it. The electrical [fish] gymnotus and torpedo, if they do not render the theory very probable, make it at least possible, and this principle may be believed to follow the most common laws of electricity.[22]

Priestley and Fontana are forerunners of the view that the electrical principle found in the electric fish is the same principle of action present in the nerves and muscles in all animals. This was to be given experimental support only a score or so years later by Galvani's studies on frog nerve and muscle, as is recounted later.

THE THERAPEUTIC USE OF ELECTRICITY

The unusual properties of electricity, the shocks produced by electric fish, were put to use by the early Romans and Greeks as a cure for gout and headaches. In a passage written in the first century B.C., it was said that:

> For any type of gout a live black torpedo should, when the pain begins, be placed under the feet. The patient must stand on a moist shore washed by the sea and

---

[20] Ibid., pp. 92–93.
[21] (Priestley, 1966), p. 102. This refers to the shock given when applied from a charged Leyden jar.
[22] Ibid., p. 102.

he should stay like this until his whole foot and leg up to the knee is numb. This takes away present pain and prevents pain from coming on [if], it has not already arisen. In this way Anteros, a freedman of Tiberius, was cured.[23]

Scribonius was also said to have described the use of the electric fish to cure headaches. So also did Galen, who ascribed its curative power to its "frig-orific principle," probably because a cold application could have a similar numbing effect.[24] Kellaway relates that, in modern times, "many primitive African tribes still use the shock of the *Malopterurus electricus* as a medic-inal agent, a practice probably of very ancient origin. A Jesuit missionary reported in the sixteenth century that the Abyssinians told him that

in these Rivers and Lakes is also found the Torpedo [referring actually to the electric catfish], which if any man holds it in his hand, if it stirre not, it doth produce no effect; but if it move itself ever so little, it so tormenteth him which holds it, that his Arteries, Joints, Sinew and all his members feel exceeding paine with a certain numbness; and as soon as it is let go and out of hand all the paine and numbness is also gone. The superstitious *Abassines* believe that it is good to expel the Devils out of the body.[25]

Kellaway also found instances in the literature up to about the year 1850 in which Europeans used electric fish for medicinal purposes until artificially generated electricity, more ready available and capable of being better con-trolled, replaced their use.

The use of artificial electricity as a therapy advanced to the point where Priestley, in his treatise on electricity, observed that the entry for medical electricity in the *materia medica* of the eighteenth century had become of considerable size.[26] He noted that a professor of medicine named Kratzen-stein, had in 1744 reportedly cured a woman with a contracted little fin-ger and relieved a person who had two lame fingers. In 1747, Jallabert, a professor of philosophy, cured a locksmith whose arm had been para-lyzed for fifteen years; although the Abbé Nollet later found the condition to have relapsed. Sauvages of the Academy in Montpelier had consider-able success with some but not all paralytics. Dr. Bohadtch, a Bohemian, found that "hemiplegiacs" were most amenable to the treatment. In 1757, Patric Brydone claimed to have completely cured a hemiplegiac within three days, and John Godfrey Teske "very nearly cured" a young man of a par-alytic arm of five years standing. Tetanus in a young girl was reported to be cured in 1762 by a Dr. Watson, to the point where her rigidity was re-lieved and she could walk and even run. Other reports of cures were given, including that by Benjamin Franklin who gave an account in 1758 of the application of electricity in a number of paralytics that were brought to him. They appeared to have gained relief, but "finding the shocks severe,

---

[23] (Kellaway, 1946), p. 130.   [24] Ibid., p. 123.   [25] Ibid., p. 133.
[26] (Priestley, 1966), Volume 1, Section 14, pp. 472–489.

became discouraged and in a short time relapsed; so that he never knew any permanent advantage from the palsy." Others reported cures were of headaches, sciatica, cramps, violent pains, toothache, quotidian and tertian ague, deafness, bruises, running sores, dropsy, kidney stones, consumption, rheumatic pains, St. Vitus' dance, obstructed menses, "mucous apoplexy," and so forth.[27]

MAGNETIC AND ELECTRIC CURES THAT FAILED

Applications of electricity and magnetism to medicine as therapeutic agents extended well into the nineteenth century, though questions of their effectiveness were raised early on. A prominent figure in the medical use of magnetism was *Franz Anton Mesmer* (1734–1815), who practiced what he termed *animal magnetism*, soon known as *mesmerism*. He theorized that the planets have a magnetic influence on physiological and pathological processes, the celestial magnetic influence entering the body to become insinuated into the substance of the nerves to enhance their power of action. In one arrangement to channel this celestial influence, the patient sat in a tub containing a weak acid with iron bars poking out the sides. By grasping the bars, the magnetic fluid from the atmosphere could be drawn into the patient and thereby have the desired therapeutic effect.[28]

Mesmer's claims of cures were so extraordinary and had aroused so much public interest that a French commission with Benjamin Franklin at its head was appointed to investigate them.[29] Sadly, the commission concluded that animal magnetic fluid "is capable of being perceived by none of our senses, that no action either upon themselves or upon the subjects are assured. That it is the imagination which is involved and that there is no proof for a magnetic fluid."[30] Mesmer and some of the other practitioners could very well have been believers themselves, making them more convincing than if they were outright charlatans. The renowned French neurologist Charcot later concluded that the effects were from exaggerated suggestibility, a manifestation of hysteria. He induced hypnotic states in these subjects and was able to show that much of what Mesmer claimed to caused by magnetism, could be seen in the hypnotic state.[31]

An interesting example of suggestibility in relation to the supposed beneficial effects of electricity was that excited by Perkins and his "tractors." Inspired by the findings of Galvani and Volta, the tractors were made of two different metals which, according to the specifications described by Perkins in his English patent, were about the thickness of lead pencils, either three,

[27] (Bonnefoy, 1782).   [28] (Amadou, 1971).
[29] (Hunter and MacAlpine, 1963b) pp. 483–486.
[30] Yet, amputations of limbs have been successfully carried out under its influence.
[31] (Charcot, 1878). Charcot influenced Freud to investigate the phenomenon of suggestion and hypnotism, with Freud later going on to develop the school of psychoanalysis.

four, or five inches in length joined together at one end with the opposite ends spread and tapering to blunt points. In the process of healing (e.g., of lower back pain or sciatica), the extended points of the tractor were drawn down over the skin of the back and lower limb in a continuous sweeping motion. A miraculous relief of pain followed after a certain number of such applications. These reports, heavily advertised, attracted a flood of patients eager for cure by the tractor practitioners. Unfortunately, tractors made of wood and covered with metal paint to disguise their nature when similarly applied produced the same "cures." When this was revealed, the use of Perkins' tractors underwent a rapid collapse, and the tractors entered the lists of failed cures.[32]

GALVANI AND ANIMAL ELECTRICITY

The electricity produced by electric fish was for the most part held to be a special case until the concept of electricity as the nerve agent was given a new spark of life by *Luigi Galvani* (1737–1798). He found what he called *animal electricity* when experimenting on frogs in 1780 and found contractions of their muscles to coincide with sparks generated by an electrical machine in the room a distance away when a scalpel was held that touched the frog (Figure 7.2). Studying the phenomenon further, he later found that when frogs were suspended on an iron balcony by a brass hook with their legs hanging down, contractions occurred when they touched the iron balcony. Imitating the effect of the two metals, he fashioned from them what he referred to as a "metal arc." By means of the arc, muscle contractions were produced when they completed a circuit between the muscle and the rest of the animal. At first Galvani inferred that the contractions were caused by the metallic arc, but soon after he theorized that they were caused by animal electricity produced by the animal, with the arc serving only as a means to reveal its presence.[33] He equated animal electricity with animal spirits whose source was the brain. He hypothesized that animal electricity carried down from the brain into the spinal cord and then into the hollow nerves innervating the muscles invoked the contractions in them. Galvani analogized muscle to a Leyden jar, wherein electrical charge in the muscles is in equilibrium at rest and discharged when its nerve is excited. The nerve innervating the muscle was compared with the wire passing down into a Leyden jar. When the nerve was cut short or pulled from the muscle, a

[32] (Haygarth, 1801) and (Walsh, 1923). An interesting account of "tractorizing" was given by John Greenway at www.uky.edu/~engjlg/research/perkins.htm. Among those who tried it, he noted that George Washington had bought a pair.

[33] Two English translations appeared in the year 1953. That by Foley (Galvani, 1953b) contains a valuable introduction on historical aspects by I. B. Cohen and that by Green (Galvani, 1953a) biographical information on Galvani by his nephew Aldini.

7.2. The basic observation made by Galvani. When an electric machine in the form of a disk of glass was turned to generate static electricity and a spark drawn at (**B**), a knife blade touching the nerve of the spinal cord of the frog causes the leg muscles to contract. Note that tissue has been removed leaving only the nerves (**N**) connecting the cord to the muscles, indicating that the nerves conduct the animal electricity. Modified from (Galvani, 1953), Plate 1.

failure or weakness of muscle contractions was seen, as would be the case if the Leyden jar's central wire were removed.

In response to the theoretical objection to electricity being the nerve agent earlier raised by von Haller, Galvani provided an explanation as to how the nerve channels were able to retain the animal electrical fluid flowing down within them: "so that it [the electric fluid] is not diffused and spread to adjacent parts, with a great diminution of muscular contractions." Galvani anticipated the modern view of nerve fibers when he postulated that:

> the nerves [fibers] are so constituted that they are hollow internally, or at any rate are composed of some material adapted to carrying electric fluid, and that externally they are oily [a substance already known not to conduct electricity], or have some other similar substance which prohibits the dissipation and effusion of this electric fluid flowing through the nerves. Indeed such a structure and composition of the nerves will admit their functioning in both ways, namely in conducting the electric nerve fluid and at the same time, in preventing its dissipation, and will be completely consistent with the economy of the animal as well as the experiments, since indeed the economy of the animal seems to require that its life force be confined within the nerves and experiments reveal that the nerves are composed of a particularly oily substance....
>
> This idioelectric [property of a non-conducting] substance in the nerves,[34] however, which seems to prevent the electric nerve fluid from being dispersed with great loss, will not prevent this fluid, coursing through the inner conducting substance of the nerves, from leaving these same nerves, when it has become necessary, to produce contractions and to be transferred to the muscles as quickly as possible through an arc in its own characteristic way.[35]

Enlarging on what the nature of animal electricity in nerve might be, and its source, he continued:

> I should think this [animal electricity] would be identical with that indicated by physiologists as being the source of animal spirits, namely the cerebrum. For although we have stated that electricity is inherent in the muscles, we are not of the opinion, however, that it emanates from them as from its proper and natural source [the cerebrum].
>
> For since all nerves, not only those which extend to the muscles, but also those which project to other parts of the body, seem to be identical in appearance and in their natural qualities, who will rightly deny that all carry a fluid of a like nature? Now inasmuch as we have already shown that electric fluid is carried through the nerves of the muscles, it must therefore be transmitted through all of the nerves. Furthermore all these nerves must draw it from a single common source, namely the cerebrum. Otherwise there would be as many sources as there are part[s] wherein the nerves end, and since these parts are completely different in character and composition, they do not seem to be adapted to activating and secreting one and the same fluid.
>
> We believe, therefore, that the electric fluid is produced by the activity of the cerebrum, that it is extracted [there] in all probability from the blood, and that it enters the nerves and circulates within them in the event that they are hollow and empty, or, as seems more likely, they are carriers for a very fine lymph or other

---

[34] What we now recognize as the lipid bilayers forming the myelin sheaths of the nerve fibers (Chapter 8).

[35] (Hoff, 1936), and (Galvani, 1953b), p. 76.

similarly subtle fluid which is secreted from the cortical substance of the brain, as many believe. If this be the case, perhaps at last the nature of animal spirits, which has been hidden and vainly sought after for so long will be brought to light with clarity.[36]

Galvani proposed that an impediment to the flow of animal electricity could account for certain neuropathies as in:

> severe rheumatic afflictions and particularly in sciatica, when, as Cotunius points out the humor ceases to move between the covering and the surface of the nerve, very sharp pains and as well as extremely violent and persistent muscular contractions of the afflicted membrane take place. This frequently results in a permanent contraction of the limb or one that persists for a long time . . . [the condition of paralysis accounted for by] . . . a stoppage in the circuit of nerveo-electric fluid flowing either from the muscle to the nerve or from the nerve to the muscle . . . [as by] an oily substance occupying the innermost part of the nerve.[37]

DISPUTE OF VOLTA WITH GALVANI OVER THE REALITY OF ANIMAL ELECTRICITY

Alessandro Volta (1745–1827) at first accepted Galvani's concept of animal electricity, but then later rejected it. He came to the conclusion that the different metals composing the metallic arc were not just acting to complete the circuit allowing conduction of animal electricity, but that the metals themselves were responsible for exciting muscle contractions. Volta went on to convincingly demonstrate this by connecting couples of dissimilar metals in series and showing that the electrical strength increased in proportion to the number of such couples. With these "batteries" generating constant well-defined currents, physicists could more satisfactorily measure the properties of electricity flowing in different materials, and physiologists could determine the strength and duration of current flow needed to excite nerve and muscle: such studies were much more definitive than those made using static electrical discharges. Volta's success with the discovery of batteries seemed to have completely undermined Galvani's concept of animal electricity, relegating it to experimental error.

If the nerve agent was not animal electricity, then how does that account for the remarkable effectiveness of electricity to excite the nerves? *Francois-Achille Longet* (1811–1871) evoked some as yet unknown principle in nerve which, although not electrical in nature, could be initiated by electricity: "To explain the phenomenon of life, we must admit the presence of a special

---

[36] (Galvani, 1953b), p. 79.
[37] Ibid., p. 83. The reference made was to *Dominico Cotugno* (1736–1822), who was the first to recognize sciatica (*ischialgo postica*) as a clinical entity (Cotugno, 1764).

agent which has been variously designated the nerve principle, agent or fluid or active principle of nerves."[38]

Volta's position was so fully accepted that interest in pursuing studies of animal electricity flagged in France, Britain, and Germany, though still carried on in Italy, notably by *Leopoldi Nobili* (1784–1845) and his pupil *Carlo Matteucci* (1811–1868). Nobili found that when a skinned and decapitated frog was placed with its feet in a glass containing a saline solution, its trunk in another glass containing saline, and the circuit between the glasses then completed with a saline-soaked cotton thread, a contraction of the muscles was evoked. Here, contra Volta, the contraction was effected by a nonmetallic arc, convincing evidence for the reality of animal electricity.[39]

Nobili went on to construct a sensitive galvanometer, and with it he and Matteucci found a potential difference in muscle between a site that had been cut or injured and an intact part of the muscle – a phenomenon they termed the "injury potential" or "demarcation potential."[40] The injured site was negative with respect to the intact part of the muscle. The potential difference so produced causes a flow of current, the "injury current," between the two sites on the muscle. When a number of such injured muscles were placed in series, the injured portion of one muscle in contact with the intact part of another muscle and its injured portion in contact with the intact part of yet another muscle, and so on, a larger voltage could be produced – one that was related to the number of muscles in the series. This was just like the increase in voltage that Volta had obtained by adding metallic couples in series to form a battery with greater voltages, strong evidence for the reality of animal electricity.

Matteucci, however, could not directly record an electrical sign of a propagated nerve or muscle potential. Nor could others, including the well-known investigators Prevost and Dumas. This was because of the inadequacy of the slow galvanometers in use at the time, but it was effective as an argument against accepting the reality of animal electricity. It led Matteucci to reverse his stance and to side with the opinion of Longet that the nerve principle had a special nature, a fluid or perhaps an ethereal vibration as Newton had hypothesized (Chapter 5).

MEASUREMENT OF NERVE CONDUCTION VELOCITY

That the nerve impulse had a special property and was not electrical was a view at first shared by the influential Johannes Müller.[41] However, he thought that a closer investigation was warranted, and he had his student

---

[38] (Longet, 1842), p. 30.   [39] (Hoff, 1936).
[40] (Matteucci, 1844), and (Moruzzi, 1963).
[41] (Clarke and Jacyna, 1987), pp. 204–205.

Emil du Bois-Reymond (1818–1896) make a further examination of the nature of nerve and muscle responses. Du Bois-Reymond enthusiastically took up the challenge, feeling that by Matteucci's reversal of his original position on the electrical nature of the nerve impulse, he now had the field to himself. He devised a new sensitive galvanometer, and in his first studies confirmed Matteucci's finding of an injury potential resulting from current flowing between the site of injury and an intact region of muscle. He also found a similar injury potential in nerve. Unlike Matteucci, he accounted for the phenomenon by the fibers of muscle and nerve having a continual flow of current, one that was normally present. This concept was later shown by Hermann to be erroneous, no current flows until a potential difference is created by damage of a site as Matteucci had earlier found.

Du Bois-Reymond went on to determine a most important characteristic of the nerve impulse: its conduction velocity. He stimulated the nerve of a frog nerve-muscle preparation at a point close to the muscle and then again at a distance from it. On recording the different times of the resultant onset of the muscle contractions with a sensitive galvanometer, and taking into account the distance between them, he calculated a nerve conduction velocity of 30 to 40 M/sec, a value quite close to that found using modern instrumentation.

The nerve impulse was too rapid to be measured directly by the slow response times of the galvanometers then in use, but this was accomplished by a clever method using a *rheotome* (from the Greek for flow and cutting) devised by du Bois-Reymond. A rapidly rotating set of switches was so arranged that a brief stimulus could be given to the nerve and then at a recording point further along the nerve later connecting it to a galvanometer for a short time to record a brief time slice of the nerve's action potential. By systematically increasing the time between the stimulation and recording, a graphical representation of the action potential could be constructed from the time slices.[42] This principle was incorporated into an advanced version of the rheotome by *Julius Bernstein* (1839–1917), a pupil of du Bois-Reymond, who found the nerve impulse to consist of a negative wave having a duration in frog nerve of 0.7 M/sec – one quite similar to that measured today with modern techniques. He also determined its conduction velocity by stimulating at a distant point and recording the time of onset of the propagated negative wave. From the delay and the distance between stimulation and recording sites, he determined the conduction velocity of 29 M/sec, again a value close to that found using present-day instrumentation.

[42] (Hoff and Geddes, 1957).

THE IONIC BASIS OF MEMBRANE AND PROPAGATED
ACTION POTENTIALS

At the end and turn of the nineteenth century, the relatively slow conduction velocity demonstrated by du Bois-Reymond and Bernstein, compared with the vastly greater velocity of electrical conduction in metals, posed a difficulty. The nerve response was classified as a "special nervous current" or "nervous energy," with vacuous terms such as *neurorheuma* to distinguish it from a "true" flow of electricity in metallic wires.[43] The electrical nature of nerve and muscle potentials became better understood when the properties of ions in electrolyte solutions were demonstrated by Arrhenius and other physical chemists. Electrical potentials were produced in chambers divided by a semipermeable membrane with different concentrations of a salt, such as potassium chloride, present on the two sides. Membranes permeable to one ion of the pair (e.g., to potassium and not to chloride) allowed the positively charged potassium ions to be driven through the membrane from the side with a higher concentration of the salt to the lower one, leaving the side with the higher concentration negatively charged. The negativity then acts to attract the positively charged potassium back to it, thus creating an equilibrium situation in which the potential across the membrane is balanced by the potassium concentration difference on the two sides. The potential so produced is given by the *Nernst equation*.[44]

Nerves and muscles were found to conform to the membrane potentials predicted by the Nernst equation. They have a higher potassium concentration in them in comparison with the relatively low concentration present in the outside medium and an inside potential negative to that in the outside medium. Bernstein postulated that nerve and muscle fibers are bounded by a semipermeable membrane with the properties that produced potentials on the basis of the Nernst equation.[45] In his theory, when nerve or muscle is excited and threshold reached, the semipermeability of its membrane is temporarily lost. This depolarizes the membrane, the resting potential falling to zero. The resulting current flow into that region from the adjoining polarized membrane constitutes a local current that acts to depolarize it and initiate the same fall in potential to zero. The process repeated over and over at successively farther distances down the fiber constitutes the propagated action potential. The action potential at any point lasts for only a matter of a millisecond or so before the original membrane semipermeability is restored and the resting membrane potential is recovered. The speed of propagation of the action potential is the result of the relatively slow conduction of ions in electrolyte solutions and the time taken for the channels in the membrane to fully open when excited.

[43] (White, 1886).   [44] (Nernst, 1908).
[45] (Bernstein, 1902), and (Bernstein, 1912).

RECORDING OF COMPOUND AND SINGLE FIBER ACTION POTENTIAL RESPONSES

It was the introduction of tube amplifiers in the early part of the twentieth century that allowed for the revelation of the properties of the nerve action potential.[46] This was noted by Adrian, who wrote that: "Valve [tube] amplifiers for detecting wireless signals were developed during World War I and were applied to physiological research soon after."[47] A three-stage Audion tube amplifier was used by Adrian[48] and by Erlanger and Gasser[49] in their studies to amplify the small voltages developed by nerve impulses. They showed that the action potentials were compounded of the smaller voltages produced by the individual nerve fibers. These summed to give different groups of waves of the compound action potential which depend on the sizes and number of the individual nerve fibers contributing to the response. Erlanger and Gasser used an oscilloscope for recording with photographic paper set against the face of the oscilloscope tube to preserve the traces whereas Adrian used a more primitive capillary electrometer and a slit camera. Both these methods were later supplanted by cameras to photograph the oscilloscope traces and in our day by digital processing and computers.

Adrian and his collaborators recording from single fibers were able to determine the properties of the receptors. Spindle receptors of muscle generated a repetitive discharge of action potentials – the rate of discharge signaling the degree of stretch – this constituting the *sensory code*. A similar relation between the intensity of stimulation and rate of discharge of the single fibers was seen for pressure receptors, optic nerve fibers in response

---

[46] (Harlow, 1936). This development was initiated after the demonstration by Hertz in 1888 of the transmission of radio waves, followed by Lodge and Marconi who developed practical instrumentation to receive radio messages. High-frequency generation of radio waves were produced by sparks and picked up by receivers using galena crystals to rectify the received signals. These were given in Morse code originally designed for telegraphy. A much more sensitive reception was afforded when Fleming developed a vacuum tube using the "Edison" effect to rectify the signals. In it, electrons passed from a heated filament to a plate when a positive voltage was placed on it, this acting as a rectifying valve. The device was vastly improved by Lee de Forest, who inserted a grid between the filament and the plate. By means of a small voltage placed on the grid of these *Audion* tubes, a much larger current flowing from the filament to the plate was altered, in effect producing an amplification of the grid voltage. By coupling the output of the plate of one tube to the grid of another and this again to yet another, the degree of amplification was greatly increased. These tubes allowed modulating voltages of speech to be transmitted. Radio broadcasting was used by ships at sea and carried out by a multitude of radio amateurs who built their own sets, greatly spreading knowledge of the technique.

[47] (Adrian, 1928), p. 42.   [48] (Adrian, 1928), and (Adrian, 1932).

[49] (Erlanger and Gasser, 1924) and (Erlanger and Gasser, 1937).

to light, chemoreceptors, and so forth, the rate increasing logarithmically for many receptors (the Weber-Fechner law), but not for all. The repetitive discharge arises from a slow, long-maintained depolarization of the receptor terminals that acts as a *generator potential* producing a repetitive discharge in the sensory fiber.[50] Further development of electronic techniques made during World War II became available for use in physiological investigations soon afterward. Such instrumentation included precisely timed pulse generators for nerve stimulation and improved amplifiers and oscilloscopes with cameras for recording electrical potential responses.[51]

The sodium hypothesis
Just before World War II, the Bernstein hypothesis was tested in the giant nerve fibers of the squid.[52] The fibers of these invertebrates have unusually large diameters, as much as 0.5–1.0 mm, allowing a fine electrode to be inserted inside them from one end, and with another electrode placed in the electrolyte medium on the outside, the potential across the membrane could be measured directly. Using this preparation, Hodgkin and Huxley in the United Kingdom and Cole and Curtis in the United States found a voltage of 60–70 mV with the inside of the axon negative to the outside. This *resting membrane potential* was close to that predicted by the Nernst equation, given the concentrations of potassium found inside the fiber and that in the media outside the fiber.[53] However, when the fiber was stimulated to generate a propagated action potential, the action potential did not simply fall to zero as expected by the Bernstein model, but showed the inside to become 30 to 40 mV more positive than at rest. This unexpected and puzzling finding was termed the *overshoot*. Further analysis of the phenomenon had to be put aside until after the war when the overshoot was examined using a modification of the clamping device originated by Marmont in 1949.[54] The idea of clamping was that the whole of the membrane could be simultaneously brought to a determined level of potential and held there by a feedback circuit so that the resulting current through the whole of the membrane could be measured. When the membrane was depolarizing to the point where it would excite an action potential, a brief inward flow of

[50] (Ochs, 1965b) pp. 205–213.
[51] Many of these techniques were developed by the Radiation Laboratory during WWII in conjunction with the development of radar. Such instrumentation has developed to the point where studies can be readily carried out by students in teaching laboratories (Ochs, 1965b), and (Kandel, Schwartz, and Jessell, 2000). At present, much of the instrumentation is based on digital processing and the use of computers.
[52] The giant axons were rediscovered by Young (Young, 1936a) and (Young, 1936b).
[53] (Hodgkin and Huxley, 1939), and (Curtis and Cole, 1940).
[54] (Marmont, 1949).

current across the membrane was found, followed by an outward flow. The inward current was traced by Hodgkin, Huxley, and Katz to a rapid change in the semipermeability of the membrane to the sodium ion allowing the positively charged sodium ion to enter the fiber bringing the interior to a positive potential of as much as 40 mV, the overshoot.[55] Then, after a short delay, the permeability to sodium decreases and that of potassium increases, the loss of the positively charged potassium restoring the membrane potential to its normal resting value of 60 to 70 mV. The sequential movement of ions through the membrane – first sodium and then potassium, accounting thereby for the action potential and its overshoot – is termed the *sodium hypothesis*.[56] The resulting small gain of sodium ions and loss of potassium ions from the axon is reversed by a "sodium pump" in the membrane. The pump requires ATP as the source of energy to eject excess sodium from the fiber, thus keeping the level of this ion low in the fiber and the level of potassium high in the face of repeated action potential discharges.[57]

Further evidence that the membrane potentials are determined by ion concentration differences was given by expressing axoplasm from the giant fibers and replacing it with solutions of ions either similar to or different from those it normally contains.[58] Their study showed that the resting membrane potential followed the expectation of the Nernst equation: Lowering the concentration of sodium in the external medium decreased the amplitude of the action potential without much effect on the resting membrane potential, whereas altering potassium changed the resting membrane potential according to the Nernst equation.

The isolated nerve preparation *in vitro* cannot maintain its potentials indefinitely. The ion channels and the sodium-potassium ion pump in the membrane on which the potentials depend are proteins; and, like all proteins in the cells, they have a short life and must be constantly replenished. These membrane proteins are manufactured in the nerve cell bodies, and are continually moved out within its fiber by axoplasmic transport to supply it with its needed components. The characterization and molecular basis of the transport mechanism will be described in the following two chapters.

MICROELECTRODE RECORDING OF MEMBRANE POTENTIALS

The large size of the giant fiber was instrumental in establishing the ionic basis of membrane potentials. The question raised was whether this same mechanism holds for the smaller nerve fibers and muscle cells of vertebrates.

---

[55] (Hodgkin, Huxley, and Katz, 1952).
[56] (Hodgkin and Huxley, 1952). Curtis and Cole also found the overshoot but they interpreted it in a formalistic manner as being due to an inductive element in the membrane.
[57] (Baker et al., 1969). [58] (Baker, Hodgkin, and Shaw, 1962).

The fibers of the sartorius skeletal muscle of the frog are relatively large, 70 μm in diameter, allowing them to be impaled with fine tip microelectrodes. With microelectrodes made of glass capillaries drawn out to a tip diameter of 5 μm, Graham found resting membrane potentials to average 62 μm.[59] This was close, but fell short of conforming to the Nernst equation in which resting potentials of 85–95 mV were expected from their concentrations of potassium. This was most likely because of damage to the membrane, which depolarized the cell to some extent. The problem was largely overcome by Gilbert Ling, who was able to draw out microelectrodes by hand with tips well under 1 μm. With them, he found resting membrane potentials averaging 78.4 mV with, in one small group, an average of 97.6 mV.[60]

With further advances in the circuitry of the input amplifier and with microelectrodes drawn out by a mechanical needle puller to tips of 0.5 μm, Nastuk and Hodgkin succeeded in recording active potentials from muscle fibers.[61] They found an average value for the resting membrane potential of 88 mV, with action potentials measuring 119 mV – the difference giving an overshoot of 30 mV, a value similar to that found in the giant axon. And, in accordance with the sodium hypothesis, lowering the concentration of sodium in the external medium decreased the amplitude of the action potential, whereas altering potassium changed the resting membrane potential.

The microelectrode technique opened the way for Katz and his colleagues to examine the mechanism of neurotransmission in frog muscle.[62] With microelectrodes inserted close to the end-plate, at the site where the motor fiber terminates on the muscle, an *end-plate potential* (EPP) was seen to be generated in response to an action potential in its motor nerve. It occurred after a delay of a millisecond or so. This *synaptic delay* clearly showed that synaptic activation does not occur by the passage of an electric current from nerve to muscle. It occurs by the release of a *neurotransmitter* from the motor nerve terminal acting on the end-plate – the agent identified as *acetylcholine*. The EPP depolarizes the adjoining muscle fiber membrane

[59] (Graham and Gerard, 1946). These microelectrodes had been used earlier by Gelfan to localize the simulation of the muscle fiber surface (Gelfan, 1930), and (Gelfan and Gerard, 1930).

[60] (Ling and Gerard, 1949). Gilbert Ling went on to develop a theory of cell potentials that depend on selective ion accumulation in the cytoplasm and not on the selective ion channels in the membrane. Although his association-inductance theory has not stood up to the evidence for membrane potentials, there is much of interest in the wide-ranging presentation of physical-chemical information given in his book.

[61] (Nastuk and Hodgkin, 1950).   [62] (Katz and Miledi, 1965), and (Katz, 1966).

setting off a propagated action potential in it, one similar to that found for nerve.

Eccles used microelectrodes to record from the large motoneurons of the cat spinal cord. By stimulating the afferent nerve fibers entering the lumbar spinal cord terminating directly on the motoneurons, he found a similar synaptic delay of 1 to 2 M/sec between the action potential in the nerve terminals and the onset of an *excitatory postsynaptic potential* (EPSP), a depolarizing response in the motoneuron that is similar to and plays the same role as the EPP does in the muscle fiber. The synaptic activation of the motoneuron was clearly shown not to be caused by an electrical current flowing from the afferent terminals to the motoneurons. Such an electrical theory had earlier been theorized by Eccles for the neuromuscular junction and motoneurone, but it was contradicted by the evidence, and, as in the case in the neuromuscular junction, the action of a neurotransmitter was indicated for the motoneurone.[63] The synaptic delay represents the release time of the neurotransmitter from the presynaptic terminal, its diffusion to the postsynaptic membrane, the binding of the neurotransmitter to receptors for it on the postsynaptic neuron, followed by a flow of ionic currents in the postsynaptic neuron seen as the excitatory postsynaptic potential that sets off a propagated action potential in the motoneuron fiber. In addition, on stimulating afferent nerve input from an antagonist muscle acting on the motoneuron, an *inhibitory synaptic potential* (IPSP), a hyperpolarization, was found. This is the basis of the reciprocal innervation that Descartes had proposed, placing it in the periphery, at the muscles (Chapter 5, Figure 5.3), instead of centrally in the spinal cord. The use of the microelectrode technique was extended by Eccles and his colleagues and subsequently by many other investigators to a variety of neurons in the central nervous system, including systems in which cell changes were recorded that were related to behavior (Chapter 15).

[63] (Eccles, 1953) and (Eccles, 1964). The reversal by Eccles of his electrical theory to one in which neurotransmitters effect synaptic transmission, caused some of his former colleagues and supporters of the electrical theory to feel betrayed. Eccles was bolstered in his reversal by the philosopher Popper, who viewed the advance of science to come about by refutation of theories and their replacement with other theories, these in turn also to be tested by refutation (Popper, 1959), pp. 40–42.

8

NERVE FIBER FORM AND TRANSFORMATION

In the previous chapters, the concept of channels in nerve through which animal spirits are conveyed was an inference made from the empty blood vessels seen in optic nerves. When, starting in the seventeenth century, microscopes became available, they were eagerly taken up in the search for them. Despite the difficulties in handling the soft nerve tissue and imperfect lenses used in early microscopic studies, they did show nerve fibers that were cylindrical in form. Their internal composition, however, was a matter of dispute, whether fluid as the concept of moving spirits demanded, or solid as called for by vibratory theories. A resolution of this point was of major importance. When microscopes with achromatic lenses and with reduced spherical aberration became available in the nineteenth century, their greatly improved resolution showed the contents of the fibers to contain fluid and filamentous structures. The nerve fibers were seen to be extensions of the cell bodies, parts of the same entity, the concept expressed as the *neuron doctrine*. With the advent of electron microscopy, the filamentous structures within the fibers were resolved and shown to consist of several species of longitudinally organized protein polymers: neurofilaments, microtubules, and microfilaments that are collectively referred to as the cytoskeleton. The functions of these different protein structures were related to the shape of the fiber and the means by which materials are carried out into the fibers to maintain their structure and functions, by the mechanism known as *axoplasmic* or *axonal transport*, to be discussed in detail in Chapters 11 and 12. In the first part of this chapter, the history of the recognition of the form of the fiber and its constituent components is given, and, in the second part, the transformation of fiber form known as *beading*. This alteration in the form of the fiber seen in a variety of different situations, is characterized using stretch to evoke it.

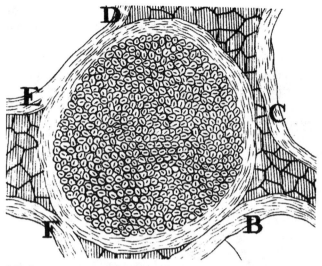

8.1. Transverse section of optic nerve pictured by Leeuwenhoek. Fascicles with sheaths are indicated by the letters **B** to **F**. The fascicle in the center shows nerve fibers as flattened with slits inside them. Leeuwenhoek describes how, when the section was first mounted in his microscope, the fibers were seen to be circular, but soon after a mound of fluid appeared over each fiber that evaporated leaving behind the flattened form. From a letter of Leeuwenhoek sent in 1719 (Leeuwenhoek, 1968), Figure 2.

EARLY MICROSCOPIC STUDIES OF THE NERVE FIBER

With the use of simple, single lens, microscopes that he had constructed for himself, *Antony van Leeuwenhoek* (1632–1723) was able to visualize individual nerve fibers and observe something of their contents. In a letter written in 1717,[1] he describes how he cut thin cross-sections of fresh bovine optic nerve and, quickly transferring the sections to his microscope, observed that they contained small circular structures with a clear interior and an outer sheath around it. However, within a minute or two, a mound of pearly fluid appeared over the middle of the fibers that soon passed off as a vapor, as the sheaths of the fibers collapsed to form flattened bands (Figure 8.1). Van Leeuwenhoek concluded that the fluid normally present within the fibers exerts a pressure distending their walls to give them their cylindrical form.

[1] (van Leeuwenhoek, 1968). This drawing, often shown, is of a nerve fascicle containing myelinated fibers that are drawn with their axonal regions collapsed to slits, a shift from their circular form he saw to occur as a result of the loss of their axoplasmic fluid.

The abbé *Felice Fontana* (1720–1805) gave further evidence for the cylindrical form of the nerve fibers.[2] Examining a longitudinal length of nerve under high magnification, he found the sheath around the individual fibers to have a wrinkled aspect, whereas their inner portion appeared to be a solid, smooth, homogeneous thread. He was able to separate the contents of the nerve fiber from their tubular sheath:

> The extremities of the nerves were placed in water, and I ran the point of the needle along them to break the cylinders, or deprive them in some way of their irregularities. I succeeded at length in meeting with one that had the form [where] ... about half of this cylinder was formed of a transparent and uniform thread; the other half was almost twice as thick, less transparent, irregular, and rugged. I then suspected that the primitive nervous fibre was formed of a transparent cylinder, smaller and more uniform, covered with another substance, the nature of which was perhaps cellular. The observations I made afterwards confirmed me invariably in this hypothesis, which at length became an established fact. I have very often seen the two parts that compose the primitive nervous cylinder. The exterior one is unequal and rugged; the other, a cylinder which seems formed of a particular transparent homogeneous membrane that appears to be filled with a gelatinous, consistent humour.[3]

To better study the individual nerve fibers, Fontana devised a *compressorium* that consisted of a pair of crystals (glass lamellae) between which a piece of nerve could be compressed (Figure 8.2).[4] By fine variation of the amount of pressure applied, the fibers could be moved apart and their individual features more readily observed:

> Having decomposed a very small nervelet into its ultimate nervous filaments [fibers], composed of different primitive nervous cylinders, of which I have spoken all along in my book, I succeeded in stripping off from the last internal envelope or tortuous filaments, several of these primitive nervous cylinders; they were transparent, homogeneous, not empty, as I had found them on other occasions.
>
> Examining thus a little mass of different primitive nervous cylinders with my instrument, I noticed that in proportion as I approximated the two sheets of crystal, there ran out of these crushed filaments a glutinous, elastic, transparent material, which the water in which the filaments bathes could in no way dissolve. When two or more neighboring cylinders were compressed together, the glutinous materials of the one did not mix at all with those of the other although these different materials were mutually compressed, and the one pushed aside the other.

---

[2] (Fontana, 1787).

[3] Ibid., p. 236. Although the description is clearly that of nerve fibers, the observation of their gelatinous content contrasts with the fluidity observed by van Leeuwenhoek.

[4] (Hoff, 1959).

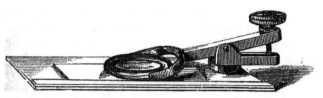

8.2. Compressorium first used by Fontana to spread and separate the individual fibers in fresh nerve. Turning the knob causes graduated pressure to be exerted on top of the glass lamellae pressing on the bottom glass sheet between which the nerve is interposed. The illustration is taken from a nineteenth century instrument supply house catalog. Referred to in (Hoff, 1959).

Continuing to compress the cylinders more and more, I came to notice that the glutinous material decomposed itself into little round grains of a diameter four or five times less than a red blood globule. I saw on this occasion that many of the little grains flowed with great celerity in the center of the primitive cylinders, and ran out at the cut ends of these cylinders. In this state, the water of the slide transported them with great rapidity from one place to another, and they did not reunite to form the glutinous material. This glutinous material, examined with the strongest lenses, at the moment when they (the grains) leave the primitive nervous cylinders, seemed to be formed of granular filaments, tenacious and elastic, which the water would neither dissolve nor separate. I confirmed this observation several times on several kinds of nerves, in several animals, and even in man, so that I do not fear to certify to its verity. It is thus necessary at present to consider the nervous cylinders as true canals, containing an elastic, glutinous, graniform material; that at least is where observation leads us.[5]

Fontana advanced a hypothesis of nerve action based on his observations of the actions of these grains:

I do not know if the Physiologists would like to regard the little grains that I have observed as the animal spirits, and as the mechanical principle of all movements. In such hypotheses one would have difficulty in explaining the instantaneous velocity of animal motions; for these little grains seem too slow to move when they are in the nerve, where they form more a viscous and inert glue than a subtle and very mobile fluid, as it seems there ought to be. Animal movements could be explained more easily by supposing that the graniform material is elastic, and continues throughout the whole length of the nerve canal, as observation demonstrates. The measurement [movement] could be transmitted in the very moment there is a mechanical alteration, or a disturbance in any part whatever of the nerve. This kind of nervous movement is indeed different than the movement attributed to animal spirits; it differs also from the hypotheses invented by the Solidists, who make the whole nerve enter into vibration.

Thus the existence of these imaginary animal spirits flowing from one part to another does not seem to be compatible with the observations that I have just reported, and the vibration of nerves is contrary to experience and to the

[5] (Fontana, 1787).

structure of the nerve itself. But it would not be at all absurd to think that the elastic, gelatinous material which fills the primitive nervous cylinders, could have insensible vibrations similar to those of the air in the formation of sound; that is to say, without there being any transport of particles from one place to another.[6]

STUDIES USING MICROSCOPES EQUIPPED WITH IMPROVED ACHROMATIC LENSES

The early microscopes of the seventeenth and eighteenth centuries, hindered by the chromatic and spherical aberration of their lenses, could give rise to fallacious images. With the development of achromatic lenses corrected also for spherical aberration, microscopes using these compound lenses became available in the early part of the nineteenth century allowing for better resolution and higher magnification. One of the first biologists to make use of the improved microscopes was *Christian Gottfried Ehrenberg* (1795–1876). In his studies of brain, spinal cord, and peripheral nerves, he saw what could have been cell bodies and confirmed the observations of van Leeuwenhoek and Fontana that nerve fibers are cylindrical.[7] In addition, he found nerve fibers having the form of a series of expanded and constricted regions along their length, a shape he called *varicose* or *articulated* (Figure 8.3). These varicose (beaded) forms were seen more frequently in sensory than in motor fibers, suggesting that this shape could be related to the function of the fiber.

Robert Remak (1814–1850), who studied microscopy with Ehrenberg and later was a member of the renowned group of investigators associated with Johannes Müller in Berlin, undertook an examination of varicose fibers. He soon found that both sensory and motor fibers could become beaded when subjected to pressure, as was the case when using the compressorium to separate the fibers, with the sensory fibers more prone to bead. Other investigators at the time also concluded that the varicosities were artifacts of preparation. In addition to pressure, exposure to air, the use of water as a medium (with, as a consequence, hypotonic swelling), and other unphysiological media were seen to produce the varicose appearance.[8] Nevertheless, Müller suggested that varicosities could be normally present in nerve, or at least that normal fibers are readily caused to bead:

> The fibres of the brain, spinal marrow, and all nerves, are, in the perfectly fresh and uninjured state, quite uniformly cylindrical, without any enlargements, but that the varicose appearance may be produced by pressure. Notwithstanding the proneness of the fibres of the brain and spinal cord to present the varicose

[6] Ibid. Note the affinity to Newton's theory (Chapter 5).
[7] (Ehrenberg, 1838).   [8] (Müller, 1840), (Todd, 1847), and (Longet, 1842).

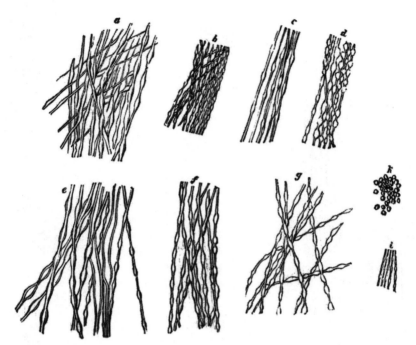

8.3. Beaded fibers present in various central nervous system regions and sensory nerves shown by Ehrenberg: (a) medulla oblongata, (b) middle medullary substance of brain, (c) spinal marrow, (d) olfactory nerve, (e) and (f) not identified, (g) from decussating optic nerves (i) retinal fibers, and (k), globules of blood and their *nuclei* (?). This last item is most likely an error. From (Ehrenberg, 1838), Plate 1.

appearance, I have frequently succeeded in separating a lamella of cerebral substance with so little violence that the fibres preserved their uniform cylindrical aspect; and I, like other observers, have seen the fibres of this form in the optic nerve and retina likewise. The fibres seemed to me to suffer most injury when it was attempted to cut too thin [a] lamina from the soft cerebral substance.... It is however, a characteristic property of the fibres of the brain, and the nerves of special sense, that they are exceedingly prone to become varicose. No other tissue has this property; and, in enumerating the characters of the cerebral fibres, it cannot be omitted. It is not quite certain on what this property depends.[9]

Although emphasizing the readiness with which the slightest compression could cause the fibers to form a bead-like appearance, Müller noted that the immediate environment also had an effect to promote beading. This was shown by fibers in which that part of it in the spinal canal is beaded and not so outside of it (Figure 8.4).

[9] (Müller, 1840), pp. 649–650.

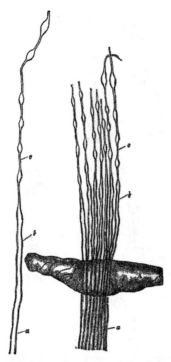

8.4. Fibers showings variation in their form. The fibers on the right are cylindrical outside the spinal canal (below a transverse bar of tissue) and beaded within it. A single such fiber is shown on the left. Constricted regions (α) and cylindrical regions (**a**) of beaded fibers. From Valentin in (Todd, 1847), Volume 3, Figure 330.

REMAK'S FUNDAMENTAL OBSERVATIONS ON THE RELATIONSHIP OF THE NERVE FIBER TO THE CELL BODY: HIS "EARLY NEURON" VIEW AGAINST THE "TWO-NEURAL ELEMENT" CONCEPT OF PURKINJE AND VALENTIN

Remak found that the fibers to be composed of a solid or gel-like core or band surrounded by an investing sheath that was rougher and apt to appear wrinkled.[10] These observations were made without knowing the work of Fontana; but, when he became aware of it, he gave due credit to his predecessor. However, Remak found that the band-like interior was not composed of small granules as Fontana had proposed, but that it had a gel-like consistency with a finer fibrillary structure contained in it. This view of Remak's was recognized by the eminent Czech physiologist *Johannes Evangelista Purkinje* (1787–1869), who gave the term *axis cylinder* to the interior core, the term used to this day. However, Purkinje and his younger colleague *Gabriel*

[10] (Remak, 1838).

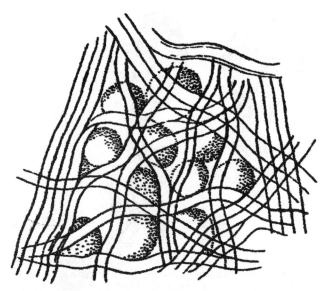

8.5. Fibers and cells in the otic ganglion of a sheep. These were considered by Valentin to be interlaced, closely apposed but distinct and separate neural entities. From (Steida, 1899), Table 1 and Figure 10.

Gustav Valentin (1810–1883), along with the noted histologist *Jakob Friedrich Henle* (1809–1885), regarded the interior of the fiber as being fluid. Müller commented on the difference of opinion between the two conflicting positions, writing that:

> Both Valentin and Henle regard Remak as incorrect in describing the contents of the nervous cylinder to be solid. In the fresh state, when no reagent had acted on the nerves, they found the fibres to contain a perfectly transparent and fluid matter, which the addition of water, however, caused to coagulate. The flattened pale filet which remains after a nervous fiber has been subjected to pressure, and which Remak supposed to be the solid fibre freed from the investing tube, appeared to Henle to be the tube itself emptied of its contents.[11]

The question as to whether the core was gel-like or fluid was of fundamental importance with regard to the theory of how the fiber functioned and also how the fibers were related to the cells that Ehrenberg had observed in the central nervous system. Purkinje and Valentin verified that they truly were cells on the basis of the presence of a nucleus and nucleolus within them.[12] At this point in time, the relation between the cells and fibers became a defining issue. Purkinje and Valentin thought of cells and fibers as two separate neural elements that, in some places such as the ganglia, are in close apposition but not connected (Figure 8.5). They presumed the cells to

[11] (Müller, 1840), p. 648. [12] (Purkinje, 1838).

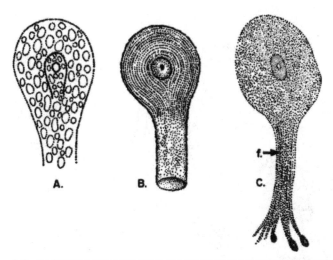

8.6. (**A**) Globular structures within the cell body shown by Valentin. (**B**) Fibrillary structures in the cell seen by Remak. (**C**) Fibrillary structures appear also in the fiber as shown by Remak (**f**), indicating that the fiber is an integral extension from the cell, making him the first to propose such an "early neuron doctrine." Redrawn from Remak (Remak, 1838) in (Ochs, 1975d), Figure 2.

be "energy generators" causing a circulation of the fluid within the hollow nerve fibers. The fibers were considered to have the form of long continuous loops, descending to innervate a peripheral organ such as a sensory organ or muscle, looping to ascend to the brain then looping back down to form a closed path for the circulating fiber fluid.[13] The energy for the movement of the nerve's fluid was gained in passing through a ganglion wherein the cells approximated to the fibers communicates its energy for the movement of nerve fluid. This Purkinje-Valentin theory requires a highly mobile fluid able to move within the looped fibers, and Remak's claim that the inner core of the fibers was solid or gel-like constituted a major obstacle to their theory. An even greater obstacle was Remak's observation that the fibers arise from the cells, appearing to constitute a single neural entity rather than that they are approximated (Figure 8.6). This fundamental view of Remak's was called an *early neuron doctrine.*[14]

Support for the unity of cells and fibers in the sense of Remak was given by the eminent physiologist *Hermann Ludwig Ferdinand von Helmholtz* (1821–1894). For his doctoral thesis, under the direction of Müller, he used invertebrates with large cell bodies and fibers where the origin of nerve fibers from the cells was clearly evident.[15] Other prominent investigators also

[13] (Van der Loos, 1967), and Ochs, 1975c).   [14] (Ochs, 1975c).

[15] (Helmholtz, 1842). Interestingly, in his thesis, Helmholtz did not give credit to Remak for priority in this regard, although as a fellow member of Müller's circle

supporting Remak's findings were Hannover, Koelliker, and Deiters, whereas among other noted neurobiologists who gave support to the Purkinje-Valentin theory were Prevost and Dumas[16] and Burdach.[17] They pictured fibers in the form of long loops, with the ends of the loops contacting muscles and sensory organs similar to those endings that Valentin had presented.[18] Although Müller pictured such loops in his authoritative book on physiology,[19] he also showed fibers terminating in the corpuscle of Pacini,[20] on muscle fibers, at the specialized structure of the muscle end-plate[21] – findings that contradicted the Valentin theory of a continuous looping of fibers. He wrote that:

> It would be important to know whether the large globules of the grey substance in the brain and in the ganglia have, or have not, any connection with other parts or with one another. Certain processes, which are, under favourable circumstances, seen issuing here and there from the globules, suggest the probability that they are connected with each other by fibres. I saw these tooth-like processes first on the club-like bodies in the medulla oblongata of the *petromyzon*. Remak observed them *soon afterwards* [my italics] on the globules [cells] of the grey substance of the brain and ganglia. He not only saw fibres coming off from the surface of the globules of the ganglia, but succeeded even in isolating them [showing their length] to the extent of several or many times the diameter of the globules.[22]

This passage seems to suggest that Müller was advancing his own priority in the discovery of the connection of fibers and cell. However, in a later edition, he more clearly gave credit to Remak – enlarging on the confirmation of Remak's view given by Helmholtz, Hannover, and Wagner.[23]

Nevertheless, Valentin held to his position that cells and fibers in the ganglia are separate, but contiguous neural elements – later conceiving of the cell body as lying within a widened expanse of the nerve fiber with a membrane around it that supposedly separated it from the fiber proper. However, the evidence for that supposition was unsatisfactory and the influential Koelliker concluded that "the contents of the nerve cells cannot be understood as being contained in a dilated nerve tube."[24]

he must have been very familiar with Remak's work, perhaps even having learned some microscopy from him. See (Kisch, 1954).
[16] (Prevost and Dumas, 1823).   [17] (Burdach, 1838).   [18] (Valentin, 1841).
[19] (Müller, 1840), pp. 51, 52, and 520, showing figures given by Burdach.
[20] Ibid., p. 519.
[21] (Clarke and O'Malley, 1968), p. 77 and Figures 18–20. The plate-like structure is due to the folds of the muscle membrane under the terminating nerve fibers where acetylcholine receptors are densely represented. The neurotransmitter acetylcholine was later shown to act at this site (Chapter 7).
[22] (Müller, 1840), p. 657.   [23] (Müller, 1845), pp. 525–526.
[24] (De Reuck and Knight, 1966), and (Kolliker, 1853).

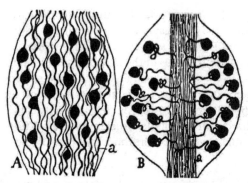

8.7. (A) Bipolar neurons shown in the ganglion of a fish on the left. In (B), the section taken from a mammalian ganglion shows its T-shaped neurons. From Cajal, redrawn in (Ochs, 1975d), Figure 4.

Some of the evidence that seemed to have supported Valentin's view had to do with the apparent lack of connection of fibers and cells in the mammalian dorsal root ganglia. In fish, it was clearly seen that the dorsal root ganglia of fish are bipolar, with fibers originating from the cell body; whereas in the ganglia of mammals and birds, no such obvious connection

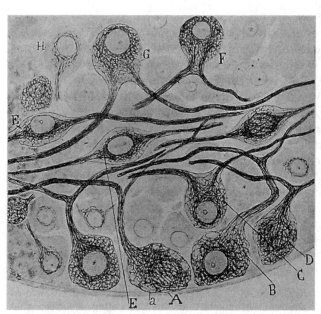

8.8. Early development of spinal ganglion neurons in a chick embryo on the 14th day of incubation showing their transformation from bipolar (E) to (C) (F) (G) and then to T-shaped neurons (A), (B). The fibers from the poles come together and join for a short stretch to form the initial segment, with its outer two projecting branches constituting the T-shaped neuron. From (Ramón y Cajal, 1995), Volume 1, Figure 268.

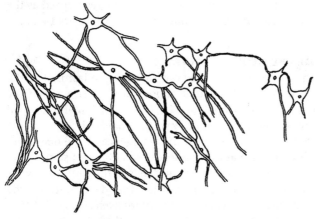

8.9. Direct interconnectivity of cell bodies and fibers in the ventral horn pictured in (Schroeder van der Kolk, 1859). Redrawn in (Steida, 1899), Plate ii, Figure 21.

appeared evident (Figure 8.7). This can be understood because, from their ganglion cell bodies, the fiber arises at a relatively small place on the large cell body and can easily be missed. Clear evidence for their connection was later provided by *Louis-Antoine Ranvier* (1835–1922). He fixed the ganglion *in situ* with osmium and then teased out the individual cells with their fiber extensions still intact.[25] From the cell body, an initial fiber emerges that then divides into two branches: one ascending in the dorsal root to enter the spinal cord, and the other descending as a sensory fiber in the peripheral nerve. The cell with its stem and branches forms a T-shaped neuron. Cajal[26] then showed that, in the early stage of development in birds (and also in mammals), the ganglion neurons are at first bipolar like those in the fish; but, then the fibers from the poles move together and merge to form the initial stem segment of the T-shaped neuron of the adult (Figure 8.8).

In the spinal cord, and elsewhere in the central nervous system, numerous tree-like fiber projections from the cell bodies were found, the *dendrites*, which *Otto Freidrich Karl Deiters* (1834–1863) was able to differentiate from the axon.[27] As later shown by Cajal, axons contact the dendrites and the cell bodies of cells that in turn send its axons to contact dendrites and cell bodies of other cells. The inability to visualize the endings of axons on the dendrites and cell bodies with optical microscopy led to the hypothesis that cells were directly interconnected in a reticulum or syncytium (Figure 8.9).[28] The interconnecting fibers were supposed to modulate the flow of excitation from cell to cell through the network and so provide for the integration of function. With the acceptance of the Neuron Doctrine, the individuality

[25] (Ranvier, 1875).   [26] (Ramón y Cajal, 1995), Volume 1, Figure 268, p. 521.
[27] (Deiters, 1865).   [28] (Schroeder van der Kolk, 1859).

of neurons, their synaptic interconnectivity has become recognized as the basis of the nervous system's integration and behavior (Chapters 14–16).

CONFIRMATION OF THE NEURON DOCTRINE

The Remak-Valentin controversy dragged out through the last half of the nineteenth century until incontrovertible support for the concept of the neuron was advanced: *Wilhelm His* (1831–1904) (a pupil of Remak's) showed the growth of neurites from their cell body in the course of embryogenesis[29]; *Fridtjof Nansen* (1861–1930) showed the emergence of the fiber from its cell body in invertebrates[30]; and *Auguste Forel* (1848–1931) showed degeneration of fibers after transaction from their cells in the central nervous system,[31] similar to the distal amputated portion that degenerates in the peripheral nerves (Chapter 9), as detailed by *Santiago Ramón y Cajal* (1852–1934).[32] Acceptance of the neuron doctrine was further strengthened by evidence that the nerve net in the jellyfish consists of contacts between individual neurons rather than a reticulum, as had generally been thought. At the end of the nineteenth century, *Wilhelm Waldeyer* (1837–1921) summarized the evidence for the origin of fibers from the cell body giving the concept its present name: the *neuron doctrine*.[33] The neuron was accepted as the basis of the organization of the nervous system by Barker,[34] Morat and van Gehuchten[35] in their comprehensive monographs on the nervous system published near the turn of the nineteenth century. The great neurophysiologist *Charles Scott Sherrington* (1857–1952) based his studies of the reflex behavior of the nervous system on the neuron doctrine, introducing the term *synapse* for the junction at which the axon of one neuron contacts another to influence its action.[36]

However, an objection to the neuron doctrine was raised by *Camillo Golgi* (1843–1926). He proposed that the dendritic arborizations of neurons served a metabolic function for the cell body and that it lacked the power of conduction.[37] It was the inability to resolve the synaptic junction of fibers on the dendrites microscopically at the time that gave room for Golgi's belief. Another opposition to the neuron doctrine was advanced by Apathy[38] and Bethe,[39] who used stains that appeared to show fibrils passing from one neuron directly into that of another, making them act as a

---

[29] (His, 1887).
[30] (Nansen, F., 1887). The famous Norwegian biologist and arctic explorer was given a Nobel Peace Prize for his humanitarian work in feeding the Russians starving as a result of WWI.
[31] (Forel, 1887).   [32] (Ramón y Cajal, 1954).   [33] (Waldeyer, 1891).
[34] (Barker, 1899).   [35] (Morat, 1906), and (van Gehuchten, 1900).
[36] (Sherrington, 1906). French gives the history of the term "synapse" (French, 1970).
[37] (Golgi, 1967).   [38] (Barker, 1899).   [39] (Bethe, 1904).

syncytium. These fibrils were shown not to be present,[40] evidence later verified by use of the electron microscope that fully established the autonomy of the neurons and confirmed the neuron doctrine as the basis on which the nervous system is organized.[41]

With the period of controversy lasting most of the latter half of the nineteenth century until the neuron doctrine became accepted, the priority of Remak in proposing its early formulation was overlooked, or only grudgingly recognized.[42] This led Kisch to refer to him as one of the "forgotten leaders of medicine."[43] Only in the latter part of the twentieth century has Remak's great pioneering contributions to neuroscience been more justly recognized.[44]

THE SCHWANN CELL IN RELATION TO THE AXON OF THE NERVE FIBER

Theodore Schwann (1810–1882), who had been associated with Müller, was famous for showing, with *Matthias Jakob Schleiden* (1804–1880), that the tissues of animals and plants consist of cells. In his embryological studies of muscle, Schwann found that individual cells coalesce to form the muscle fiber. Nuclei were also seen distributed at intervals along the length of developing nerve fibers, and Schwann undertook an investigation of these cells that we now refer to as the *Schwann* cells. He specifically questioned Remak's view that the axis cylinder is the functional part of the nervous fiber, with the myelin sheath subsequently laid down around it to form the nerve fiber.[45] Schwann referred to Remak's work on the embryological

---

[40] (Ramón y Cajal, 1971).   [41] (Shepherd, 1991).   [42] (Steida, 1899).

[43] (Kisch, 1954). Kish refers the difficulties experienced by Remak to the widespread antisemitism present in Germany at the time and to Remak's "unbending Judaism." Some of the great leaders of science at the time who were liberal recognized Remak's worth – Müller and, above all, the eminent *Alexander von Humboldt* (1769–1859) who tried to gain a full professorship for him in Germany, but to no avail. Remak became a lecturer and a professor *ordinarius*. These positions, without the funding connected with a full professorship, forced him to practice medicine to gain a livelihood and hampered his research. His scientific opponent, Valentin, though also Jewish, gained a professorship in the more liberal environment of Switzerland. Their opposite professional careers had roots in their personalities. Remak was little inclined to make allowances for his Jewish background and resented the gratuitous slights given him for his "Jewishness," whereas Valentin, with a much more congenial nature, won friendship and advancement. However, Valentin suffered from stubborn pride, which made him unable to alter his position until, when he finally did accede that Remak was right, it was too late to help him achieve the recognition and position he deserved.

[44] (Meyer, 1971), (Clarke and O'Malley, 1968, p.46), and Clarke and Jacyna, 1987). A book on the life and achievements of Remak, based on recently examined German sources, has been published (Schmiedebach, 1995).

[45] (Schwann, 1969).

development of nerve in the rabbit, quoting him as saying that "the substance of the cerebrospinal nerves of the rabbit, in the third week of embryological existence, consists of corpuscles, some of which are irregularly spherical, others slightly elongated, having a very delicate filament adhering to them; they are most transparent, and arranged in rows."[46] Schwann decided that the nerve fiber is in reality formed by the union of these cells. In his studies of the fetal pig fibers, he designated the cells as "the primitive structure of nerve for the younger the foetus the greater is their relative quantity,"[47] analogizing these cells to the muscle cells coalescencing to form fibers.

This view was controverted by the observations made by Harrison in the first decade of the twentieth century when he showed the outgrowth of nerve fibers from the cell bodies without the Schwann cells, using the technique of tissue culture that he had developed.[48] In a letter written by the eminent neurophysiologist *Walter Holbrooke Gaskell* (1847–1914) he commented favorably on Harrison's finding, stating "the idea of a chain of cells forming a nerve fibre has always seemed to me absurd and I imagine is universally discredited although at the same time the sheath cell may assist in providing pabulum for the young nerve."[49]

THE STRUCTURE AND FORMATION OF THE MYELIN SHEATH
The functional relationship of the Schwann cells to the axon was suggested by Galvani, who at the end of the eighteenth century had supposed that the nerve fiber to be sheathed with lipid – acting as an electrical insulator – that constrains its animal electricity to flow inside the fiber without being dissipated in the surrounding tissue (Chapter 7). The view that the lipid was present in fluid form was held well into the twentieth century. Young, in his theory of axonal flow (Chapter 11), viewed the myelin sheath as a fluid that shows the property of surface tension to account for the beading of nerve fibers, much as drops of dew appearing on spider webs, viscous fluids on thin glass fibers, and so forth.[50] As evidence, Young showed the beaded appearance of fibers in the early stage of Wallerian degeneration, the phenomenon to be discussed later in Chapter 9.

---

[46] Ibid., p. 143. In addition to his fundamental work on the nervous system, Remak was a pioneer in embryology and is given credit for first defining the three germ layers as described by Papez in (Haymaker and Schiller, 1970), pp. 66–69. As we know, the axon is of ectodermal origin, and the Schwann cell is of mesodermal origin.

[47] Ibid., pp. 143–144.   [48] (Harrison, 1910).

[49] Gaskell, W.: Autograph letter signed "to E. H. Starling. Cambridge, August 8, 1910." The letter is in the possession of the Royal College of Physicians of London. Referred to in (Ochs, 1975c), p. 253.

[50] (Plateau, 1873). See also (Thompson, 1942).

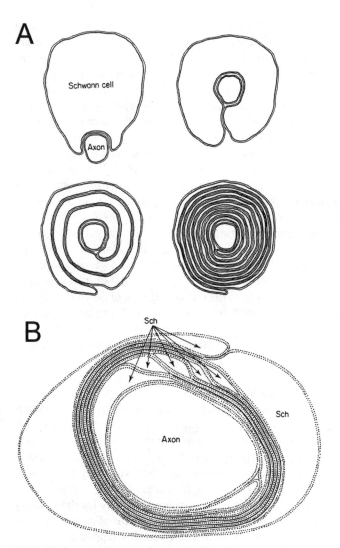

8.10. (A) Genesis of the myelin sheath shown by an axon invested by the Schwann cell membrane. The membrane of the Schwann cell engulfs the axon and meets to form a double membrane known as the mesaxon. The mesaxon then becomes wrapped around and around the axon to form the multilayered myelin sheath. Larger fibers may contain as many as several hundred lamellae. This representation shows a much larger relation of the myelin sheath to the axon than is usually present in the well-fixed nerve. From (Geren, 1954). (B) Diagram of myelin lamellae in a cross-section of a small myelinated nerve. The merging of the Schwann cell (Sch) membranes to form the mesaxon is shown at the top, where the double-membraned lamellae can be followed as it wraps around the axon. At the lower right, the membrane surrounds the axon joining to form the *mesaxon*. In this figure, separation of some of the myelin lamellae by the cytoplasm of the Schwann cell is seen at the top with arrows from Sch, the incisures of Schmidt-Lantermann. From (Robertson, 1960). Redrawn in (Ochs, 1965), Figures 6.5 and 6.6.

However, examination of the myelin sheath by polarized light indicated that the myelin sheath has a highly organized structure. X-ray diffraction studies revealed that it is composed of a series of layered membranes wrapped around the axon, with the myelin lipid molecules in the membranes arranged radially in an ordered structure.[51] The X-ray analysis showed a *fundamental period*, a spacing between the layers of close to 170 Å. This brought up a problem. Electron microscopic studies of cell membranes stained with osmium showed them to consist of a *unit membrane* of approximately 75–80 Å in thickness, with the membrane formed of lipids oriented radially in the membrane as a bilipid layer.[52] How the bilipid membranes of the Schwann cells form the myelin sheath was shown by Betty Geren in her electron microscopic study of the embryonic development of the nerve fiber.[53] She found that axons first grew out and then Schwann cells moved down along them to become positioned at intervals along their lengths. The Schwann cells then expand circumferentially to engulf the axon, the Schwann cell membranes meeting around the axon forming a double-membrane *lamella* (of twice the thickness of the unit membranes, making the lamellae 170 Å in thickness) that is then laid down in jelly-roll fashion around the axon (Figure 8.10). The adult myelin sheath so formed may contain as many as a hundred or more such lamellae – the larger the axon, the greater the number of lamellae and the thicker the myelin sheath.

THE FUNCTIONAL ROLE OF MYELINATION
Each Schwann cell forms an independent segment of the myelin sheath 0.5 to 1.5 mm in length (the length proportional to the size of the axon), with a narrow gap left between the territories of the neighboring Schwann cells spaced along the fibers that constitute the internodes. The electrical resistance of multiple layers of membrane is high, with only the narrow gap between the territories of the myelin, the *nodes of Ranvier*, allowing ion movement through the axon membrane at this site to serve for the conduction of the action potentials (Chapter 7). The constraint of current to flow only through the nodes between the myelin sheath wrapped around the axon at the *internodes* acts to greatly speed up the conduction of the action potential. Lillie demonstrated this with his iron-wire model. It consisted of short segments of glass threaded along the wire to represent myelin insulation, the small spaces between them the nodes causing the slow spread of the surface change seen when placing an iron wire in an acid medium to be considerably speeded up.[54] The flow of current takes much less time that the surface change in the iron of the model. A similar faster ion current flow

[51] (Schmitt, Bear, and Palmer, 1941), and (Finean, 1958).   [52] (Robertson, 1960).
[53] (Geren, 1954).   [54] (Lillie, 1925).

occurs within the myelin covered part of the axon, with slower channel changes at the nodes. The relatively fewer such slower sites in myelinated fibers accounts for their faster conduction in comparison to unmyelinated fibers of similar dimension. The process is called *saltatory conduction*, from the Latin *per saltum* (by leaps).

At first, the myelinated fibers in the spinal cord and elsewhere in the central nervous system seemed to pose a problem in that their conduction velocity was fast, similar to that of peripheral myelinated fibers. But, they did not appear to have the same distribution of Schwann cells along their lengths. They were, however, shown to be similar to peripheral nerve fibers, with glial cells in the central nervous system taking the place of Schwann cells, the difference being that, in the central nervous system, a single glial cell, the oligodendrocyte, has cytoplasmic extensions to several axons – each of the extensions wrapping around the axons to form the jelly-roll layering of their myelin sheath.[55]

Yet another early finding that seemed to threaten this view of the propagation of the action potential was that of the apparent openings or slits in the myelin sheaths seen at intervals along the internodes, these known as the *clefts* or *incisures of Schmidt-Lantermann*.[56] If such openings were truly present, a saltatory conduction through the nodes could not be sustained. One explanation offered was that the incisures were artifacts produced in the course of histological preparation. However, electron microscopic preparations showed all the lamellar layers to be present – the appearance of incisures seen with optical microscopy caused by some degree of separation between the lamellae and the interposition of Schwann cell cytoplasm (cf. Figure 8.10B). This is considered to be the means by which materials formed in the Schwann cell nucleus are sent into the sheath to replenish their structural components.

In addition to the myelinated fibers that attain a maximal diameter of 24 μm or so in mammals, smaller unmyelinated fibers of the order of several microns were discovered by Remak, the *fibers of Remak*. These do not have nodes, but have cell bodies distributed along their length in an irregular fashion. Electron microscopy was needed to resolve their fine structure. The fibers were found by its means to be composed of a group of small axons, from a few to several score or more, each less than 1 μm.[57] The axons are partially engulfed by the membrane of the Schwann cells investing them (cf. Figure 8.19).[58] They give the appearance of an arrest of the myelination seen in the larger fibers at an early stage of myelination, often before the membranes meet to form lamellae (Figure 8.10A).

---

[55] (Bunge, Bunge, and Ris, 1961).   [56] (Hiscoe, 1947).
[57] (Gasser, 1958).   [58] (Thomas, Berthold, and Ochoa, 1993).

FIBRILLARY ELEMENTS WITHIN THE AXOPLASM

As discussed previously, on examination of fresh nerve fibers before adequate fixation and staining was available – as was the case in the early part of the nineteenth century – it was hard to be certain that fibrillary materials were present in them as had been reported by Remak.[59] Valentin pictured globular structures rather than fibrillary material, but Remak pointed out that the fibrillary structures were fragile and may well have disintegrated in the course of preparation to form the globular inclusions that were pictured by Valentin (Figure 8.6). Valentin inverted the argument saying that the globular structures are labile and fibrillary artifacts appear rapidly after death or are generated by the fixatives then in use, which included such substances as alcohol, ether, acids, and metallic salts[60] – agents that could readily produce artifacts generating faulty interpretations.[61]

Fibrillary elements in the axons were later more fully established when using silver and other more reliable stains.[62] However, there still remained the suspicion that the stains could generate the fibrillary structures seen.[63] Evidence for the reality of fibrils was advanced by *Sigmund Freud* (1856–1939).[64] Using the large neurons in the crayfish, and examining fresh unstained tissue, Freud demonstrated a large number of fibrils present in the cell body that passed directly down into their fibers.[65] It is noteworthy that, at the time when Ringer solutions were still in the future, Freud used the blood of the animal to keep neural tissues in a normal physiological state when preparing them for microscopic examination. Although such observations would appear to have settled the matter, reservations regarding the reality of fibrils persisted. *Sir William Maddock Bayliss* (1860–1924), in his influential textbook *Principles of General Physiology* (published in 1914), referred to the so-called "neurofibrils" calling them artifacts produced by the methods used in histological preparation "and in no sense were they to be considered a constituent of the living nerve cell."[66] On the other hand, Bozler gave strong evidence in support of their presence in the fresh fibers of the jellyfish *Rhizostoma*.[67]

DIFFERENTIATION OF THE FIBRILLARY STRUCTURES IN THE AXON AND THEIR PUTATIVE FUNCTIONS

An early hypothesis advanced for the function of fibrils was that they were responsible for the form of the fiber. This was rejected by Bozler, who thought they were not firm enough to serve that function. By exclusion, he supported the view that had earlier been advanced by Schultze, namely that

[59] (Müller, 1840), p. 596. [60] (Valentin, 1843). [61] (Liddell, 1960).
[62] (Schultze, 1870). [63] (Burnside, 1975).
[64] (Triarhou and Del Cerro, 1987). Freud started his career as a neurobiologist, undertaking a study of the nerve cells and fibers of invertebrates before gaining fame later on as the Father of Psychoanalysis.
[65] (Freud, 1882). [66] (Burnside, 1975). [67] (Bozler, 1927).

their role was to conduct the action potential. Parker also thought this at first, but he later reversed his opinion and advanced the hypothesis that the fibrils served to carry components made in the cells out into their fibers to support their function.[68] This is basically the concept of axoplasmic transport to be discussed at length in Chapters 11 and 12.

The fibrillary material in the axons, seen when stained with silver and other stains, was resolved with the used of electron microscopy into several classes of protein polymers. One such species is that of the *neurofilaments* measuring 100 Å in diameter (Figure 8.11).[69] Using gel filtration, they were later found to consist of a triplet of three proteins with molecular weights of 200,000, 160,000 and 68,000, respectively.[70] The rod-like core consists of coiled polymers formed by the 68,000 proteins, which is longitudinally oriented in the fibers with their other proteins projecting laterally from it as side-arms.

A thicker structure was also fleetingly seen with osmium tetroxide, which was generally used in early studies for fixation and staining. It was approximately 250 Å in diameter and appeared to be hollow-walled.[71] When glutaraldehyde and similar aldehydes were introduced in 1963 for fixation,[72] these structures were revealed to indeed be tubular and were named *microtubules* or *neurotubules*. In a review published only a few years later, Porter described the microtubules as being present in many, if not all, of a wide variety of cell types: plants, protozoa, and animal cells.[73] They are prominent in the long spicule-like extensions of the protozoan *Tokophyra*, in which particle movement was seen, and in cilia and flagella that move by a sinuous bending motion – suggesting early on that they would be responsible for its form. This appeared to be the case for their presence in neurons: Porter also conjectured that "they provide the motive force and directional guidance for such flow of particles and metabolites as take place in the dendrites (and axon), and we would suggest further that here, as in other instances, they may process form-building and form-maintaining functions."[74] Porter also noted the presence of ATPase associated with the microtubules, a prescient observation that has turned out to be a key component of the mechanism of transport, as will be described in Chapters 11 and 12.

The microtubules oriented longitudinally in the fibers are composed of a heterodimer of two tubulin proteins, α and β, having molecular weights of

---

[68] (Parker, 1929a), and (Parker, 1929b).
[69] (Shelanski and Liem, 1979), (Willard, 1983), and (Marotta, 1983).
[70] (Shelanski and Liem, 1979).   [71] (Fernández-Morán, 1952).
[72] (Sabatini, Bensch, and Barrnet, 1963), (Karnovsky, 1965), and (Porter, 1966). The superiority of aldehydes for preservation of the microtubules and other axonal structures is due to the ease with which the monomer rapidly enters the fibers to quickly cross-link the proteins. This reduces the formation of regions of coagulation and shifts of fluid that give rise to the distortions and destruction of organelles seen with other fixatives.
[73] (Porter, 1966).   [74] Ibid., p. 325.

A

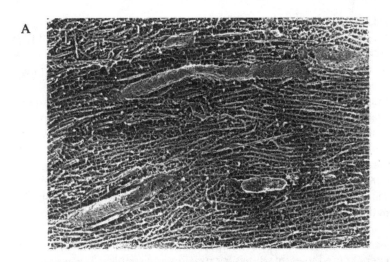

B

8.11. (A) Neurofilaments shown in a longitudinal section of nerve. A long, sausage-like mitochondria is seen with cross-arms of the neurofilaments connecting to them. (B) At higher magnification, several thicker microtubules are seen with cross-arms appearing to interconnect microtubules and neurofilaments. From (Willard, 1983), Figure 4.1.

57,000 and 53,000 (Figure 8.12B), respectively.[75] The dimer tubulin subunits join end to end to form a linearly organized polymer, the *protofilament* – with thirteen of the protofilaments aligned longitudinally side to side to form the wall of the microtubules (Figure 8.12A).[76] Other proteins, the *microtubular-associated proteins*, extend laterally from the wall as sidearms,[77] which appear to cross-link with those of other microtubules and neurofilaments. Such cross-linking appears evident in the preparations produced by

---

[75] (Dustin, 1984).
[76] (Amos, Linck, and Klug, 1976), (Amos et al., 1976), and (Kim, Binder, and Rosenbaum, 1979).
[77] (Kim et al., 1979).

A

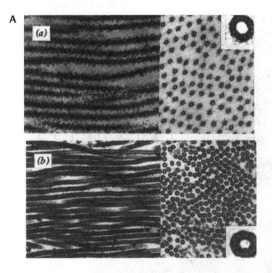

B

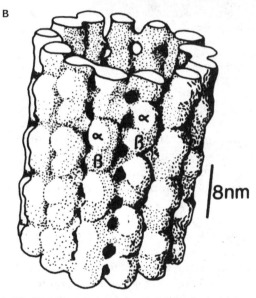

8nm

8.12. (A) Microtubules shown in longitudinal and in cross-section. In the upper panel, microtubule-associated proteins is seen as side-arms of the microtubules **(a)**. In the inset, on the right, the ring-like structure of a microtubule contains thirteen protofilaments joining laterally to form its wall. In the lower panel, microtubules assembled from tubulin without microtubule-associated protein are smooth-walled **(b)**. From (Kim, Binder, and Rosenbaum, 1979), Figure 9. **(B)** Wall of a microtubule formed by tubulin dimers composed of α and β proteins that are linked end to end to make the protofilaments, with thirteen protofilaments joined side to side forming the microtubular wall. Note staggering of tubulins in the neighboring protofilaments. From (Amos, Linck, and Klug, 1976), Figure 3.

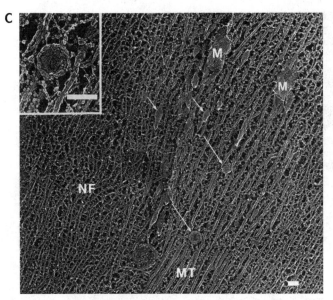

8.12. (*cont.*) (**C**) Domains of microtubules (**MT**) and neurofilaments (**NF**) in fibers prepared by quick-freezing and deep-etching showing cross-linking of the side-arms. Short arrows point to tubular structures, long arrows to globular structures. **M** indicates mitochondria. Inset shows details of a globular structure with connecting side-arms to the microtubules. From (Hirokawa et al., 1989), Figures 8 and 9.

Hirokawa using the technique of quick-freezing and deep-etching (Figure 8.12C).[78] A still more extensive interconnectivity of side-arms was seen using the technique of "critical point drying." Such connectivity appeared not only between the cytoskeletal organelles, but also to mitochondria, endoplasmic reticulum, and other structures in the axon.[79] However, Ris showed this picture to be an artifact brought about by traces of water present during the critical point drying process.[80]

Of these two organelles, is one more responsible in conferring form on the fiber than the other? The proportion of neurofilaments to microtubules varies considerably in the different types and sizes of nerve fibers. In the dendrites, unmyelinated axons and smaller myelinated fibers, microtubules are the predominant structure; and, on that basis, they would appear to determine form. However, in the larger myelinated fibers, much the greater proportion of the cytoskeleton is of neurofilaments.[81] These differences in their proportion in the various fiber types suggest that some other factor(s) is responsible for fiber form. Does the apparent cross-linking of the side-arms of cytoskeletal organelles confer form on the fibers by increasing their rigidity –

---

[78] (Hirokawa, 1982).   [79] (Ellisman and Porter, 1980).   [80] (Ris, 1985).
[81] (Friede and Samorajski, 1970), and (Hoffman and Griffin, 1993).

a concept that we may call the *rigid matrix* hypothesis? This possibility is discussed in the following sections of this chapter dealing with the form change characterized as beading.

A third class of polymer proteins found in the axon is that of the *microfilaments*, 80 Å in thickness, that are composed of actin subunits.[82] Unlike the axially directed microtubules and the neurofilaments, the actin filaments are irregularly oriented within the axon and are more densely concentrated near the axolemma. The microfilaments play an important role with respect to the molecular components closely related to the axolemma, the *membrane skeleton*, which has been held responsible for cellular form. How it contrasts with the axially oriented microtubule and neurofilament cytoskeleton organelles in relation to fiber form change of is described in the following sections.

CHARACTERIZATION OF THE TRANSFORMATION OF FIBER FORM SEEN AS BEADING

As indicated in the beginning of this chapter, Ehrenberg and other early pioneers of optical microscopy of the nerve had observed the marked variation in their form seen as varicosities or beading, which is characterized by alternating constrictions and expansions along the fiber rather than it appearing as a relatively uniform cylinder. Beading was ascribed to a pathological change, either to the pressure entailed in the use of the compressorium, of an unphysiological medium used in its preparation for microscopic examination, or as the result of neuropathy.[83] However, the beaded form can readily be produced at will in the fresh nerves of the cat, rat, and other animals by subjecting them to a mild stretch.[84] This allowed the properties of beading to be studied. But the beaded form is labile, and not ordinarily preserved by the fixatives commonly used for histological examination. The technique that preserves beading was that of freeze-substitution.[85] In unstretched

---

[82] (Goldman, Pollard, and Rosenbaum, 1976), (Kuczmarski and Rosenbaum, 1979), and (Matsumoto, Tsukita, and Arai, 1989).

[83] (Ochs et al., 1997).   [84] (Ochs and Jersild, 1987).

[85] The beaded form change is held in place while under stretch by the use of freeze-substitution (Ochs et al., 1994). In this technique, the nerve is rapidly frozen in freon 12 or isopentane at a low temperature of about $-140°C$, and the frozen nerve tissue is transferred to a substitution medium at a low temperature of about -50 to $-60°C$ containing acetone or alcohol and osmium tetroxide. Over a period of days to weeks, the ice of the tissue is gradually replaced by the solvent, the replacement occurring gradually in laminar fashion as the solvent moves deeper and deeper into the tissue until the fixative has entirely replaced the ice of the tissue. By this means, an almost instantaneous snapshot of the original form of the fibers at the time of fast-freezing is obtained. The technique has been described by a number of investigators (Harvey, 1981), (Ornberg and Reese, 1981), (Benshalom and Reese, 1985), and (Van Harreveld and Crowell, 1964).

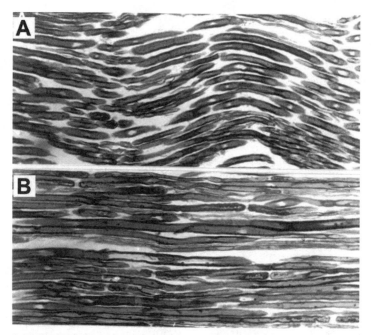

8.13. Longitudinal sections of quick frozen and freeze-substituted nerves showing myelinated fibers with their dark staining myelin sheath. **(A)**. Unstretched nerves have a wavy disposition. A very small stretch of less than a gm suffices to straighten them **(B)**.

peripheral nerve, its fibers are disposed of as sinuous long waves (Figure 8.13A).[86] On elongation of some 15 percent using very little force, the sinuous waves disappear as the fibers are straightened (Figure 8.13B). The fibers are cylindrical, but with very little further tension applied; with tensions of as little as 2–5 grams, the fibers become beaded (Figure 8.13C). Beading appears quickly, within some score of seconds on applying tension and quickly disappears on relaxation.

When the nerves are stretched to the point where the fibers become beaded, a greatly increased resistance to further elongation occurs. This is because of the collagen fibrils oriented in the longitudinal direction between the nerve fibers, and in the epineurium and perineurium sheaths.[87] When that restraint is decreased by treating the nerves with collagenase, further elongation can occur, the nerve fibers then showing an exaggerated form of beading termed *hyperbeading* (Figure 8.13D).[88] In the hyperbeaded

[86] (Pourmand, Ochs, and Jersild, 1994).

[87] (Thomas, 1963). These are seen in EM [(Thomas et al., 1993) and (Peters, Palay, and Webster, 1991)] and in scanning electron micrographs (Friede and Bischhausen, 1978).

[88] (Pourmand et al., 1994), Figure 5.

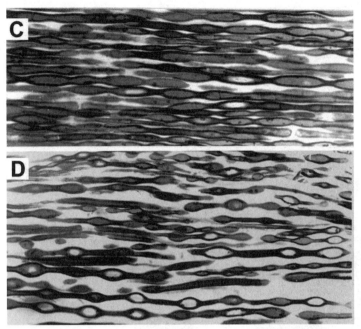

8.13. (*cont.*) At this point an additional stretch of only a few grams, 2–5 gm, evokes beading in the fibers seen with their typical series of enlargements and constrictions (**C**). In the fibers of nerves treated with collagenase to reduce the restraint to stretch due to collagen, the beading constrictions are narrower and lengthened at the expense of the expanded regions that appear more bulbous, the resulting form change termed *hyperbeading* (**D**). From (Ochs and Jersild, 1987), Figures 2 and 3.

fibers, the constrictions are longer and narrower, and the expansions have a more bulbous form than is ordinarily seen.

That beading is not generated by the use of freeze-substitution was shown by its retention when using glutaraldehyde fixation at temperatures close to 0°C.[89] The success of this *cold-fixation* in retaining beading is due to the rapidity of the entry of the monomeric form of the aldehyde and its cross-linking of axonal proteins before the fixative itself undoes the beading mechanism. The low temperature retards the disabling action of the fixative. Some nerves do not require these specialized methods. Beading was seen with standard methods of fixation in the smaller fibers of human fetal nerves aged 19 to 24 weeks[90] and in the nerves of the newborn or very young animals.[91] Presumably, the fixative can enter these smaller fibers rapidly enough to fix their beaded form before the fixative undoes the beading.

[89] (Ochs et al., 1994).   [90] (Tome, Tegner, and Chevallay, 1988).
[91] (Sokolansky, 1930).

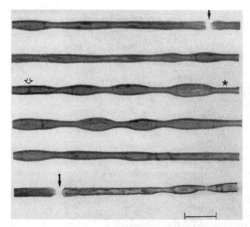

8.14. A single beaded myelinated fiber teased from a nerve that had been stretched and quick-frozen. The arrows point to the nodes of Ranvier; the length of fiber between the nodes is termed the "internode." The swellings and constrictions of beading are more prominent in the midregion of the internode, where the nucleus of the Schwann cell is present (*). Little or no such beading occurs in the region of the internodes adjoining the nodes. The bar indicates a length of 50 μM. From (Ochs, Pourmand, and Jersild, 1995), Figure 1.

SPATIAL FEATURES OF BEADING IN ISOLATED FIBERS

To better determine the pattern and degree of beading, single fibers were teased from freeze-substituted or cold-fixed nerves (Figure 8.14).[92] The most heavily beaded part of the fibers is seen in the middle part of the internodes flanking the nucleus of the Schwann cell at or near the center of the internodes. The degree of beading varies from fiber to fiber. Those with the greatest degree of beading have the largest expansions and the most narrow constrictions; and those with the least degree of beading have small enlargements and shallow depressions. Differences in the degree of beading can be seen in the same fiber between adjoining internodes, with one internode showing a great degree and little or none in the adjoining internodes. Some fibers may even show only half an internode beaded. These observations indicate that the beading process is not caused by simple physical changes on stretching, but that some mechanism in the fiber is responsible for the beading evoked by stretching.

Despite the marked differences in diameter in the constrictions and expansions, the myelin sheath retains the same thickness in both regions as can be seen in a longitudinal section taken through the axis of the fiber (Figure 8.15A). What has occurred is that the axoplasmic fluid and soluble

[92] The technique of isolating single fibers has been described by Dyck and his associates (Dyck and Lofgren, 1968), and (Dyck, Gianni, and Lais, 1993a). Their use of this method has been of great value in describing the various altered forms of nerve fibers in peripheral neuropathies.

A

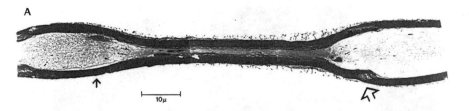

B

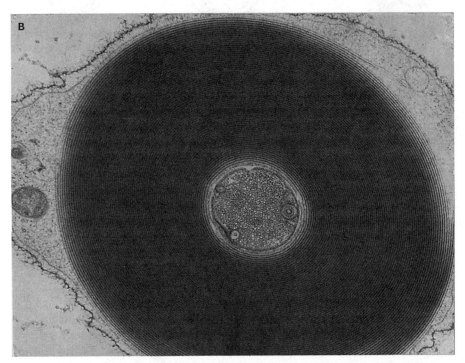

8.15. (**A**) Longitudinal section taken axially through the center of a myelinated fiber showing the similarity of the myelin sheath thickness in the constricted and expanded regions.
The lightly stained cytoskeletal organelles in the expansions have become compacted in
the constriction causing it to appear darker stained. The open arrow indicates an incisure
of Schmidt-Lantermann located outside the constricted region. Small arrow shows some
shrinkage in expansion from myelin sheath. From (Ochs et al., 1987), Figure 3. (**B**) Cross-
section of a myelinated fiber taken through a beading constriction showing in the electron
micrograph details of the sheath and axonal contents. The axon diameter is much smaller
with respect to the myelin sheath than that of a normal fiber (cf. the larger relative diameter of the axon in a normal fiber shown in (**C**)); Microtubules and neurofilaments are
closely compacted in the axon. The larger circular structures in the axon represent a portion of the inner layers of the myelin sheath that had intruded in the course of beading to
become reconfigured into these spherical structures. From (Ochs et al., 1987), Figure 8.

C

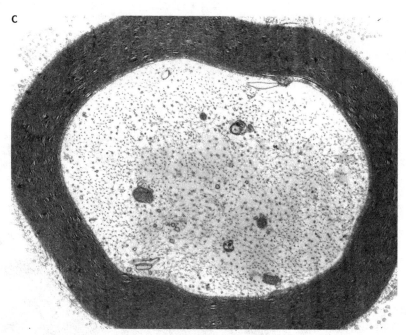

8.15. (*cont.*) (**C**) Cross-section of a nonbeaded myelinated fiber showing its normally relatively larger axon size in proportion to the overall diameter of the fiber (cf. relationship shown diagrammatically in Figure 8.10A). The microtubules, neurofilaments and other cytoskeletal organelles are widely dispersed in the axon. From (Ochs et al., 1987), Figure 7.

components of the axon are expressed from the constricted regions and driven into the adjoining expanded regions increasing their diameter. Left behind in the constrictions are the cytoskeletal organelles, principally the microtubules, neurofilaments, and microfilaments in a closely compacted state, into a region that can be as small as 5 percent of its original area (Figure 8.15B). The reduced ratio of the diameter of the axon to that of the fiber is to be compared with the normal myelinated fiber, where the axon takes a greater share of the total fiber diameter (Figure 8.15C). The myelin sheath remains unaltered in thickness in the expansions and constrictions of the beaded fibers, as is clearly seen in the series of consecutive cross-sections taken through a beaded fiber (Figure 8.16).

THE LOCUS OF THE MECHANISM BRINGING ABOUT BEADING
Is the myelin sheath responsible for beading? Ramón y Cajal, in showing a myelinated nerve fiber that displays constriction and expansions, wrote of "the axon where it is *squeezed* [my italics] at the level of a strangulation

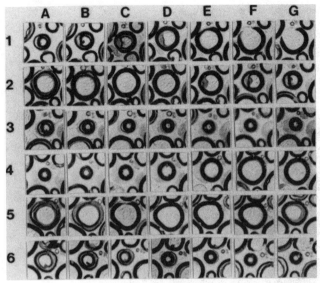

8.16. Serial cross-sections taken successively through constrictions and expansions of a beaded myelinated fiber. The thickness of the myelin sheath remains unchanged despite the great changes in the overall diameter of the fibers. Each section is 6 μm in thickness and set in order in row 1 from columns **A** to **G** and continued on in row 2 column **A** to **G**, and so on through row **6**. The fiber, seen in the center of the frames, starts in **A** row 1 with a constriction and goes on to an expansion that is continued in row **2**, where it decreases to the constriction throughout row **3**. Row **4** continues the constriction that goes on to an expansion at the end of the row. The expansion is continued through row **5** and then in row **6** goes on to another constriction. From (Ochs et al., 1987), Figure 5.

of the myelin,"[93] suggesting an active role of the myelin in producing the constriction. However, axons in unmyelinated fibers also undergo beading on stretch.[94] They show the same succession of constrictions and enlargements in longitudinal section typical of beading in myelinated fibers (Figure 8.17).[95] In cross-section, the constrictions of the beaded axons are seen to contain closely compacted cytoskeletal organelles, mainly microtubules, the predominant organelle, while they are widely spread out in the expansions (Figure 8.18). This picture is to be compared with the unbeaded axons in unmyelinated fibers that are generally more similar in size and have a similar distribution of the cytoskeletal organelles in them (Figure 8.19).

[93] (Ramón y Cajal, 1968), Volume 1, Figure 32.
[94] (Ochs, Pourmand and Jersild, 1996), Figure 2.
[95] See (Markin et al., 1999) for an analysis of the beaded form of unmyelinated fibers.

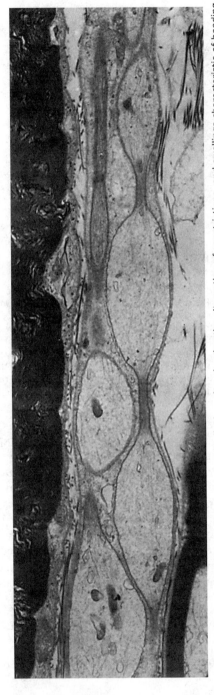

8.17. Axons of an unmyelinated fiber cut in longitudinal section showing the alternation of constrictions, and swellings characteristic of beading in its axon. One of the axons shows a series of three beads. In their constrictions, the cytoskeleton (mainly of microtubules) is seen to be darkly stained because of their compaction. From (Ochs et al., 1996), Figure 2.

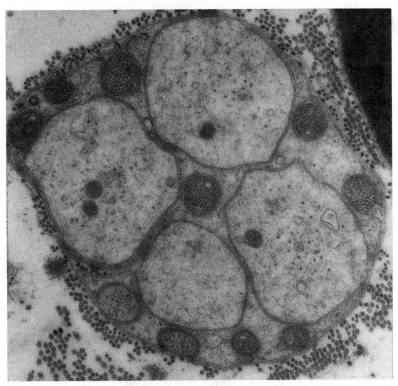

8.18. Cross-section of a beaded unmyelinated fiber showing expansion and constriction in its axons because of beading. In the small constrictions, the microtubules are closely compacted. They are dispersed in the expanded axons that are not only larger, but also more circular than is normally seen in unbeaded axons (cf. Figure 8.19). The larger dark staining structures in the axons and in the cytoplasm of the Schwann cell around the individual axons are mitochondria. The membrane of the Schwann cell fiber partially surrounds the individual axons of the fiber. The solid profiles around the fiber are collagen fibrils c. 60–75 nm thick. From (Ochs et al., 1996), Figure 3.

Beading in the unmyelinated axons was taken as *prima facie* evidence that the myelin sheath is not the agent responsible for beading. The possibility was entertained that beading could be brought about by an alteration in the arrangement of the cytoskeletal organelles within the axons. Such a concept, known as the *tensegrity theory*, has been advanced to account for form changes in other cells, where, following a change in a signal molecule in the membrane (e.g., β-integrin), the cytoskeletal organelles undergo a change in their organization.[96] This hypothesis was tested by exposing nerves *in vitro* to β,β'-iminodiproprionitrile, which was found to cause a complete dissaggregation of microtubules and neurofilaments (Chapter 14). In fibers

---

[96] (Ingber, 1993), and (Ingber et al., 1993).

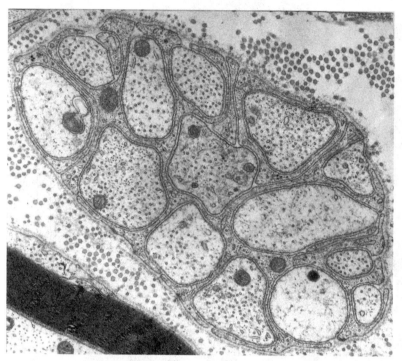

8.19. Cross-section of a nonbeaded unmyelinated fiber. The range of axon sizes in it is much less than that seen in beaded axons, and they are not as circular (cf. Figure 8.18). The microtubules are dispersed in all the axons, with some variation in their number from axon to axon. Larger dark profiles are mitochondria. The solid round profiles ringing the fiber are collagen fibrils. As in (Ochs et al., 1996), Figure 4.

so treated, beading was readily produced by stretch, and it was even enhanced.[97] Because beading constrictions do not arise from a constriction of the myelin sheath, nor by a reorganization of the cytoskeleton within the axon, the hypothesis advanced was that the mechanism responsible is one associated with the *membrane skeleton*.[98] This is the subaxolemmal network consisting of α-spectrin/fodrin, actin microfilaments, ankyrin connected to ion channels and to integrins, and other transmembrane proteins that is present in many cells.[99] Stretch could activate the membrane skeleton mechanism through a transmembrane signaling molecule, such as β-integrin, which is associated with it.[100]

[97] (Ochs, Pourmand, and Jersild, 1996).   [98] (Ochs et al., 1997).
[99] (Byers and Branton, 1985).
[100] (Arvidson, 1992), and (Wang, Butler, and Ingber, 1993).

IMPLICATIONS OF BEADING FOR THE MATRIX THEORY
OF FIBER FORM

In the previous section, where preparations were pictured indicating that the cross-arms of the cytoskeleton organelles appear to be linking (Figure 8.12C), suggesting that the cytoskeleton exists as a rigid matrix that determines the form of the fiber. However, the readiness with which beading is set in motion shows that the cross-arms between the cytoskeleton organelles do not act as rigid struts keeping the organelles at fixed distances apart in the constrictions. Nor do cross-arms serve to tether the cytoskeletal organelles to keep them at a fixed distance, as shown by their greater than normal separation from one another in the expansions of the beaded fibers. A lack of restraint by the cross-connections on the lateral movement of the neurofilaments was also reported in the extruded axoplasm of the squid giant axon that swells when its axolemma is removed.[101]

SHIFT OF MYELIN SHEATH COMPONENTS IN BEADED FIBERS

It was noted previously that, despite the dramatic narrowing of the fiber diameter seen in the constrictions and the enlargement in the expansions of the beaded myelinated fiber, the thickness of the myelin sheath in these regions remains the same (Figures 8.15, 8.16). The decrease in the circumference of the fibers in the constrictions can amount to as little as one-fifth of its normal dimension, without a change in the thickness of the myelin sheath, in the number of its lamellae, and their thickness with respect to the expanded regions.[102] This means that the total amount of myelin sheath material must be reduced in the constrictions. It compels the conclusion that lipids, and most likely other mobile components in the myelin sheath, must be shifted from the constrictions into the expansions to account for the similarity of the sheath thicknesses.[103] For this to occur, the mobilized lipids and other components of the sheath membranes constrained within the plane of their individual lamellar membranes

---

[101] (Brown and Lasek, 1993).

[102] Despite the sizable shift of its materials, the myelin sheath retains not only its characteristic thickness, but also its lamellar fine structure (Ochs and Jersild, 1987). The periodicity of the main dense lines of the myelin lamellae averaged 14.0 nm within both the constrictions and enlargements, a value similar to that found for freeze-dried cat nerves (Elfvin, 1963) and freeze-substituted rabbit nerves (Malhotra and Van Harreveld, 1965). The periodicities are somewhat reduced by the dehydration incurred in the further processing of the freeze-substituted tissues (Kirschner and Hollingshead, 1980). But, in general, the differences from fresh tissue are less than those measured in osmium-fixed (Fernández-Morán and Finean, 1957) or glutaraldehyde-fixed nerves (Karlsson, 1966), where a greater shrinkage occurs in the course of preparation.

[103] (Ochs and Jersild, 1987).

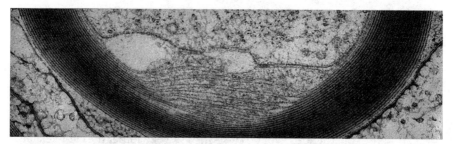

8.20. Leafing seen as a separation of myelin sheath lamellae in a region of a beaded myelinated fiber between constrictions and expansions. Leafing is due to accumulated myelin water squeezed from the constrictions to settle in the intermediate region between the constrictions and expansions (cf. Figure 8.16 and see text). From (Ochs et al., 1987), Figure 12a.

would have to move longitudinally. The movement envisioned is one occurring in a concatenated fashion – the molecules first moved displacing their neighbors, those in turn displacing their neighbors further along, and so on.[104]

INTRUSIONS, INFOLDING, AND LEAFING OF MYELIN SHEATH LAMELLAE IN THE BEADED FIBER

Some local inhomogeneities of the myelin sheath occur circumferentially in the course of beading. These may be seen as narrow, finger-like intrusions of axolemma, along with several inner layers of the myelin sheath into the axon; as dome-like "infoldings" of axolemma and inner layers of the myelin sheath; or as circular, small, multilamellated bodies that appear to have arisen from the myelin sheath (cf. Figure 8.15B). Those bodies appear to become separated from the sheath. Some such intrusions appear to occur normally, passing into the axon where they undergo a progressive degradation as they are carried by retrograde transport to the cell body to presumably undergo further degradation and their components reutilized.[105] The observations of Antal and Szekely, indicating a phagocytosis of myelin sheath fragments by dendrites, could be a related phenomenon.[106]

---

[104] The rapidity of movement indicates a high fluidity of lipid in the myelin membranes, a fluidity that has been shown for the membranes of a variety of cells by various techniques (Quinn, 1981), (Finean and Mitchell, 1981), and (Kates and Manson, 1984). Material shifts in the circumferential direction by a slippage of the lamellar layers, a mechanism invoked for much slower axonal diameter increases [(Friede, 1972), and (Friede and Bischausen, 1982)], and decreases in diameter (O'Neill et al., 1984) that would result in a change in the overall thickness of the myelin sheath. Such was not found to be the case in either the constrictions or expansions of the beaded fibers.

[105] (Ochs and Jersild, 1990).   [106] (Antal and Szekely, 1987).

8.21. In leafing, a separation of the bilipid layers of lamellae can be seen, as well as that between the lamellae (cf. Figure 8.20). From (Ochs et al., 1987), Figure 12b.

Another type of myelin sheath alteration seen in beaded fibers is that of *leafing*, lucent regions of the myelin sheath appearing as a result of the separation of the lamellae (Figure 8.20) and their bilipid layer membranes (Figure 8.21). A considerable amount of water is present in the myelin sheath, estimated to be as much as 30 to 50 percent of the myelin by weight.[107] Leafing appears to result from the movement of water expressed from the myelin sheath in the constricted regions. The excess water separates the lamellar membranes and the bilipid layers of its membranes. Leafing occurs between the constrictions and expansions because the water squeezed from the constriction is unable to enter the expanded regions where tension there would prevent its entry. This intermediate location of leafing can be seen by the lighter regions in the serial sections of a beaded fiber (Figure 8.16: row 1, B–D; row 2, E–G; and row 6, A–B).

Leafing is to be distinguished from the incisures of Schmidt-Lantermann, places where some excess of Schwann cell cytoplasm appears between the membranes forming the lamellae (Figure 8.10B).[108] The incisures are found at intervals in the internodes, where the lamellae are staggered circumferentially around the sheath giving them a cone-like appearance in longitudinally sectioned fibers. In cross-sections, they appear as a light-staining region within the myelin, where the separations of the lamellae takes place (Figure 8.16: row 3, A–C; row 5, A–B; row 6, A–C). The incisures are not causally related to the beading constrictions, because they are seen in the expansions and in the constrictions of beaded fibers.[109] The incisures appear to represent a spread of cytoplasm from the Schwann cell nuclear region serving for the turnover of myelin membrane components.

---

[107] (Schmitt et al., 1941) and (Kirschner, Ganser, and Caspar, 1984).
[108] (Ghabriel and Allt, 1981).   [109] (Ochs and Jersild, 1987).

EFFECT OF BEADING ON NERVE FUNCTION

When nerves are stretched in vivo, while recording compound action potential responses from them, potential amplitudes were seen to fall off in amplitude – with a complete failure appearing within a half-hour to an hour after applying stretch.[110] The failure of responses was traced to the closing off of the *vasa nervorum*, supplying the nerve with oxygen with, as a result, the production of an ischemia.[111] This same time of failure was seen in chambers in which mammalian nerves in vitro were stimulated to give compound action potentials. On replacing the oxygen with nitrogen a similar failure and loss of the responses appeared within 20 to 30 minutes in the anoxic nerves.[112] The recording of responses in vitro allowed an examination of the effect of stretch-induced beading to be made without the confounding effect of vascular shut-off of circulation and anoxia seen in the in vivo studies.[113] Action potentials were maintained for hours in the face of tensions 10-fold greater than that needed to maximally bead their fibers. Axoplasmic transport in vitro (Chapter 11) was also maintained in the face of much higher tensions.[114] These results show that beading per se had no direct effect on these basic nerve functions.

INITIATION OF BEADING BY AGENTS OTHER THAN STRETCH: BEADING A GENERALIZED PROPERTY OF THE FIBER

Beading can appear without stretch. It is seen in the early stage of Wallerian degeneration, appearing within a day or so before the formation of ovoids and spheres characteristic of degeneration.[115] It first appears in the midregion of the internodes, at just the sites where the greatest degree of beading is seen in fibers when stretched. This early onset of beading characterizes the onset phase of degeneration (Chapter 9). Beading has been initiated in fibers of the sciatic nerve of the mouse by injection of a neurotoxin obtained from the spider *Phoneutria nigroventer*.[116] Unlike Wallerian degeneration, however, the toxin-induced beading reverses spontaneously within twenty-four hours, the fibers resuming their cylindrical form. Anoxia and a decreased supply of ATP can initiate beading, this seen as somewhat irregularly shaped swellings and constrictions in cultured myelinated fibers exposed to cyanide.[117] Injected into mouse nerve, cyanide caused a series of swellings and constrictions of unmyelinated nerve axons that Hall referred to as beading.[118] A metabolic change was proposed to cause large

---

[110] (Lundborg and Rydevik, 1973), and (Rydevick, 1990).
[111] (Kwan et al., 1992).   [112] (Leone and Ochs, 1978).
[113] (Ochs et al., 2000).   [114] (Ochs and Jersild, 1986).
[115] (Ramón y Cajal, 1968), (Young, 1945), and (Williams and Hall, 1971).
[116] (Cruz-Höfling et al., 1985).   [117] (Masurovsky and Bunge, 1971).
[118] (Hall, 1972), Figure 6.

focal swellings seen in axons in a culture medium depleted of glucose. This was considered to cause the loss of semipermeability of the axolemma, allowing ions to enter the fibers accompanied by water to swell the fibers.[119] On reintroduction of glucose with recovery of a metabolic energy of supply to the sodium pump, the ions are pumped out and along with it the excess fluid to account for the reduction of the swellings. As support for their hypothesis, the authors pointed to studies made on the cerebral cortex, where – following anoxia by arresting circulation – sodium and chloride along with water were shown to move from the extracellular space into the dendrites, causing them to swell at the expense of the extracellular space.[120] However, this mechanism cannot account for the rapid form change of the beading induced by stretch, which takes place in a matter of seconds and is as quickly reversed on relaxation. The total amount of axoplasmic fluid in the fibers remains unchanged when they are beaded, the form change resulting from a redistribution of axoplasmic fluid as shown by the theoretical analysis given by Markin et al.[121]

Dendrites of central nervous system neurons in the spinal cord were found to bead when stimulating dorsal root afferents terminating on them, the release of *substance P* from the afferents held responsible for the form change.[122] The beading was reversible, spontaneously disappearing within an hour. Dendrite form changes similar to beading have been seen in various pathological conditions using Golgi staining. Ward found varicosities in apical dendrites in a site of focal epilepsy.[123] The Scheibels reported that the apical dendrites in the hippocampal dentate complex taken from the temporal lobe of epileptics showed a form change they referred to as "string-of-beads."[124] In children with progressive behavioral failure due to an undefined cerebral pathology, form changes similar to that of beading were found in the apical dendrites of their pyramidal cells.[125]

This raises the question of what function beading may serve. One possibility is that it is a mechanism that serves to close off axons when they are transected, the process termed "resealing." Fiber narrowing seen at the cut end of fibers can be seen in the fibers described by Lubińska in frog nerves,[126] and a similar narrowing at the cut end of giant axons has been observed by Flaig[127] and by Gallant.[128] Fishman and his colleagues have shown it in both the giant axon of the squid and in the earthworm,[129] and

---

[119] (Mire, Hendelman, and Bunge, 1970).
[120] (Van Harreveld and Ochs, 1956), and (Van Harreveld, 1966).
[121] (Markin et al., 1999).   [122] (Mantyh et al., 1995).   [123] (Ward, 1969).
[124] (Scheibel and Scheibel, 1973).   [125] (Purpura, 1974), and (Purpura et al., 1982).
[126] (Lubińska, 1956).   [127] (Flaig, 1947).
[128] (Gallant, 1988), and (Gallant, Hammar, and Reese, 1996).
[129] (Krause et al., 1994) and (Fishman et al., 1995).

a similar resealing process was reported in the cockroach giant axon.[130] Sealing could prevent the loss of ions and allow the injured fibers to regenerate (Chapter 10). How the mechanism responsible for beading appears to be controlled by a factor supplied by fast axoplasmic transport is discussed in the following chapter.

[130] (Yawo and Kuno, 1985).

9

WALLERIAN DEGENERATION: EARLY AND LATE PHASES

The loss of sensation and muscle power after a nerve transection has been known from antiquity. In the nineteenth century, when the microscopic structure of the nerve fiber became known, the amputated stump of transected nerves was seen to undergo the characteristic breakdown called *Wallerian degeneration*. The phenomenon led to a major advance in understanding the different functions of the spinal cord roots; sensory fibers carried in the dorsal roots, motor fibers in the ventral roots, the *Bell-Magendie law*. On cutting a root, degeneration was seen only in that portion of its fibers separated from the cells in the ganglia. The inference of these results was that the cell bodies are required to maintain viability of their fibers. The pursuit of how this comes about led to the recognition of the neuron doctrine and the need for some means by which materials from the cells are carried out into their fibers, the mechanism of axoplasmic transport, which will be discussed in detail in Chapters 11 and 12. In this chapter, the analysis of Wallerian degeneration is presented and shown to be a two-stage process in which the earliest phase is a beading of the fibers.

THE BELL-MAGENDIE LAW

In a paper he had privately printed in 1811 and that was privately circulated, and only much later publically revealed, the famous English anatomist *Charles Bell* (1774–1842) reported that injury to the anterior (ventral) portion of the spinal cord marrow caused convulsive muscular movements in vivisected animals, more so than an injury to the posterior (dorsal) portions of the cord.[1] When he cut the dorsal root, the muscles were not convulsed. But when the anterior root was even only touched with his knife, the muscles immediately did so. Bell considered that the anterior roots are

---

[1] (Cranefield, 1974). A complete analysis of the history of the findings leading to the establishment of the law is given in this book, one that allows a reasonable judgment to be made of the priority claims for its discovery.

connected to the anterior columns of the cord, leading to the cerebrum where sensation and willed movements are controlled, whereas the posterior roots are connected via the posterior columns to the cerebellum from which the "secret" operations of the body are maintained. It is not entirely clear what the secret operations might be. Bell wrote that they are required for the "integration of vital function." This term could be an allusion to experiments in which experiments aimed at destruction of part of the cerebellum to assess its function – which in fact caused damage to the underlying brain stem where autonomic centers are located – appeared as an alteration of autonomic function.[2]

The great French physiologist *Francois Magendie* (1783–1855) clarified the role of the roots when he found on cutting the dorsal root that the animal became insensible to painful stimulation of the body below the root's level of entry to the cord, whereas animals were able to move about normally. When the ventral roots were cut, the limbs were immobilized, while sensibility retained via the dorsal roots. Magendie concluded that the ventral roots convey motor influences to the muscles whereas sensation is carried into the cord by the dorsal roots.[3] There was a bitter controversy over the priority to be given for this discovery of the separation of function in the roots that was resolved by its designation as the Bell-Magendie law.[4] A curious side note to this episode is the priority claimed by Walker for making the suggestion that different functions are subserved in the roots.[5] Unfortunately, he picked the dorsal root as the path for motor outflow and the ventral root for sensory input.

WALLER'S LAWS OF DEGENERATION
The name of *Augustus Volney Waller* (1816–1870) is used today to eponymously designate the degeneration of nerve fibers distal to a transection or crush, *Wallerian degeneration*. Although he was not the first to observe the microscopic changes in nerve fibers distal to a transection, priority goes to Nasse (Nasse, 1839).[6] However, the study carried out by Waller on the

---

[2] (Neuburger, 1981) and (Ochs, 1989).   [3] (Magendie, 1822).
[4] Cranefield lays before the reader the original relevant papers and other documents, along with correspondence relating to the dispute, with Bell's copy of his 1811 paper containing his own extensive notations (Cranefield, 1974). In the reprint of A. D. Waller's paper (the son of A. V. Waller) reprinted by Cranefield (1974, 1911m and 1912b), A. D. Waller concluded that the priority for the discovery should go entirely to Magendie, while earlier his father, A.V. Waller, gave credit entirely to Charles Bell (Waller, 1861) p. 11.
[5] (Walker, 1973).
[6] (Stannius, 1847). He had noted degenerative changes down to the "finest terminals." See (Nasse, 1839), (Lent, 1856), and (Gunther and Schoen, 1840). Schiff found the phenomenon present in all nerves, as was quoted in (Wedl, 1855), p. 180.

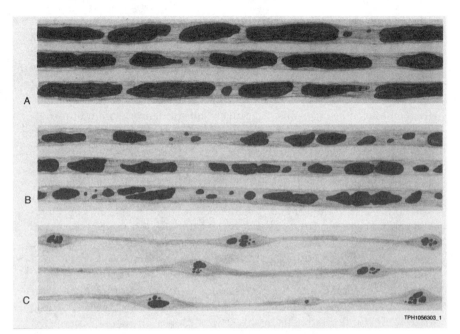

TPH1056303_1

9.1. Wallerian degeneration shown by myelin sheath irregularities in single myelinated nerve fibers teased from the distal stump of transected nerve; ovoid segmentation (**A**), and clusters of ball-shaped myelin figures (**B**). Fibers with little myelin remaining are found in the late stage of degeneration (**C**). From (Dyck, Giannini, and Lais, 1993a), Figures 30–34.

glossopharyngeal nerve innervating the taste buds of the tongue of the frog remains the exemplar of the phenomenon.[7] After a lag of several days, the fibers distal to a transection lose their normally tubular form as the myelin breaks up into ovoids and then spheres, which later appear as rough granules all along the length of the fiber (Figure 9.1). The progress of this transformation appeared to occur simultaneously along its length. Waller recognized that this change, designated the *first law of Waller*, could be used as a means to trace the course of the degenerated fibers intermingled with normal nerve fibers in the nervous system. In the citation of the Montyon prize awarded to him by the Academie des Sciences of France in 1856, it was referred to as Waller's *method neurotrophique*.[8]

Waller drew another, most important, theoretical inference from the phenomenon, namely that the cell body acts as a "trophic center" to maintain

---

[7] (Waller, 1850).
[8] (Ochs, 1975c), p. 253. Waller had close connections to France. Born in England, his childhood was spent in France, where he later obtained his M.D. in Paris and where he became interested in the histology of the frog nerve, publishing many of his papers in french journals (Denny-Brown, 1970), pp. 88–91.

the viability of the nerve fiber – that it was the loss of an influence from the cell bodies on their fibers that resulted in the fiber's degeneration. This concept constituted the *second law of Waller* for which he was awarded the Medal of the Royal Society of England for the year 1860.[9] The citation read, "for the discovery of an important relation between the ganglia of the spinal nerves and the nutrition of their sentient fibers." Making use of the Bell-Magendie law, Waller selectively cut the sensory and motor roots of the cervical cord segments in a group of kittens (Figure 9.2).[10] On cutting the dorsal roots, the portion of the fibers central to the transection was found to have degenerated, whereas that still connected to the ganglion remained unchanged. On cutting the sensory fibers distal to the ganglion, the part below the section degenerated, whereas that portion of the fiber still connected to the ganglion maintained its normal appearance. When the ventral root was cut, only that part of the fibers distal to the transection degenerated, and the fibers above the transection still connected to the spinal cord remained normal. From these observations, Waller concluded that "the ganglion corpuscles [cell bodies] present in the dorsal root ganglion exert a trophic influence necessary to maintain the form and function of the sensory fibers ascending in the dorsal root fibers and as well on those descending in the peripheral nerve." The portion of the fiber disconnected from the cell bodies degenerates because "the fibers are deprived of a sustaining influence emanating from the cells."[11] Similarly, Waller made the inference that motor cell bodies are present in the spinal cord that have a trophic influence on their fibers emerging in the ventral roots.[12]

Waller viewed the trophic influence of the cells on their fibers as a dynamic process:

> As long as the influence of the ganglion over the nerve fiber occurs, the equilibrium [forces of renewal as opposed to those of degeneration] is maintained, but as soon as the connection of the ganglion corpuscle with the nerve fiber is destroyed, its peripheral [severed] end is subjected to forces of degeneration.[13]

Waller pointed out that "nerve [fibers] when separated from their trophic center [cells] degenerate" show that "the direction of degeneration is independent of that of nerve conduction," for conduction in the sensory nerves passes centrally to and through the ganglion to the cord, whereas

---

[9] (Ochs, 1975c), p. 253.
[10] (Waller, 1861) Figure on p. 21. Redrawn in A. D. Waller's textbook (Waller, 1891), Figure 149, and (Ochs, 1975c), Figure 5.
[11] Ibid., p. 20.
[12] At the time, the origin of the motor fibers from cell bodies in the cord had not yet been established.
[13] (Waller, 1852a). See also (Waller, 1852b), (Waller, 1852c), (Waller, 1852d), and (Waller, 1852e).

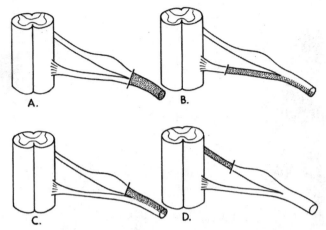

9.2. Roots and nerves cut in the cervical cord region of a group of kittens. **(A)** Nerve cut below the ganglion. **(B)** Ventral root cut. **(C)** Sensory nerves cut just distal to the ganglion. **(D)** Dorsal root cut. The portion of the fibers separated from their cell bodies undergo Wallerian degeneration; this is shown by stippling. From Waller modified in (Ochs, 1975c), Figure 5.

the trophic influence from the cells passes in both directions. The importance of Waller's trophic concept was discussed by Ramón y Cajal,[14] among others, but has not received as much attention as Waller's first law of degeneration. In the middle of the nineteenth century, it was not universally accepted that the nerve fiber took issue from the cell body, the two-neural element concept of Valentin still having its influential adherents. It is not clear what position Waller took in this debate. His scientific work was interrupted in 1856 by a serious angina attack. However, he did write reviews that indicated his later views. Of particular importance was a lecture he gave at the Royal Institution in London in 1861, which was privately printed. A rare copy of the full lecture was fortunately preserved.[15] It gives more of an insight into his thinking than the resumé of the lecture later printed in the Institution Proceedings.[16]

In the full paper of his lecture, Waller reprinted a figure of the dorsal root ganglion (Figure 9.3) taken from Leydig.[17] It shows the cells as bipolar, with fibers exiting from its poles. With that figure in mind, Waller wrote that

[14] (Ramón y Cajal, 1968), pp. 4 and 5. Ramón y Cajal gave much credit to Waller's clear-sightedness in assessing the implications of his findings, more so than some of those who came later to the subject.

[15] (Waller, 1861). In the Medical School library of the University of Newcastle, England.

[16] (Waller, 1862). In this paper it was noted that in consequence of Waller's illness, the paper was read by Dr. F. Bond.

[17] (Denny-Brown, 1970), p. 57.

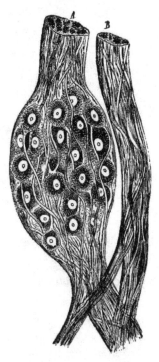

9.3. (A) Dorsal root ganglion of a fish showing multipolar cells. Most have two fibers exiting from their poles, several with more. (B) Alongside the ganglion and its fibers at the right is a band of motor fibers. From a figure of Leydig's given by Waller and reproduced in (Ochs, 1975d), Figure 6.

the ganglion cells maintain and regulate the vitality of the fibers that are attached to them: "A nerve-cell would be to its effluent nerve-fibres what a fountain is to the rivulet which trickles from it – a centre of nutritive energy."[18] Such an analogy would require a true union between the cell body and its fiber. In discussing the ganglion pictured in that figure, Waller states: "the points at which the fibres are attached to the cells are called their poles; and according as a cell has one, two, or more poles, it is called unipolar, bipolar, or multipolar." In an additional figure, he shows bipolar and multipolar cells and states: "the cells of which this ganglion is composed are all bipolar, that is, each of them is attached to two fibres. In that case each of the fibres in the posterior root in passing through the ganglion may be considered to pass, as it were through a cell."

But the pictures shown are those of the fish. In the mammalian dorsal root ganglion, the connection of the fiber to the cell is difficult to see. An

[18] (Waller, 1861), p. 20.

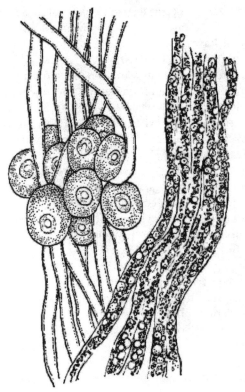

9.4. Preparation of a mammalian dorsal root ganglion in which the cell bodies and fibers are seemingly separate entities that are closely apposed. On the right are fibers showing Wallerian degeneration. Drawn from a preparation of Waller's by his daughter-in-law, Mary Waller. From (Stirling, 1902). Reproduced in (Ochs, 1975d), Figure 7.

indication of what he did see was a pencil drawing made from one of his preparations by Mrs. Mary A. Waller, the wife of Waller's son, Augustus Desiré Waller.[19] This found its way into Stirling's biography of A. V. Waller (Figure 9.4).[20] It shows a band of degenerated nerve fibers and at the left a normal ganglion and fibers with the connection between them not visible, allowing for the impression that they could be separate elements apposed to one another in Valentin's sense. This view was furthered by Waller in his Croonian lecture of 1870, when he wrote: "The nervous system, central and peripheral, is an immense and intricate series of nerve-tubules and of ganglion cells, and by the method I have laid before you we have already

[19] Waller's son, Augustus Desiré Waller, gained recognition as an electrophysiologist who introduced electrocardiography into medicine. Often confused with his father, he said of himself that, "he was the original Wallerian degeneration."
[20] (Stirling, 1902), Figure 7 in (Ochs, 1975c).

recognized in these *elements* [my italics] a great degree of material dependence."[21] I have emphasized the term "element" because of the implication that Waller at the time could well have thought of the cells and fibers as separate neural entities in Valentin's sense (Chapter 8).

Waller's apparent leaning to the Valentinian view was certainly not unique to him. This may be seen in the writings of other authorities, including the eminent French physiologist *Claude Bernard* (1813–1878). Bernard seemed to have accepted the Valentinian view when he wrote of fibers and cells as elements, referring to fibers as "conductors" of the nervous system and cells as "elaborating or collecting agents."[22] Yet, despite the uncertainty regarding the true relation of the cell to the fiber, Bernard clearly recognized the fundamental significance of Waller's findings, namely that the cell body exerts a trophic control over the fiber. He repeated Waller's fundamental experiment of selectively cutting dorsal and ventral roots, with essentially the same results. In a phrase reminiscent of Waller's, he wrote that, "the sensitive nerves are distinguished from motor nerves by the presence of the ganglion which plays in their nutrition the same role that the spinal cord plays in the nutrition of the motor nerves."[23] But Bernard could not see how it comes about that, "in the actual state of our knowledge, the nutrition of one nervous element [is subserved] by another nervous element." Bernard's dilemma shows the remaining influence of the concept that the cell and fiber are different neural elements. However, he went on to say that our "ignorance of how nerves receive their nutrition renders the conclusion of Waller difficult to explain in present terms but they will later receive their interpretation."

THE PROXIMO-DISTAL PROGRESSION OF WALLERIAN DEGENERATION

It appeared to Waller, and to many others who followed him, that once degeneration starts it develops simultaneously all along the whole length of the nerve fibers. However, by electrical recordings made at successively greater distances along the length of the amputated portion of nerve, a centrifugal progression of degeneration was indicated.[24] Yet, still other studies gave opposite results.[25] In large part, the conflict was due to taking nerves too late after their transection and thus missing a faster spread of degeneration than had been anticipated. The study made by Parker and Paine in the lateral-line fibers of the fish clearly demonstrated a proximo-distal spread of

---

[21] (Waller, 1870). As had been noted, Waller was quite ill at the time, dying later that same year.
[22] (Bernard, 1867) and (Bernard, 1858).
[23] (Bernard, 1858) Volume 1, pp. 235–245. See (Ochs, 1975c) Figure 5.
[24] (Rosenblueth and Dempsey, 1939), and (Rosenblueth and Del Pozo, 1943).
[25] (Joseph, 1973).

degeneration – its revelation aided by the lower temperature at which the study was made:[26] The rate of degeneration is slowed at low temperatures in this poikilothermic animal as Waller had shown earlier.[27]

Until recently, microscopic studies of fibers undergoing degeneration appeared to support the view that degeneration occurs simultaneously all along the length of the amputated nerve. This was the position taken by Cajal in his studies of nerve degeneration.[28] After nerve transection, the myelin was said to contract and break up, leaving the old Schwann cell free of debris:

> The process of medullary fragmentation ... occurs simultaneously along the entire length of the peripheral stump ... near the wound the disorganization of the myelin and axon occurs earlier than in the rest of the stump ... the fibres suffer not only the effects of Wallerian degeneration, but also the mechanical effects and the chemical effects of the inflammatory exudation. One must, however, reject the idea of Bethe, who believes that the degeneration of the peripheral stump progresses centrifugally, that is, from the wound to the nervous termination. Erb, Tizzoni, Neumann, Büngner, and some others, also adopted this erroneous conception.[29]

Some early changes precede the fragmentation. The progression pictured by Young indicates that at first a widening of the nodes occurs along with some depression in the middle of the internode (Figure 9.5).[30] These then pinch off to form two long ovoids on either side of the center of the internode, which then subdivide to form ovoids with shorter lengths and then spheres. A somewhat different sequence was seen by Lubińska in single degenerating fibers teased from the distal amputated stump of rat phrenic nerves.[31] Following a lag period of twenty-six hours after cutting the nerve, fibers in the amputated portion first show an indentation of the myelin sheath in the middle of the internode under the Schwann cell nucleus, soon followed by a series of constrictions and enlargements embracing the middle region of the internode (Figure 9.6). This has the appearance of the beading produced in the fibers of lightly stretched nerves (Chapter 8). Then, the fibers go on to a later stage, where ovoids form – these appearing first in the midregion of the internode, where beading is greatest and then throughout the internode.

The time of appearance of beading and then ovoid form changes was related to the fiber size: occurring first in the thinnest fibers, next in larger fibers, then in still larger ones, and finally the largest fibers – the last to exhibit the form change (Figure 9.7). By taking pieces of nerves at different distances from the site of transection, at 5 and 15 mm, Lubińska was able to calculate the rate of the proximo-distal spread of degeneration and show that it also was dependent on fiber size. The most rapid rate of degeneration

[26] (Parker and Paine, 1934). [27] (Waller, 1852a). [28] (Ramón y Cajal, 1968).
[29] Ibid., Volume 1, p. 74. [30] (Young, 1949). [31] (Lubińska, 1977).

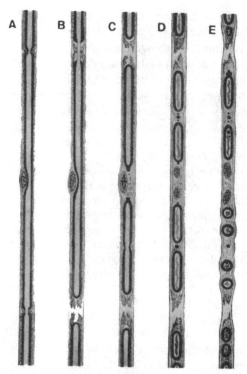

9.5. Classical sequence of morphological changes occurring in myelinated fibers in the course of Wallerian degeneration. On the left, **(A)** shows a normal fiber with nodes near the top and bottom and a slight depression at the middle of the internode under the Schwann cell nucleus. At **(B)**, a day or two after transection, a widening of the nodes is pictured. This is followed by closure of the myelin at the nodes and in the middle of the internode to form long ovoids. These then further subdivide **(C–E)** to form shorter ovoids and then myelin spheres that are subsequently resorbed. More recent studies show the earliest changes constituting the first stage of Wallerian degeneration to occur first in the midregion of the fiber with beading and ovoid formation (cf. Figure 9.6). From (Young, 1949). Modified in (Ochs, 1965b), Figure 8.3.

spread was found for the thinner fibers (Figure 9.8). The calculated rate in the thinnest fibers, 252 mm/day, approaches that of fast axoplasmic transport (Chapter 11). Lubińska inferred that some factor(s) preventing degeneration is continually being supplied to the fibers, the loss of which leads to ovoid formation and degeneration. The lag between transection and the appearance of degeneration was ascribed to the limited store of the factor in the amputated fibers: When the supply is used up, degeneration ensues. The longer lag period seen in the larger fibers, and the lower apparent rate of its proximo-distal spread, is because of the greater amount of the factor stored in them, compared with the smaller fibers.

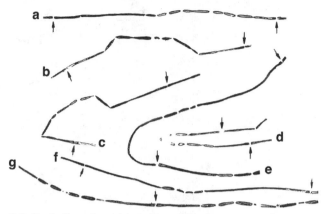

9.6. Single fibers teased from nerves distal to a transection. The early stage of degeneration is seen as beading and ovoid formation in the midregion of the internodes. The different fibers are marked off by arrows pointing to their nodes in the individual fibers labeled **a–g**. Ovoids are seen in the mid-region of the internodes. Modified from (Lubińska, 1977), Figure 2.

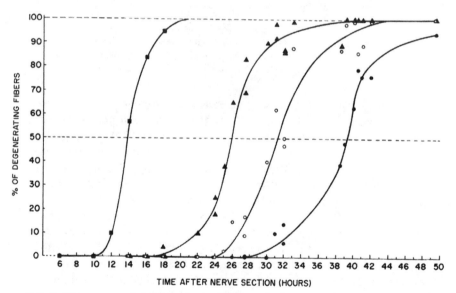

TIME AFTER NERVE SECTION (HOURS)

9.7. The time course of degeneration in the fibers shown by the onset of ovoids in the internodes (cf. Figure 9.5) depends on fiber size. The smallest fibers (■) show the earliest appearance of ovoids, at just ten hours after transection of the nerve. Fibers somewhat larger (▲) show the onset of ovoids at eighteen hours. Ovoids appear in the larger fibers (O) starting at 24 hours and in the largest fibers (●) at twenty-eight hours. From (Lubińska, 1977), Figure 3.

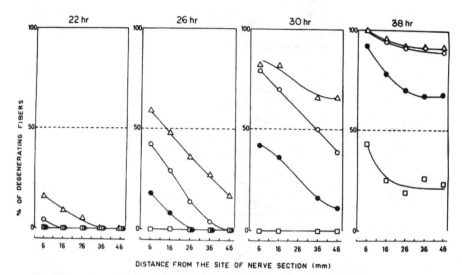

DISTANCE FROM THE SITE OF NERVE SECTION (mm)

9.8. The proximo-distal spread of degeneration in fibers distal to a nerve transection seen as a function of time and fiber size. At twenty-two hours, the thinnest (Δ) fibers first show degeneration appearing closest to the transection. At twenty-eight hours, the medium (o) and then the larger (•) fibers show greater evidence of degeneration, the change appearing first proximally then distally. At thirty hours, degeneration is more pronounced. The thickest fibers (□) do not show signs of degeneration until thirty-eight hours. Distances along the nerve shown in millimeters on the abscissa. From (Lubińska, 1977), Figure 4.

TWO STAGES OF WALLERIAN DEGENERATION; BEADING FOLLOWED BY CYTOSKELETON DISINTEGRATION AND OVOID FORMATION

Rather than ovoid formation as the earliest change in Wallerian degeneration, the earliest stage is that of beading. As had been described in Chapter 8, when normal fibers are stretched to initiate beading, it is the presence of the cytoskeleton in the constrictions that presents a barrier to the merger of the axolemma and myelin sheath to form ovoids. In Wallerian degeneration, the earliest stage is seen as beading, wherein the presence of the cytoskeletal organelles hinders closure of the membranes to form ovoids.[32] But then, after a short delay, the cytoskeleton becomes disrupted[33] allowing the axonal membranes and the myelin sheaths in the constrictions to merge to form the ovoids and spheres typical of Wallerian degeneration.

A dramatic example of the separation of Wallerian degeneration into these two stages is seen in the nerves of the C57BL/6/Ola mouse (Ola).

[32] A bead-like early stage of Wallerian degeneration is seen in (Gutmann and Hnik, 1962), Figure 1. It is alluded to by Ramón y Cajal (Ramón y Cajal, 1968). The presence of compacted cytoskeleton in the early beaded stage of degeneration was seen in electron micrographs (unpublished observations).

[33] (Donat and Wisniewski, 1973).

This is a mouse strain in which ovoid formation in Wallerian degeneration is much delayed, taking weeks instead of the several days normally seen after transection.[34] Beading, however, occurs rapidly in the fibers of amputated segments of their nerves, but it remains in this stage because of the longer delay of the disintegration of the microtubules and neurofilaments than that usually seen. That the cytoskeleton remains functional was determined by the continued transport of organelles within the amputated fibers shown by AVEC-DIC microscopy and by use of isotopes as tracers (Chapter 12).[35] The delayed disintegration of the cytoskeleton in these Ola genetic animals while beading remains present, clearly demonstrates that two processes are involved in Wallerian degeneration: the first stage, the initiation of beading, and the second stage, the disintegration of the cytoskeleton. As indicated previously, a factor normally supplied to the fibers suppressing the beading mechanism is lost on transection of the nerve, whereas some other factor causing the disintegration of the cytoskeleton is further delayed in the Ola genetic mouse strain.

DEGENERATION OF UNMYELINATED AXONS

In the analysis of beading presented in Chapter 8, the axolemma and the membrane skeletal structures associated with it[36] were pointed to as the locus of the beading mechanism rather than the myelin sheath: Beading is seen in unmyelinated axons. We would therefore expect beading to also appear as the earliest stage of degeneration in unmyelinated axons before the disintegration of their cytoskeleton. Analysis of degeneration in unmyelinated fibers is technically more difficult without the guidance of myelin sheath to show the early course of beading and degeneration. Nevertheless, electron microscopic study of unmyelinated fibers has indicated a similar sequence of changes in their axons after amputation. The early stage of beading was indicated by Kappeller and Major, who reported "alternating swellings and narrow regions" in unmyelinated axons distal to their interruption made twenty-four hours beforehand.[37] Similar beading changes were found to develop distally to transection of cultured unmyelinated axons.[38] The frank degeneration of the later stage has been reported to occur at times that have varied greatly. Friede and Martinez found that most unmyelinated axons undergo degeneration with a time lag of 2 to 3 days after nerve transection.[39] Thomas et al. reported the appearance of degeneration

[34] (Lunn et al., 1989) and (Brown et al., 1991).
[35] (Smith and Bisby, 1993), (Watson, Glass, and Griffin, 1993), and (Glass and Griffin, 1994).
[36] (Alberts et al., 1994), pp. 491–493.   [37] (Kapeller and Mayor, 1969).
[38] (Friede and Martinez, 1970).   [39] Ibid.

as early as twenty-four hours,[40] as did Williams and Hall.[41] Still, earlier degenerative changes were reported by Dyck and Hopkins.[42] They found some of the unmyelinated fibers showing degenerative changes as early as 9 to 11 hours after crushing the cervical sympathetic trunk in the rat, with such changes seen more regularly twenty-two hours after injury. A proximo-distal gradient of degeneration in the unmyelinated fibers was suggested by the finding of "empty spaces" at sites where unmyelinated fibers were ex-pected to be present. These empty regions appeared in sections taken at progressively more distant sites from the injury, as would be the case of a proximo-distal gradient of degeneration. The very early appearance of de-generation in these fibers is in accord with the hypothesis of the limited reserve of a factor inhibiting the beading mechanism – one that would be in lesser amounts in the smaller unmyelinated fibers, compared with that in the myelinated fibers.

SCHWANN CELL PROLIFERATION COINCIDENT WITH THE PROXIMO-DISTAL SPREAD OF NERVE DEGENERATION

Although axons and the Schwann cells that invest them have different cellular origins, a mutual influence exists between them.[43] A transfer of phospholipids from the axon to the myelin has been indicated by Droz and his colleagues.[44] Some factor transmitted between axons and Schwann cells was indicated by the increased uptake of [³H]thymidine, a marker of DNA synthesis, and the proliferation of the Schwann cells coinciding with axon degeneratation. The early uptake of [³H]thymidine by the Schwann cells showed a proximo-distal spread down the amputated nerve fibers at a rate of 200 mm/day,[45] a rate of spread remarkably close to that of the spread of Wallerian degeneration found by Lubińska. This could be from the loss of some signal that normally passes from the axon to the Schwann cells to prevent Schwann cell proliferation. Alternatively, the degenerating axons could give rise to a signal initiating Schwann cell proliferation.[46]

EFFECT OF NERVE DEGENERATION ON MUSCLE

That muscle atrophy (from the Greek *trophos* for nutrition) follows from loss of its innervation has been known since ancient times. *Soranus*, in the second century, wrote that after nerve lesions "the affected part [muscle] becomes pale and thin because of want of nourishment and there is a con-traction of the affected parts."[47] In later times, experimental investigation

---

[40] (Thomas, King, and Phelps, 1972).   [41] (Williams and Hall, 1971a and 1971b).
[42] (Dyck and Hopkins, 1972).   [43] (Bray, Rasminsky, and Aguayo, 1981).
[44] (Droz et al., 1979).   [45] (Oaklander and Spencer, 1988).
[46] (Bray, Rasminsky and Aguayo, 1981), and (Bray and Friedman, 1999).
[47] (Caelius Aurelianus, 1950).

of the effect of nerve transection on muscle was carried out by Fontana.[48] He cut the sciatic nerves of frogs and found that later, when stimulation of the nerve could no longer elicit contractions of their leg muscles, they could still contract when stimulated directly. This indicated that either muscle itself has the independent property of excitability, or that some nerve fibers remaining are required to maintain the muscle's contractions. This point was studied by von Humboldt, who used a magnifying glass to examine the nerves innervating muscles in the upper part of the frog's thigh and in the fins of fish.[49] When the nerves were cut and nerve fibers were no longer visible in the muscles, their stimulation could no longer produce a contraction. If, however, traces of nerve filament were seen left in the muscle, a contraction could be elicited. This was taken to indicate that muscle contractility depended on the viability of its remaining innervation. However, von Humboldt's finding was likely the result of his using too weak a stimulation, one that reached the threshold for nerve but not that of muscle which requires a stronger stimulus. Longet found that after transection of a motor nerve, by the fourth day, when the nerve below the transection was stimulated by mechanical, chemical, or electrical means and has lost every trace of its excitability, the muscle it innervates could be directly excited to contract.[50] He concluded that muscular irritability is independent of the motor nerves, that it depends on the arterial blood that maintains the nutrition of the muscular tissue and imparts the property of irritability to it.

John Reed added support to Longet's conclusion that muscles can be directly stimulated to contract. He then went on to show that direct galvanic stimulation could have an ameliorating effect on denervated muscle.[51] After cutting the nerves to the muscles of frogs so that both hind legs were completely deprived of their nervous input, he stimulated the denervated muscles of one leg daily with galvanic currents, while those on the other side were left to atrophy. Stimulation was carried out for 2 months. At the end of that time, the galvanized muscles were seen to be close to their original size and firmness, and were able to contract vigorously while the muscles of the limb that had been kept quiescent shrank to half their former bulk.[52] Further evidence in support of the ameliorating effect of direct stimulation of denervated muscle was forthcoming,[53] with programs of muscle stimulation used clinically to aid in recovery when the muscles may later become reinnervated (Chapter 10).

---

[48] (Althaus, 1859), p. 124.   [49] (Rothschuh, 1960).   [50] (Longet, 1842).
[51] (Reed, 1886).   [52] (Gutmann, 1962), and (Gutmann and Hnik, 1963).
[53] (Duchenne, 1855), (Eccles, 1941a), (Eccles, 1941b), (Gutmann and Hnik, 1963), and (Card, 1977).

CHANGES OF MUSCLE MEMBRANE PROPERTIES AFTER DENERVATION
In addition to the muscle atrophy after denervation, other changes occur in its membrane following nerve transection. A striking example is the hypersensitivity of the denervated muscle to the motor nerve neurotransmitter acetylcholine (ACh) (Chapter 7). This was shown by the greatly increased contractions recorded when ACh was injected into a blood vessel supplying the denervated muscle, compared with normally innervated muscle.[54] The increased contractions were later traced to an increase in ACh receptors in the muscle membranes that normally in the mature muscle are localized to the end-plate region (Chapter 7). In the denervated muscle fibers, the receptors were found to spread out over its length. This was shown by recording potential changes of the muscle fibers with microelectrodes when minute amounts of ACh were released from a second pipette. When ACh was released close to the end-plate of normal muscle fibers, a depolarization was seen that was similar to the end-plate potential developed when the motor nerve is stimulated to release the neurotransmitter, with little or no such potentials seen in the muscle fiber distant from the end-plate. In the denervated muscles, the responses to ACh were found all along the length of the fibers[55]; they show in effect a reversion to the immature condition of the muscle fibers at birth when ACh receptors are distributed all along their lengths. Soon after, in the course of early development, they become restricted to the end-plate region in the normal mature pattern.

The question arises as to how these muscle changes come about. Is it the loss of some trophic factor supplied to the muscle by the nerve? Or is it the result of the inactivity of muscles when they are no longer stimulated to contract by nerve impulses. The answer is that both factors contribute. Direct electrical stimulation to activate the denervated muscle does prevent atrophy and the spread of receptors in the membrane outside the end-plate region, but not entirely.[56] Some factor supplied to muscle by its nerve supply must also be operating. It is known that normally a small number of discharges of ACh from the nerve continually take place, these seen as "miniature" end-plate discharges too small to cause contractions.[57] The small packets of ACh released could perhaps be preventing muscle atrophy and membrane receptor changes in its membrane. However, supplying ACh to denervated muscle did not prevent the onset of atrophy and membrane changes in them. Most likely, some trophic agent released along with the miniature ACh discharge prevents muscle from degenerating.

Some other factor(s) appears to control the contractile properties of specific muscles, those that have a slow contraction or a fast one. Taking a pair

---

[54] (Brown, 1937).   [55] (Axelsson and Thesleff, 1959), and (Miledi, 1960).
[56] (Ochs, 1982) and (Ochs, 1988). See also Chapter 14.
[57] These are the miniature end-plate potentials (Katz, 1966).

of these two kinds of muscle in the leg of cats, the fast contracting *flexor hallucis longus* and the slower contracting *soleus* of the hind limb, cutting their nerves and cross-connecting them so that on regeneration (Chapter 10), they innervate their new targets, the muscles take on the contraction characteristics of their new innervation.[58] The different rates of motoneuron discharge to slow and fast muscles could also be a factor in causing the changes seen in the cross-innervated muscles.

NEUROTROPHIC CONTROL OF TISSUES OTHER THAN MUSCLE
In addition to the atrophy of skeletal muscles when denervated, atrophy of other tissues has been reported (skin, hair, nails, etc.) after denervation. Special trophic nerve fibers have been invoked by Duchenne who remarked that if "we had no knowledge of such [trophic] nerves we would be forced to invent them."[59] Samuel supported the reality of trophic nerves, considering that they act by controlling the nutritive state of the tissues.[60] Another explanation offered for the atrophic tissue changes seen after nerve interruption was that the lack of sensation after denervation impairs guarding against damage. Examples given are the ulceration of the cornea by foreign objects after denervation and ulceration of the feet by trauma after loss of sensation in diabetic neuropathy.

The possibility that tissue degeneration after interruption of their nerve supply results from a the loss of sympathetic innervation of blood vessels was examined by S. Weir Mitchell (1829–1914). Mitchell gained prominence as a leading American neurologist by his studies of nerve lesions in a large number of Civil War patients.[61] He concluded that an interruption of the sympathetic nerve supply to tissues was unlikely to be the cause of atrophic changes: "If, in fact, we exclude vasomotor influences as capable alone of explaining the pathological changes which follow nerve wounds, we are forced to fall back upon the nerves of motion and sensation [themselves as trophic nerves], or to believe in a system of independent trophic nerves."[62] Mitchell pointed out that there was no physiological or anatomical support for the reality of specialized trophic nerves. By the twentieth century, it was agreed, except for some holdouts,[63] that some trophic factor(s) is released from the motor, sensory, and autonomic nerves themselves. The involution of sensory organs (e.g., the taste buds and various sensory organs such as the lateral line organs of fish), shows the trophic control over these organs by their sensory nerve innervation. In the autonomic nervous system, the

---

[58] (Buller, Eccles, and Eccles, 1960a, and (Buller, Eccles, and Eccles, 1960b).
[59] (Duchenne, 1894).   [60] (Samuel, 1860).
[61] (Mitchell, 1965). This book, written years after his Civil War experiences, discusses the long-lasting aftermath of nerve lesions.
[62] Ibid., p. 35.   [63] (Wyburn-Mason, 1950).

effect of denervation of their smooth muscles or glands may not be seen as atrophy, but by hypersensitivity to their neurotransmitter, *noradrenaline* in the tissues of the sympathetic nervous system, and ACh in the parasympathically innervated tissues.[64] The various trophic factors are made in the cell bodies and carried down by the axoplasmic transport system in the nerve fibers (Chapter 11) to be released along with their neurotransmitters from the terminals.

[64] (Cannon and Rosenblueth, 1949).

10

NERVE REGENERATION

Just as broken bones can heal, so must it have seemed possible to the ancients that cut nerve could reunite and its function restored. Ancient authority is silent on this, but although not stating it explicitly, Galen's commentators in the Middle Ages suggested that he thought this to be so because
of the prescriptions he gave for the treatment of nerve wounds that were
aimed to bring about the "agglutination" of cut nerves. *Paul of Aegina* (seventh century) apparently followed Galen in using medications to promote
agglutination, also mentioning suturing of divided nerves:

> After the exposed nerve has been covered over, we must apply externally pledglets,
> with some of those things which are fitting for narrow wounds, such as that from
> euphorbium, or that from pigeon's dung, taking in also much of the sound parts.
> When the wound is transverse there is greater danger of convulsions, but every-
> thing relating to the cure is in this case the same, except that while the wound is
> recent some have used sutures and certain of the agglutinative applications; but
> the sutures must not be applied very superficially lest the part below remain un-
> united, but more deeply, taking care however that the nerve be not punctured by
> the needle. It is to be known once for all, that in wounds of the nerves the medicine
> which cures punctures being of a bitter nature, it is not possible to cure with it the
> division of the nerve, as the parts cannot endure pungency and inflammation.
> And neither does the medicine which cures incisions answer with punctures. For
> its strength does not reach the bottom of it, the incision of the skin being narrow.[1]

Adams, in his translation of the books of Paul of Aegina, commented that
not only did Paul closely follow Galen in this respect, but also so did other
well-known authorities, such as Avicenna, Rhazes, Lanfrancus, and Paré. In
the twelfth century, *Roger of Salerno* (fl. 1170) advised the use of Galenic
medications, such as egg albumin, to aid in the process of agglutination;
the binding properties of egg albumin were well known from its use in
cooking. In a compendium written probably around 1230, Roger's student

[1] (Paul of Aegina, 1844).

Gilbert Angelicus (fl. 1245) stated that nerves when cut and treated could be reunited (*conglutinari*) and regenerated (*consolidari*) – though he had some reservations that the latter would really eventuate.[2] In the case of a punctured nerve, it is to be divided to relieve pain and prevent tetanus, "as Galen had advised." Galen had sternly warned that a nerve puncture might lead to convulsions, and when that outcome seemed imminent, he advised cutting the wounded nerve to forestall that possibility. Perhaps he was influenced by cases in which tetanus had occurred as the basis for his proscription. *Roland of Parma* (1170–1269), who edited a later edition of Roger's work in 1264, rejected the use of egg albumin and instead favored touching the nerve with a hot iron to cauterize it, followed by placing powdered worms on the nerve wound. Presumably, this might encourage the nerve to burrow in the tissues. Regeneration, when it was accomplished, was seen by the return of motor responses.

Guy de Chauliac (c. 1300–1368), considered to be the most eminent authority on surgery in the fourteenth century, gave clear instructions for sutured nerves, observing that an excellent functional recovery can occur in the young:

> The incision of nerves... has need of three or four other particular intentions. The first is that if it is without any loss of substance it may be sutured with the flesh; the second that a drain should be put gently in the place which is lowest; the third, that some sedative and incarnative medicament proper to nerves should be placed over it; the fourth that it should be bandaged above with a compress of soft wool.... Now that such suture is useful is proved thus: because by the suture the parted ends are approximated and preserved together; beyond this by the covering of the skin of the flesh the nerve is protected from cold which injures it. And thus Avicenna, when he says in the Fourth: If the nerve is torn in its length then it is necessary to sew it, and *without that it is not agglutinated* [my italics]. William of Salicet and Lanfranc testify the same.[3]

Arguments were given suggesting that Galen may have approved of suturing nerve:

> Notwithstanding that some say that Galen does not command to suture nerves, inasmuch that they would not be consolidated and that the pricking of the needle is provocative of convulsion. But (save their reverence) Galen has not forbidden it; but if he is silent, yet he has otherwise affirmed it. What is even more he seems to consent to it in the Sixth of the Therapeutics Chapter III, when he says: 'The nerve being quite cut there is no longer any danger of convulsion, but part of it will be mutilated, and the treatment will be similar to that of other wounds.' Now it is certain that other wounds are sutured in order that one should keep their parts

---

[2] (Handerson, 1918).
[3] (Guy de Chauliac, 1890), and (Guy de Chauliac, 1924) quoted in (Ochs, 1980) p. 2.

approximated. He has signified the same thing when in the Third of the Techni, he does not make any difference in treatment of wounds in nerves from that of other wounds unless in the punctate wound alone. Nor in the Sixth of the Therapeutics unless of that same and of wounds of the exposed nerve, and of the accident of that alone which is either wholly incised or not, and of their attrition. It is by this means that according to the same author, by such suture, that nervous parts of the abdomen are agglutinated.

In suturing nerve, it is of necessity punctured by the suturing needle and this appears to be against the dire warning of Galen. The difference between puncturing and cutting the nerve completely through was commented on:

> What they say, concerning puncture of the nerve by the needle is of no value because the nerve is already pierced, seeing that the puncture penetrates all the substance. Nor what they object that nerves do not consolidate, because if they do not consolidate according to the first intention at least they consolidate according to the second as has been said before.[4] If one replies that such does not profit anything, because as well since the nerve is cut (seeing that it is only consolidated according to the second intention which is made by the intervention of some foreign substance) it loses its continuity so that vitality is not conveyed through it and thus the movement of the part is lost.

Suturing was claimed to be especially successful in restoring nerve function in the young, where the power of regeneration is at its greatest:

> Then I say that it [suturing] profits in two ways: first in children in which the nerves are consolidated almost truly and if one part loses its action all is not lost. In the young, also, when the parts of the nerves are more approximated, there is less intervention of foreign substance; and by this some vitality can travel by it and hence the limb is more useful. I have seen and heard it said that in several cases, cut nerves and tendons have been so well restored by suture and other remedies that afterward one could not believe that they had been cut.

Ambroise Paré (1510–1590), famous for his success as a military surgeon, recounts that it was the doctrine of the older physicians in treating nerve wounds to avoid their early agglutination.[5] The course Paré took was to keep the wound open so that the "filth may freely pass forth and the medicine applied to it to enter well." Medications were used to aid in the agglutination of the wounded nerves. Salves or *emplastrums* were also applied on the basis of humoral theory to "draw out" the humors. "Sharp" and more so "drying" medications were favored, particularly in nerves laid bare. Compound prescriptions were favored and might include such various substances as balsams, oil of terebinth, aqua vitae (brandy), a variety of powdered roots

---

[4] A wound was said also by the first intension when cicatrization occurs without suppuration (pus formation) and also by the second intention when suppuration occurs.

[5] (Paré, 1649), and (Paré, 1840).

and euphorbia (spurges, plants with a milky sap that are vesicant and rube-
facient). Dunglison notes in more recent times that the natives of Brazil use
a species of euphorbia as a remedy for serpent bites,[6] an interesting parallel
to the use made of such a substance by Paré to "draw humors" from an
injured nerve. Paré also lists animal fats or greases, plant oils, rosin, and
vinegar as being applied to heal nerve wounds.

Paré treated Charles the IX, who suffered a punctured nerve – the lesion
most highly feared by Galen. First, warmed oil of turpentine mixed with
aqua vitae was dropped into the wound. This was intended to penetrate
to the bottom and "exhaust and dry up the serous and virulent humor
which sweats from the substance of the pricked nerves," and so mitigate the
pain. An emplastrum was applied to dissolve the humor that had "already
fallen down into the arm." The limb was bound and later, in the course of
treatment, dissolving and drying medicines were applied.[7]

If, in dealing with such wounds, the pain increases, scalding oil was to be
used or the nerve burned with a hot cautery. *Escharotics* (powder of alum,
mercury, and so on) were also used for this purpose. Paré warned that if,
despite all such aids, when a *contumacie*, an imminent danger of convulsion
exists, it is better to cut the nerve, tendon, or membrane to preserve the
whole body, following the course of action advised by Galen.

EXPERIMENTAL STUDIES TO SHOW THE REGENERATION OF NERVE

The use of sutures fell out of favor after the Middle Ages, and by the late
eighteenth century, it was the established opinion that nerves could not
regenerate to restore function. De La Roche, in his book on nerve, stated
that, when a nerve is cut, it "loses forever its ability to transmit movement between the separated portions even though they can join and form
a cicatrix."[8] He referred to Munro, who was said to have confirmed this in
animal experiments. Nevertheless, De La Roche noted in a footnote to that
passage, that, "a medical scientist of London doubts this as a general principle and believes that regeneration may occur in large nerves at least but
probably not in the smaller branches." The reference to "a medical scientist
of London" was certainly to *William Cumberland Cruikshank* (1745–1800),
who worked for a time as a dissector for the famous anatomist *John Hunter*
(1728–1793). In 1776, Cruikshank cut the vagus nerves in the neck of several dogs and after some weeks found what he claimed to be a new growth
of nerve filling the gap. He described his experiments on nerve regeneration in a paper that John Hunter submitted for him to the Royal Society.
The paper, however, was not accepted. The editors were skeptical regarding
the nature of the regenerating material, believing that it was not truly nerve

[6] (Dunglison, 1833), Volume 1, p. 360.   [7] (Paré, 1649).   [8] (de La Roche, 1778).

substance, but rather a growth of nonneural tissue. Their reservations were later overcome when, some 15 years later, Cruikshank's paper was finally published,[9] appearing just before a similar paper on regeneration submitted by Haighton,[10] which seemed to have substantiated Cruikshank's view. Years later, Cruikshank gave what he presumed was the reason for the long delay in the publication of his paper:

> These experiments were made for another purpose, by which I discovered the independence of the heart's motion on its nerves, as well as the reunion after division, and the regeneration after loss of substance in the nerves themselves. I wrote a paper on this subject a long time since, which the late Mr. John Hunter, to whose memory and talents I am always proud to pay my tribute, presented to the Royal Society, but it was not then printed; I think Mr. Hunter gave me for a reason, that it controverted some of Haller's opinions, who was a particular friend of Sir John Pringle, then President of the Royal Society. Another gentleman [Haighton] has lately made experiments on the same subject, and has also presented them to the Royal Society. Upon hearing these read at the Society, Mr. Home, [Sir Everard Home, John Hunter's brother-in-law] with that intelligence of anatomical subjects that distinguishes his character, and the school he was bred in, remembered my experiments, though made nearly twenty years ago. The present President of the Royal Society, who, fortunately for mankind, prefers the promulgation of science to Haller or any other man, on being made acquainted with this circumstance, has caused the paper on these experiments to be printed in the Philosophical Transactions for 1794 [1795].[11]

Cruikshank's work was based on the finding that bilateral sectioning of the vagal nerves in the neck of animals leads to their death within a matter of days, a finding that had been made in the preceding century by Willis who wrote:

> We once made a tryal of the following Experiment upon a living Dog. The skin about the Throat being cut long-ways, and the Trunk of both the wandering pair [the English term for the vagus nerve at the time] being separated apart, we made a very strict Ligature; which being done, the Dog was presently silent, and seemed stunned, and suffered about the Hypochondria convulsive motions, with a great trembling of the Heart. But this affection quickly ceasing, afterwards he lay without any strength of lively aspect, as if dying, slow and impotent to any motion, and vomiting up any food that was given him: nevertheless his life as yet continued, neither was it presently extinguished after those nerves were wholly cut asunder; but this Animal lived for many days, and so long, till through long fasting, his strength and spirits being worn out, he died.[12]

Unfortunately, the survival time of the animal was not noted, other than the ambiguous reference to "many" days. Whereas bilateral vagotomy was judged to be lethal, the cutting of only one vagus nerve was not. In

---

[9] (Cruikshank, 1795).   [10] (Haighton, 1795).
[11] (Cruikshank, 1779), p. 88. Quotation from a footnote.   [12] (Willis, 1965b).

Cruikshank's experiment,[13] he cut one of the vagus nerves in the neck of a dog, removing a piece 15 mm in length. Also included was the accompanying intercostal (sympathetic) nerve, but this was recognized as having no lethal effects when cut. Soon after transection of the nerve, the animal experienced some difficulty in breathing, and an inflammation of the eye on the side cut was noted. Those changes passed off, and by the eighth day the animal appeared to have recovered fully. Cruikshank then cut and removed a section of the vagus nerve from the opposite side. Breathing soon became more labored, the animal vomited and copious salivation took place, and the pulse rose. Later, the animal ate and drank and passed stools, but seven days after the second operation, the animal died. The significant observation made by Cruikshank was that a substance of the same color as the nerve appeared to have united the two ends. The cut ends also showed swellings that Cruikshank described as rounder in form than that of a ganglion. The nerve on the side that had been cut later had a similar substance uniting the cut ends that appeared somewhat more bloody in color. Cruikshank inferred that the substance uniting the cut ends was regenerated nerve, a circumstance "never heretofore observed." He wrote: "it occurred to me that it might be objected to the reasoning, that the first two nerves [vagus and sympathetic] were doing their office before the last two were divided." Pursuing this thought, Cruikshank considered that the time he had allowed for regeneration had been too short, that "the nerves had not yet acquired the power of performing their office." Therefore, in his next experiment, he cut the right vagus nerve of a dog and waited for three weeks before cutting the left vagus nerve. Vomiting with convulsive jerks of the abdominal muscles in breathing was again seen, but soon subsided, and the animal survived for a period of two and a half weeks before it died. The nerves, which had been first divided, were now found to be firmly united with a "kind of callous" substance, which he analogized to the material found in the healed part of a broken bone:

> The divided nerves of the right side were firmly united; having their extremities covered with a kind of callous substance; the regenerating nerve, like bone in the same situation, converting the whole of the surrounding extravasated blood into its own substance. The nerves of the left side were also perfectly united; but the quantity of extravasated blood having been less, the regenerated nerves were smaller than the original; I observed too, that they did not seem fibrous like original nerves, but the recollection that the callous of bone is dissimilar to the original bone, quieted whatever doubts could arise from this circumstance.[14]

---

[13] (Cruikshank, 1795).
[14] Ibid. The use of the term regeneration here was not precise. It could have implied the outgrowth of new fibers or, as in the healing of broken bones, the reunion of the cut ends by a callus.

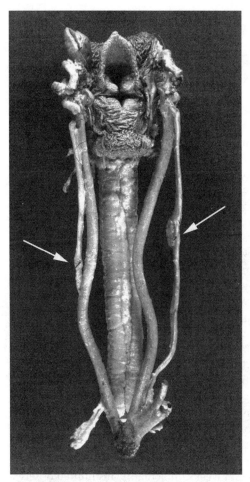

10.1. Cruikshank's preparation of transections of dog vagal nerve that he took to show regeneration. The trachea and larynx are in the center, with the vagus nerves on either side along with the thicker carotid vessels. Black threads were placed on the left, and right vagus nerves where swellings at the site of transection are indicated by arrows. The specimen is listed in Hunter's collection (Marshal and Burton, 1962), p. xlvi.

The nerves removed from this animal were stored as an exhibit in Hunter's museum (Figure 10.1).[15] In 1778, the Abbé Fontana, famous for his microscopic studies of nerve (Chapter 8), paid a visit to Hunter's laboratory. Cruikshank related his experiments on nerve regeneration to Fontana and

[15] (Marshal and Burton, 1962). Listed in the catalog as item 22.30. From a photograph kindly sent by Mr. Donald Hay, Department of Pathology at the Royal Infirmary, Glasgow.

showed him the preserved nerve preparation. When Fontana asked Cruikshank about Hunter's opinion of his experiments, "Cruikshank told me
ingenuously, that the doctor [Hunter] did not perceive a real reproduction
of the nerve in these experiments, and suspected very strongly, from the
great difference betwixt the external structure of the part cut, and that of
the other parts, that it [regeneration] was not the case."[16]

Fontana had some years earlier studied the effect of cutting the sciatic
nerve in animals without obtaining evidence of nerve regeneration, but
Cruikshank's revelations now inspired him to make a renewed study of that
possibility. In one series of experiments, Fontana removed a portion of the
vagus nerve from a dozen rabbits and some time later, on making a microscopic examination of the nerve ends, he found the nerves to have become
united with what, he concluded, was indeed regenerated nerve. The evidence supporting this conclusion was that the material bridging the ends of
the cut showed the same "spiral" or "banded" appearance he had observed
in normal nerves. Clarke and Bearn, in their historical-experimental review
of the "bands of Fontana," considered the various views that had been expressed as to its nature.[17] They concluded that it was because of the wavy
disposition of the nerve fibers within their enveloping (perineural) sheaths.
With a light source directed obliquely on nerve along its axial direction,
the raised portions of the wavy nerve fibers throw their shadows across
the shallow parts to give rise to the banded appearance. Stretching nerves
to straighten the fibers causes the bands of Fontana to disappear. This was
used in the course of studies where nerves were stretched to initiate beading
(Chapter 8).

A critical aspect of Cruikshank's studies must be considered, namely the
relatively short time allowed for regenerating fibers to reach the heart, lung,
and gastrointestinal tract to restore their function and preserve life. Cruikshank was overly impressed by the apparent prolongation of the life of the
animal from a day or so to several weeks. In actuality, his limited observations fell within the range of survival times of up to two or three weeks others reported to follow bilateral vagotomy. That still more time was needed
for regeneration was shown by Haighton.[18] He divided the vagus nerve on
one side in a dog with relatively minor symptoms resulting. Three days later,
he cut the vagus nerve on the other side and the animal died four days
later. So much was in accord with Cruikshank's findings. Haighton then
prolonged the time between the two transections. When the second vagus
nerve was cut nine days after the first, the animal died, but after a delay of
thirteen days. While this "prolongation" of the animal's life falls within the
range of lethality after a bilateral vagotomy, it was taken by Haighton as
an indication that a "partial regeneration" of the nerve had occurred rather

[16] (Fontana, 1787).   [17] (Clarke and Bearn, 1972).   [18] (Haighton, 1795).

than a complete regeneration as Cruikshank supposed. Haighton then performed his more convincing experiment. The vagus on the one side was divided, and six weeks were allowed to lapse before cutting the vagus of the other side. The effect of the second cut on the animal's physiological state was milder, the animal at first taking only milk and refusing solid food. Six months later, recovery was complete and the animal lived on for a further nineteen months. To prove that the animal's longevity was due to a regeneration of the vagus nerves rather than the possibility that another nerve had taken over the function of the transected nerve, Haighton again cut the vagus nerve above the site of regeneration, with the usual acute symptoms seen with the death of the animal on the second day. Haighton dissected out the regenerated part of the nerve and found that it was indeed united by nerve tissue.

THE LETHAL EFFECT OF A DOUBLE VAGOTOMY

The modern reader is struck by the lethality reported to follow a double vagotomy in the neck. When carried out as a treatment for peptic and duodenal ulcers, the nerve is cut subdiaphragmatically to selectively interrupt the innervation of the stomach and the duodenum.[19] When cutting the vagus in the neck, its more widespread distribution affects a variety of other organs. A number of the leading physiologists in the nineteenth century undertook to discover the organ affected that caused lethality after vagus nerve transections. Frey considered its pathological effect to be on the lungs.[20] Schiff also thought that the vagus had a trophic effect on the lungs, that cutting the nerve lead to a "nutritional" defect in the lungs, followed by inflammation and death of the animal.[21] An alternative possibility was that the lung changes seen resulted from aspiration of saliva or a regurgitation of stomach contents into the lungs. Claude Bernard provided evidence against those possibilities by inserting a tracheal cannula into animals before cutting the vagi.[22] In rabbits so prepared, death occurred as usual a few days after completing a bilateral vagotomy. Cannulas were inserted into the stomach of two dogs to divert their contents and prevent regurgitation of stomach contents into the lungs before carrying out a bilateral vagotomy. Nevertheless, those animals also died thirty-six and forty hours after bilateral vagotomy, with clear lungs in one animal and emphysema in the other. Bernard considered the emphysema to be caused by the increased depth of respiration seen after vagotomy, with a traumatization of the overextended lungs. In support of this idea, Legallois reported that the lungs of rabbits dying after vagotomy were congested and no longer able to float in water.[23] However, Ductoray de Blainville found little effect

[19] (Dragstedt, 1945).   [20] (Frey, 1877).   [21] (Schiff, 1858).   [22] (Bernard, 1844).
[23] (Legallois, 1813).

of vagotomy on the lungs of pigeons dying soon after bilateral vagotomy, and he emphasized the effect of vagotomy on the digestive system seen in mammals and birds.[24] John Reid summarized the effects of vagotomy along with his own investigations and concluded the cause of death to be paralysis of esophageal propulsion.[25]

The explanation for the early occurrence of death seen after bilateral vagotomy in the neck is still unclear. Alvarez, writing in more recent times, thought that studies made before the twentieth century were faulty, probably vitiated by infection.[26] An historical-experimental study was done in our laboratory to examine the effects of bilateral vagotomy in the neck using the approach taken by earlier investigators, except that aseptic technique was followed and an antibiotic given to prevent infection.[27] A section of the vagus nerves on one side was carried out and then the nerve on the other side cut at different times afterward. On cutting the second nerve, death occurred from a few days to within two to three weeks afterward, the longer survival times seen when the second cut was carried out with the longer delays after the first. Gastrointestinal effects were most commonly seen, an observation in accord with Reid's opinion and Ingelfinger who had pointed to a defect of esophageal function.[28] Our observations suggested that altered esophageal peristalsis, with vomiting and electrolyte imbalance leading to circulatory collapse, is the most likely cause of death.

REGENERATION OF PERIPHERAL NERVES IN THE LIMBS

Haighton's evidence for the reality of nerve regeneration was based on only one experiment. More experiments would have been required to explore the phenomenon further, but the death of animals as an endpoint is too indefinite and complicated. The study of regeneration was greatly advanced by using the sciatic nerve and other peripheral limb nerves. Fontana, as noted previously, had been unsuccessful in his early attempts to observe regeneration in the sciatic nerve. He accounted for his failure in comparison with the success met with by Cruikshank as due to "too much motion in the parts [legs] where they are situated, and it [regeneration] would in all likelihood have ensued if that motion were diminished."[29] An alternative possibility raised by Fontana was that there may be an inherent difference existing between the vagus and the sciatic nerves, with the capability of regeneration in the vagus being "a property only belonging to those nerves most essential."

---

[24] (Ductoray de Blainville, 1808).   [25] (Reid, 1838).   [26] (Alvarez, 1948).
[27] The unpublished experiments were performed on a total of 35 animals using mostly cats, along with several dogs and rabbits.
[28] (Ingelfinger, 1958).   [29] (Fontana, 1787), p. 213.

All this was set aside when he and other investigators turned to the sciatic nerve and found that it indeed was quite capable of regeneration.

Müller, after studies of nerve regeneration carried out by a number of investigators, was still left in doubt as to the reality of a true regeneration of nerve as apposed to a "mere reunion."[30] In experiments on rabbits in which the sciatic nerve was divided, he found after one month and twenty days that stimulation of the nerve below the region that had been cut caused the muscles to contract. However, electrical stimulation above the region gave rise to feeble contraction, causing Müller to conclude that regeneration had occurred to a limited degree. Alternatively, regenerating sensory fibers might have united with motor nerves.[31] This was investigated by Schwann, his assistant at the time, in frogs in which their sciatic nerves were cut. After regeneration he found that, on cutting their dorsal roots it elicited no response, while on dividing the ventral root, it caused strong motor responses as expected on the basis of the Bell-Magendie law (Chapter 9). The inference made from this finding was that when the motor and sensory nerve fibers regenerated, they did so with fibers of their own functional type.

Some further observations of Schwann's on the regenerated nerves are of particular interest. He noted that, in the previously divided distal amputated part, "nervous fibrils" were lying close together at the site of union that did not appear as white as a normal nerve, possibly because their "neurilemma was less perfectly reproduced."[32] This observation was the forerunner of later studies of regeneration made by Waller and others showing that these fibrils are regenerating neurites without myelin sheaths; these later turning white when they become myelinated.

Steinruck studied the time course of nerve regeneration by following the return of sensation and motion.[33] In all, he carried out fifty experiments, mostly on the sciatic nerves of cats, with some on the vagus, infraorbital, and hypoglossus nerves and several others on the sciatic nerve of frogs. Nerves were cut and a small piece of nerve removed, in some cases simply cut. Steinruck found that some functional recovery occurred after five weeks, but that for full restoration of function one to two years might be needed. He also found a *cross-innervation*, whereby the motor nerve of one muscle could be made to innervate another muscle by cutting and crossing the nerves and then suturing them to their new partners. This was carried out on the nerves innervating the elevator and depressor wing muscles in chickens. When regeneration was complete, stimulating the nerve that ordinarily would elevate the wing depressed it and stimulating the other nerve raised the wing.

[30] (Muller, 1840), pp. 457–465.   [31] Ibid., pp. 461–462.   [32] Ibid., p. 463.
[33] (Steinrück, 1840).

REUNION VERSUS REGENERATION

Steinruck was among those who believed the cut ends of the nerve to undergo *reunion* – the rejoining of the ends of the fibers, supposedly by an
exudate of "lymph" from the medium rather than by a true regeneration.[34]
Guenther and Schoen also believed that primitive nerve fibers in the distal
and central ends of the transected nerve join together to bring about their
reunion.[35] On the other hand, Nasse found what appeared to be new fibers
present in the distal degenerated nerve stump, smaller in diameter than
normal fibers that had a "different look."[36]

It remained for Waller to clearly describe the process of nerve fiber regeneration in the glossopharnygeal nerve in the frog.[37] He saw new fibers
passing from the central portion into the distal amputated stump that were
much smaller than their parent nerve fibers, one-quarter to one-eighth
their diameter, pale, without the double contour indicative of myelinated
fibers, and with fusiform nuclei positioned at intervals along their length.
He traced these new fibers over their full length, finding them at increasingly more distal positions, as would be expected of their slow centrifugal
growth, until they reached their fungiform papillae receptors in the tongue.
The time taken was relatively long, about nine months. Waller thus clearly
described the course of regeneration much as we would today, except for the
long time required. This was certainly because of the lower metabolism of
these poikilothermic animals and their lower body temperature, compared
with mammals.

We might suppose that after Waller's clear description of regeneration,
the concept of a reunion would be set aside. That was not to be. The "re-
unionists" received support from clinical observations. The eminent surgeon Paget[?] recounted one of his cases, that of a boy in whom the median
and radial nerves had been divided a little above the wrist. The nerves were
not sutured, but nevertheless function in the hand – which had been totally
lost directly after the injury – began to return in ten days and was nearly
perfectly restored in a month. Another case described by Paget was of a boy
whose hand had been almost severed from the forearm at the wrist, with the
median and radial nerves divided. After suturing the nerves, sensation returned in ten to twelve days. Such quick recoveries could not be accounted
for by the slow rate of regeneration described by Waller and were referred to
as clear examples of healing by "immediate union" or "primary adhesion."

Experimental support for the reunion concept was given by Schiff.[39] Using cats and dogs, he also found rapid recoveries of function after nerve
transections, at times much too soon for centrifugal regeneration to have

---

[34] Ibid.   [35] (Gunther and Schoen, 1840).   [36] (Nasse, 1839), p. 413.
[37] (Waller, 1852e), and (Waller, 1852f).   [38] (Paget, 1863).   [39] (Schiff, 1854).

taken place. Schiff believed that the axis cylinders of the fibers in the distal stump of the transected nerves remained viable and were able to reunite with those in the fibers central to the cut. The new regenerating fibers that Waller had reported were, in Schiff's view, the remaining axis cylinders of fibers that were better seen after they were deprived of their myelin sheaths. On transecting the lingual and infraorbital nerves, Schiff found fascicules of normal-appearing fibers present in the distal nerve stumps as early as three to four weeks later. This he interpreted to be caused by the reunion of the axis cylinders, followed by a fairly rapid restoration of their myelin sheaths.[40] Bruch reported a quick recovery of function after dividing the sciatic nerves of cats, adding further support to the concept of reunion. In a number of cases, he found no trace of scarring at the site of transection, with only a slight constriction of the sheath of Schwann seen at the transection site.[41] That regenerated nerve fibers had bridged the site was shown by stretching the nerve and finding the bands of Fontana to disappear. Lent supported the view that the myelin undergoes degeneration in the distal transected part of the fibers, leaving behind empty sheaths of Schwann and their nuclei that are usually hidden by the myelin. He held that the axis cylinders reverted to an embryonic form which, after reunion with the axis cylinders of the upper portion of nerve, then went on to reform the myelin.[42]

Philipeaux and Vulpian introduced a much more radical concept.[43] They considered that the axis cylinders of the nerve fibers in the stump distal to a transection not only remained viable after their myelin sheath degenerates, but also that the axis cylinders in the distal nerve stump bring about function recovery on their own *without reuniting* (my italics) with the axis cylinders in the central portion of the cut nerve. The regenerated fibers were reported to at first have a small diameter, then gradually to increase in size as regeneration proceeds. Their observations were made eighty-four days after transecting the hypoglossal nerve, and the central end moved away from the distal amputated portion away so that it would not be influenced by it. Even so, as early as forty-seven days after sectioning, young fibers were seen among the degenerated fibers of the distal part of the nerve. On stimulating the distal regenerated part four months after transection, no contraction of the muscle was seen. They took the absence of response to electrical stimulation of the amputated nerve to indicate a "lack of irritability" of the new regenerated nerve fibers that regained their power of irritability only after the myelin had fully regenerated.

---

[40] The finding could fit with the faster rate of regeneration seen in the mammal compared with the frog results reported by Waller that were carried out at lower body temperatures in the poikilothermic frog.
[41] (Bruch, 1855).   [42] (Lent, 1857).   [43] (Philipeaux and Vulpian, 1859), p. 55.

Opposition to Philipeaux and Vulpian's interpretation came quickly. Ranvier pointed out that strands of fine nerve fibers growing from the central part of the cut nerve to connect with the isolated nerve stump could very well have been missed by them. Vulpian later retreated from his position.[44] The failure of Philipeaux and Vulpian to obtain responses on stimulating the newly regenerated nerves before they had become myelinated can be accounted for by the phenomenon later studied by Erb and called by him the *degeneration reaction*.[45] In the early stages of regeneration of motor fibers, before they become fully myelinated, the reinnervated muscles could respond to voluntary commands showing that these fibers can conduct impulses, but that they were not excited by electrical stimulation. This phenomenon can be accounted for by the higher threshold of unmyelinated and small myelinated fibers to electrical stimulation, compared with the larger myelinated fibers.

WALLER'S LATER REVERSION TO THE REUNIONIST POSITION

Ten years after his clear description of regeneration by an outgrowth of new fibers into the distal degenerated part of a sectioned nerve, Waller reversed himself and accepted the reunionist view. In a lecture he gave at the Royal Institution, he stated that following its transection:

> the upper end of the nerve, that is, the portion connected with the Centres, remains unchanged, except just at its lower extremity where it swells out and an albuminous fluid, like the white of an egg, exudes around and between it and the upper end of the lower extremity [of the cut nerve], in which, after a few days, we may already detect some small new nerve-fibers. If this process of reparation is allowed to go on without interruption, the divided ends become in time united by this albuminous exudation, and the new nerve-fibers in the latter, uniting with the unchanged ones in the upper and the renovated ones in the lower end, serve to re-establish their continuity, and the nerve recovers its original structure – not as it would appear at first sight, and, as I first supposed, from the creation of new fibers [entering into] the lower [cut] end of the nerve; but simply from the formation of a fresh quantity of white medullary substance in the tubular membrane of the fibers and around their axis-cylinder, both of which remain unchanged throughout the whole process.[46]

He analogized the reunion of the cut nerve fiber to the soldering of a wire: "The central axis cylinder growing out to rejoin the axis cylinder of the amputated part becomes rejoined as in a cable where the copper wires become joined by soldering them together." He extended the analogy of an electrical conductor to the physiological properties of the nerve:

> There is another point, in which the analogy between the two agencies is close. The activity of the voltaic cell depends in a great measure upon the readiness with

[44] (Vulpian, 1866).   [45] (Erb, 1883), p. 56.   [46] (Waller, 1861), p. 14.

which the fluid element of the circuit undergoes decomposition – upon the readiness, that is, which the constituents of the fluid exhibit to take on new relations. The fluid is, in fact, the source of power in the cell. Now, although we cannot positively assert that the activity of a nerve-fibre depends upon the decomposability of its medullary substance, the readiness with which that substance undergoes decomposition indicates in it considerable activity of nutrition, however paradoxical to those who are unacquainted with physiology such a statement may appear. And activity of nutrition in the animal economy is invariably accompanied by potential energy: wherever power is most actively exhibited, there the metamorphosis of the tissues is proportionately rapid. Hence, it is only reasonable to look upon this medullary constituent of the nerve-fibers as the source, in the activity of its nutritive changes, of the peculiar powers which the fibers possess, or, at least, as the condition of the exercise of those powers. In connection with this point it may be mentioned that we have never been able to demonstrate in the grey variety of nerve-fibers, either in man or the lower animals, that power of conducting sensations and motor impulses which the white fibers possess; and in the grey fibers, it will be remembered, the medullary substance is absent. That those fibers do exercise functions of some kind there can be no doubt; but what those functions may be we are at present unable to say.[47]

Waller took the position that the medullary part of the fiber (the axis cylinder) reunites the divided fiber. What can we say of the reversal of his earlier position, of what appears to be so clear an observation of the centrifugal outgrowth of regenerating neurites? At this point in time, Waller was ill and no longer an active investigator, but a commentator. He was apparently swayed by the evidence given by those influential leaders who supported the reunionist view. Even as late as the turn of the century, Kennedy – in reviewing the history of nerve regeneration – concluded in favor of reunion, adding his own clinical experiences.[48] He cited cases in which sensation returned within two to five days after severance of the nerves in the forearm and then, a short time later by voluntary motion, though imperfectly at first. Such rapid recoveries were not in accord with a slow centrifugal growth of regenerating fibers.

We have no clear explanation for those reports of an early return of function after nerve transection that Kennedy, Laugier, Paget, and other prominent neurosurgeons had given. Collateral innervation and anatomical variants could have been involved. In any case, the extensive experience gained from the wounded in the two world wars and other military actions, the great numbers of automobile and industrial accidents, along with numerous well-controlled animal studies have shown the truth of Waller's original observation – that regeneration is due to an outgrowth of neurites from the proximal nerve into the distal portion of a cut nerve.[49]

[47] Ibid., pp. 16–17.   [48] (Kennedy, 1898).   [49] (Sunderland, 1978a).

Although regenerative outgrowth is now agreed upon by all for the verte-brate nerve, transected nerve fibers in some invertebrates can reunite. When the giant axon innervating the claw muscle of the crayfish is cut, the prox-imal and distal nerve pieces reunite, with motor responses returning much sooner than could be accounted for by a proximo-distal regeneration.[50] In electron microscopic studies, fine extensions were shown to sprout from the proximal end of the cut giant nerve fiber to make contact with its distal cut end. Some neurotrophic material supplied to the amputated fiber appears able to maintain its viability and effect a reunion.[51]

TROPHIC CONTROL OF LIMB REGENERATION BY NERVE
A number of nonmammalian species retain a remarkable power of regener-ation. The whole limb of the adult newt (*Triturus viridescens*) can regenerate when severed. That some principle in nerve regulates the process was indi-cated by the failure of such regeneration when the nerve to the limb was cut, a finding first reported by Todd in 1823:

> If the division of the nerve be made after the healing of the stump [of the cut limb], reproduction is either retarded or entirely prevented. And if the nerve be divided after reproduction has commenced, or considerably advanced, the new growth either remains stationary, or it wastes, becomes shriveled and shapeless, or entirely disappears. This derangement cannot, in my opinion, be fairly attributed to the vascular derangement induced in the limb by the wound of the division, but must arise from something peculiar in the influence of the nerve.[52]

Singer, in his review of the trophic influence of nerve on tissues in the lower amphibia, gave support to the concept that nerve is indispensable for regeneration by showing that the capability of a limb to regenerate was related quantitatively to the amount of nerve available for innervation.[53]

CELL BODY CHANGES IN THE COURSE OF REGENERATION
In early studies of Wallerian degeneration, the fibers above a nerve tran-section did not seem to differ much from the normal. Later investigations revealed a small decrease in their diameter. Most dramatic were the changes in the particles in the cell bodies described by Nissl after nerve fiber inter-ruption. The Nissl particles, staining purple with aniline dye, lose their color as the cells became swollen (Figure 10.2); this phenomenon is referred to as *chromatolysis*.[54] The Nissl particles were later shown to be composed of ri-bonucleic acid (RNA). They do not disintegrate, but become dispersed in the cytoplasm making the cell bodies appear colorless. Spectrographic analysis of the nucleic acid cell content of the cell bodies at various times after cut-ting their fibers showed, after a brief lag period, that the RNA (along with

---

[50] (Hoy, Bittner, and Kennedy, 1967).   [51] (Nordlander and Singer, 1973).
[52] Quoted in (Singer, 1952b), p. 172.   [53] Ibid., p. 172.   [54] (Nissl, 1894).

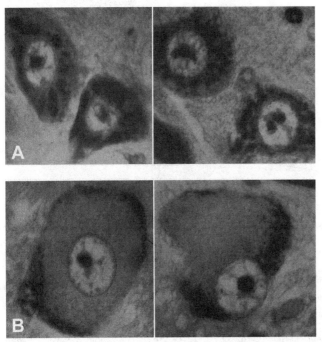

10.2. Chromatolysis shown in Nissl-stained cell bodies. The cells shown in the upper panel (A) are from a normal animal. The cytoplasm is stained dark by the dye, the nucleus is unstained, and the nucleolus is dark. The cells in the lower panel (B) are taken from an animal after transection of its fibers. They show the cell bodies to be much swollen and the staining of their cytoplasm to be greatly diminished. The nucleus is often displaced from the center, as shown in the cell on the lower right. From (Brattgård, Edström, and Hydén, 1957), Figures 1 and 5.

proteins in the cells) undergoes a large increase that lasts months before declining to control levels (Figure 10.3).[55] This was related to an increased production of proteins needed to supply the regenerating neurites. Along with the increase of those proteins, a shift in the overall program of synthesis takes place. Neurotransmitters and related components, for which there is no need before muscles become reinnervated, are down-regulated and their content much decreased in the regenerating fibers.[56]

These events in the cell bodies appear to be initiated by a signal carried by retrograde transport (Chapter 11) to the cell body from the transected site. After crushing a nerve, a second crush made above that site was used to collect proteins ascending from the lower ligation.[57] Using that technique, the

[55] (Bratttgård, Edström, and Hydén, 1957).   [56] (Johnson, 1970).
[57] (Ochs, 1975b), (Bisby, 1976), and (Bisby and Bulger, 1977).

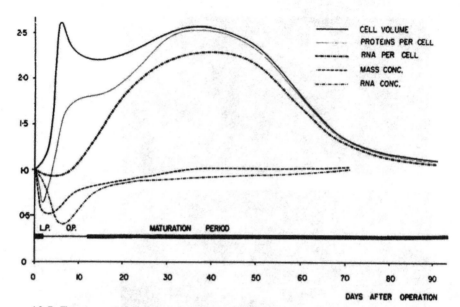

10.3. Time course of changed contents of chromatolytic cell bodies. Taking 1.0 on the ordinate to represent the normalized cell contents, a latent period (**L.P.**) is seen after transection, during which the cell volume increases by as much as 2.5× and the RNA shows minimal change. In the outgrow phase (**O.P.**), the concentrations of mass and RNA are low as a result of the volume increase. In the longer lasting maturation phase, the cell volume, proteins, and the RNA remain high and then gradually return to normal after 70 days. From (Brattgård et al., 1957), Figure 4.

proteins collected were found able to initiate chromatolysis when injected into otherwise normal cells.[58]

NEUROTROPISM – THE DIRECTION OF GROWING NEURITES

Regenerating neurites sprout from the upper end of transected rat nerve fibers as early as four hours after crushing their sciatic nerve.[59] Sprouting is in much greater evidence a day or so later, with as many as a score or more originating from the central part of a single nerve fiber.[60] Each neurite has a *growth cone* at its end that serves to guide it and the neurite elongating behind it as it moves toward the distal degenerated stump. The growth cones have an ameboid-like ability to move. Early on, Ramón y Cajal, who was the first to recognize them (Figure 10.4), characterized their action in mechanical terms – as a living battering ram pushing aside mechanically the obstacles it finds on its way. In a later resumé of their properties, he described them as being sensitive to mechanical and chemical variations in

---

[58] (Ambron et al., 1992).   [59] (Zelena, Lubińska, and Gutmann, 1968).
[60] (Perroncito, 1905), (Shawe, 1955), and (Ramón y Cajal, 1968).

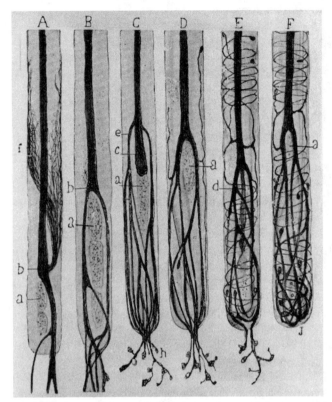

10.4. Outgrowth of fine neurites from the fibers above a nerve transection (**a–f**). The direction of neurite growth can be irregular, but for the most part it is directed toward the amputated stump of nerve. Ramón y Cajal first described the "club-like" enlargement (**h**) at their terminals, the growth cones, seen in (**C–E**). From (Ramón y Cajal, 1968), Figure 62.

their surroundings and moving with an ameboidism comparable with that of a pseudopod of a leukocyte.[61] The direction taken by the growth cone is determined by a signal(s) arising from the degenerated part of the nerve. To show this *tropic* (from the Greek *tropos*, to turn to) influence, Ramón y Cajal cut the sciatic nerve and stitched its upper cut end in place to point away from the distal amputated stump.[62] After allowing time for regeneration, he found regenerating fibers exiting from the upper part of the nerve to have turned around and then to grow down to enter the distal amputated stump (Figure 10.5). This was taken as evidence for a *neurotropic* factor released from the distal stump that attracts and orients the movement of the growth cone to it. On reaching the amputated stump, it grows between and into the

---

[61] (Ramón y Cajal, 1968), Volume 1, pp. 364–365.   [62] Ibid., Volume 1, pp. 242–244.

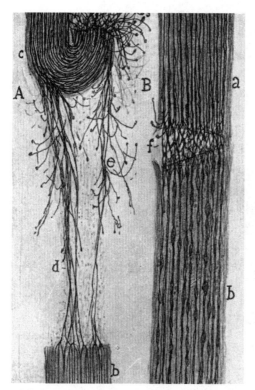

10.5. Redirection of neurite outgrowth indicating the action of a neurotropic agent. Ramón y Cajal cut nerves and stitched them so that their upper ends pointed away from the distal stump. After regeneration, some neurites (**A**) are seen to have grown directly down to the distal stump (**d**), others from the upper cut end having turned to descend down to the distal stump or ascend (**e**). In (**B**), a cut was made so that the two ends remained close, and all but a few neurites grew directly down from (**a**) into the distal cut end (**b**). From (Ramón y Cajal, 1968), Figure 100.

sheaths of the degenerated fibers, with some neurites eventually more or less reaching their target cells to reinnervate them.[63]

---

[63] The entry of regenerating neurites into the distal tubules of amputated nerve is pictured by (Ramón y Cajal, 1968) figure 90, p. 233. Not all fibers find their targets. This depends on a number of factors, including lack of responsiveness to tropic signals. Functional restitution is diminished by the entry of fibers into degenerated nerve targets other than the ones they originally innervated. A more favorable circumstance for functional restoration is seen when a nerve crush is used to interrupt the fibers. This allows many of the regenerating neurites to grow down within their original tubules to reach their original muscles. The most favorable return of function was seen when nerves were briefly frozen at temperatures below −20°C to cause degeneration. This least disturbs the normal structures of the fibers and tubules offering optimal guidance for the growth of regenerating neurites back to their original targets.

Ramón y Cajal could not define the mechanism of regeneration, but favored Heidenhain's concept of what he called by the *histodynamic impulse*. This acts "not by sending out a material substance or soluble enzyme, but by the propagation of a special impulse, different from the functional motor or sensory current."[64] Ramón y Cajal differentiated between embryonic and adult nerve cells. In the embryonic state, the neurite derives its nutritional energy from the cell body, as from a storehouse. After the embryonic stage, "the growth and ramification of the axon would occur largely at the expense, chemically, of the region traversed [i.e., by the axon], the soma then limiting itself to its function as a histo-dynamic centre."[65] As Ramón y Cajal had written, "if the trophic action depended on the diffusion of some soluble substance, capable of circulating along the axon, the absence of this substance in the peripheral stump would manifest itself in the centrifugal degeneration of the latter. But, as we have stated, this degeneration is simultaneous along the entire course of the conductor [nerve fiber]." The basis for invoking this mysterious entity was the evidence at the time that the degeneration of the distal stump of a transected nerve occurs at the same time all along its length. The present understanding that degeneration has a proximo-distal spread, at a rate much faster than had then been envisioned (Chapter 9), one consistent with fast axoplasmic transport (Chapter 11), removes the basis for a histodynamic impulse.

Ramón y Cajal's view of the growth of neurites was derived from reconstructions of fixed histological preparations. In the first decade of the twentieth century, the movement of growth cones and the growth of living neurites was directly visualized by Harrison in organ cultures, the technique he developed.[66] He saw the neurites to grow much as Ramón y Cajal had described (Figure 10.6). A similar growth of fibers was also seen by Speidel in the translucent skin of the tail fin of the developing tadpole.[67] Speidel pointed out the salient features of neurite growth:

1. Their abundance and branching early in development.
2. An apparent randomness in the direction of growth of many of the neurites with a direct orientation toward the ultimate target taken by some.
3. The achievement of the target cell by a neurite.
4. Regression and involution of aberrant branches.
5. Further growth and thickening of the successful neurites that had innervated their target cells – a process termed *maturation*.

Innervation of the target cells at first is not precise, and the target cells may be innervated at multiple sites. These become progressively reduced until a

---

[64] (Ramón y Cajal, 1968), p. 366. [65] Ibid., p. 367. [66] (Harrison, 1910).
[67] (Speidel, 1941).

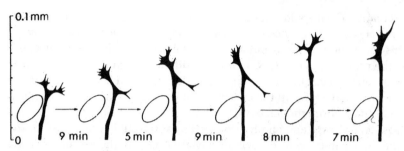

10.6. The time course of growth of an individual neurite in tissue culture is shown. At the terminals, the growth cones are seen to sprout at various points, seeming to respond to some influence from the medium. Neurite branches that sprout in an aberrant direction from the main line of growth become resorbed. The oval profile represents a stationary red blood cell that serves to indicate the neurite's growth. Time intervals in minutes. From (Harrison, 1910). Redrawn from (Jacobson, 1978) in (Ochs, 1982), Figure 15.3.

smaller number or a single neurite remains terminated on the target cell.[68] This process has been shown to be the case for skeletal muscle and for the parasympathetic and sympathetic ganglia.

The reduction of redundant neurites and further termination of other neurites on the target cells after they have been innervated by pioneer neurites suggest that a signal ascending in the successful fibers by retrograde transport (Chapter 11) acts to repress the redundant neurites. An increase of needed materials produced by the cell bodies is also sent down to the successful neurites. This channeling of the transported materials to selected neurites is an instance of *routing* (Chapter 11).

SPECIFICITY OF ACTION OF NEUROTROPIC FACTORS
To determine if the specificity of tropic factors extends to the type of muscle, Weiss and Hoag cut the peroneal and tibial branches of the sciatic nerve of rats and inserted their proximal ends into the arms of a Y-shaped sleeve of blood vessels, with the distal cut end of the tibial nerve inserted into its stem (Figure 10.7).[69] If a neurotropic factor arises from the distal amputated portion of tibial nerve to selectively attract the growth of tibial neurites to it, then more tibial fibers should regenerate down into it in preference to those from the peroneal nerve. Following time allowed for regeneration, the test of such a difference was determined by comparing the muscle responses to stimulation of the peroneal and tibial nerves. A greater muscle response on stimulating the tibial nerve would indicate that it had preferentially innervated the muscle. The result, however, was that there was no such asymmetry; reinnervation occurred on a random basis, indicating that there was no specific neurotropic signal from the two nerves.

[68] (Purves and Lichtman, 1980).   [69] (Weiss and Hoag, 1946).

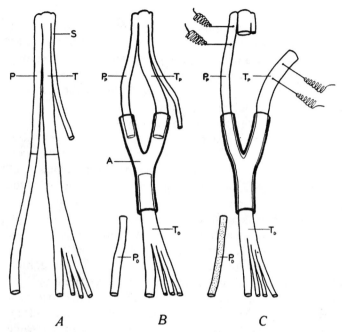

10.7. Test for a possible nerve-specific neurotropic influence emanating from the distal degenerating stump of a transected nerve, **(A)**. The tibial **(T)** and peroneal **(P)** branches of the sciatic nerve are cut and **(B)** inserted into the arms of the Y-shaped blood vessel, with the distal cut end of the tibial branch inserted into the stem **(C)**. After growth into the tibial nerve and muscle has taken place, the peroneal and tibial branches are electrically stimulated to determine if there had been a bias of tibial fibers regenerating into the tibial nerve stump. This would be indicated by a greater amplitude of its muscle contractions, but that did not occur. From (Weiss and Hoag, 1946), Figure 1.

Politis and Spencer used a similar experimental arrangement to look for a neurotropic factor.[70] Rat sciatic nerves were transected and the central end inserted into the stem of a Y-shaped silastic tube. The distal part of the cut nerve was inserted into one of the arms of the Y, while either nothing or an inert material was placed in the other arm of the Y. A preferential growth of regenerating fibers into the tube containing the degenerated part of the nerve was found, showing the action of a tropic factor. Homogenates of the degenerated nerve placed in the arm of the Y were similarly found to attract regenerating neurites. The homogenates, when heated or acted on by trypsin, were no longer active, indicating that the neurotropic factor was a protein.

[70] (Politis and Spencer, 1983).

NERVE GROWTH FACTOR

Levi-Montalcini and her colleagues found a *nerve growth factor* (NGF) present in sympathetic and sensory neurons.[71] NGF had also been found in snake venom and mouse salivary glands in large amounts and shown to consist of two identical polypeptides, each with a molecular weight of 13,259.[72] Its *neurotrophic* (growth augmenting) property was indicated by the increased density of microtubules, neurofilaments, and actin found in fibers after uptake by its terminals. In newborn rats, the marked hypertrophy of the sympathetic ganglia and sensory ganglia, to as much as 12× their normal size, was the result of a larger than normal number of cells. Normally, the oversupply of cells present at birth is decreased, with many of them dying off in the course of maturation. NGF, by inhibiting their dying off, accounts for the hyperplasia seen.

A routing of growth of fibers back into those fibers whose terminals had taken up NGF was shown using a three-chamber system. The cell body of a cultured neuron was placed in the central portion and its two neurite branches allowed to grow out into the two side chambers.[73] The addition of NGF to either one of the side chambers augmented neurite growth of that branch only. The NGF taken up by the terminals and carried to the cell body by retrograde transport results in an increased synthesis of cytoskeletal proteins and other materials that are then directed back to that same branch while the neurite in the other chamber not exposed to NGF did not participate in the enhanced growth. These findings can be explained by routing where the NGF, taken up and carried along a subset of microtubules to the cell body, also carries the increased amount of synthesized components back down along that same set of microtubules (Chapter 11).

In addition to its neurotrophic action, NGF has a *neurotropic* action. This growth-directing action was shown in chick dorsal-root explants in cultures by placing the tip of a micropipette containing NGF, from which very small amounts were released at a short distance from a growth cone, at an angle to it.[74] Within a score or so of minutes, the neurite was seen to turn and grow toward the NGF-containing micropipette. Repositioning the tip of the pipette to another site away from the growth cone caused a redirection of the growth cone toward the new position of the NGF-containing pipette.

MOLECULAR PROCESSES IN THE GROWTH CONE RELATING
TO STRUCTURAL CHANGES

The direction of advance taken by the growth cone is determined by their *filopodia*, the slender extensions from the growth cones with lengths up to

[71] (Levi-Montalcini, 1976), (Mobley et al., 1977), and (Levi-Montalcini, 1987).
[72] (Server and Shooter, 1977), and (Greene and Shooter, 1980).
[73] (Campenot, 1977).   [74] (Gundersen and Barrett, 1979).

Regeneration and Reinnervation

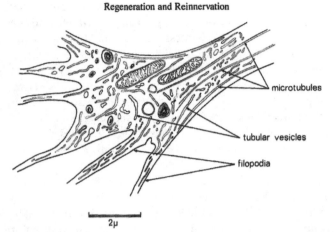

10.8. Sketch of a growth cone and its filopodia. In tissue culture, the thin filopodia are seen under the microscope to be incessantly waving about, apparently moving in response to signals from the environment. Their full length is not shown in the figure. Organelles within the growth cones are indicated: mitochondria and tubular vesicles. Additions are made to the membrane as the growth cones lay down the neurite extending behind it, with the microtubules in them elongating by assembly from the tubulin subunits brought to the growth cones along the newly laid-down microtubules. From (Bray, 1973), Figure 2.

10 to 20 μm and diameters up to 0.3 μm (Figure 10.8). Under the microscope, the filopodia show complicated motions. They extend, wave back and forth, advance in one direction, regress to move in another, all apparently in response to signal substances received from their environment that determine the direction of the growth cone's movement. The filopodia contain a filamentous matrix consisting mainly of actin microfilaments that have contractile properties and is the means by which filopodia movement comes about. The antiactin agent *cytochalasin B* disrupts the microfilaments, causing the filopodia to "wilt" and shorten with the growth cone rounding up.[75]

In addition to its action to move in one direction or another, the growth cone is the site of the assembly of the cytoskeleton elongating behind it, which enables the neurite to lengthen. Tubulin protein subunits are carried by axoplasmic transport over the microtubules to the growth cones, where they are assembled as extensions of the microtubules previously laid down.[76] As will be discussed in Chapter 11, the microtubules are the "rails" along which the transport mechanism carries materials.[77]

[75] (Yamada, Spooner, and Wessells, 1971).   [76] (Bray and Gilbert, 1981).
[77] (Letourneau, Kater, and Macagno, 1991). The papers described in this book show the impressive increase of knowledge concerning the processes carried out in the growth cone gained in recent years.

The measured rate of neurite growth has ranged from about 1 mm/day to 5 mm/day. One method used to examine the rate of growth relies on the sensitivity of the growth cones to mechanical stimulation. On tapping the skin over the site of the growing nerve ends, tingling is experienced; this phenomenon is known as the *Tinel sign*.[78] It is used to follow the growth of regenerating nerve in patients by the successive displacement of the sensation felt on tapping over the skin of the regenerating nerve endings. Clinical studies show the lowest rates, 1 to 2 mm/day, most likely because of the complications from damage to the nerve, delays in suturing, and the condition of the patient's general health that may be debilitating. Higher rates of nerve regeneration have been more consistently found under controlled conditions in animal studies in which generally rates of 3.5–4.5 mm/day were recorded using several different techniques. One such was the flexion withdrawal reflex response to a mild nociceptive stimulation of the skin near the growing end of regenerating nerve; the shift of the furthest point down the limb at which the animal responds gave a rate of 4.5 mm/day.[79] Bisby examined the rate of regenerative growth in the rat by the advance of labeled proteins (Chapter 11) to be 3.8 mm/day.[80] A similar determination using this method also gave a rate of 3.5 mm/day.[81] The precise determinations of the rate is important, for it has a bearing on the mechanism taken to account for slow axonal transport, as will be discussed in the next two chapters.

SPINAL CORD REGENERATION

The similarity of the larger fibers of spinal cord tracts with respect to peripheral nerve myelination, except that the glial myelinate – a group of fibers in the cord in contrast to the Schwann cell investment of single fibers in the peripheral nerve (Chapter 8) – led to the possibility that fibers of the spinal cord could also regenerate. This assumes the generality of the mechanisms of regeneration in the peripheral and central nervous systems, a matter of not only basic neurology, but also one of pressing clinical concern. The loss of muscular control and sensation below the level of transection leads to the pathetic situation in which the patient, often young and otherwise vigorous, becomes dependent on constant nursing care. Reflexes carried out by the local spinal mechanism below the transection still remain in force, with some reflexes exaggerated (Chapter 14), leading to the belief that if the transected fibers of the cord could regenerate and make connection with them, function could be restored.

Unfortunately, the regeneration of axons in the mammalian central nervous system (CNS) is feeble, compared with that seen in some of the

---

[78] (Konorski and Lubińska, 1946).   [79] (Gutmann et al., 1942).
[80] (Bisby, 1978).   [81] (Griffin, Drachman, and Price, 1976).

lower vertebrates.[82] Experimental studies to augment the power of their regeneration has a long history.[83] After transection of the spinal cord of the rat, cat, dog, and monkey, histological study showed nerve fibers in the spinal tracts to sprout, but then after several weeks to stop. One of the factors believed to prevent or inhibit growth was assumed to be fibrous formation occurring at the site of transection creating a physical barrier. However, pyromin, trypsin, and hyaluronidase used to suppress or decompose fibrous growth gave equivocal results at best.[84]

Examination of the spinal cord at the site of transection showed cysts or cavities near the cut ends of the cord with the formation of terminal clubs of the fibers, the clubs enlarging to form cysts and cavitations. Kao considered that, by removing these cavitations and using pieces of peripheral nerve to bridge the gap, regeneration might be enhanced.[85] The rationale for the use of pieces of peripheral nerve as a bridge was based on the increased number of Schwann cells that occurs in degenerated peripheral nerve leading to the supposition that this would promote the regeneration of CNS nerve fibers. Earlier studies using peripheral nerve implants in the CNS seemed to show an enhanced regeneration,[86] but were not confirmed by other such studies,[87] though a significant number of axons were seen traversing the graft. Similarly, Richardson et al. implanted segments of peripheral nerve into cuts made in rat spinal cords and found regenerating cord fibers to have entered and passed through the graft.[88] The fibers bridging the gap were functional, as was indicated by injecting horseradish peroxidase above and below the graft and finding the marker to be transported through the graft. Thus, it appears that some factor(s) in Schwann cells promotes regeneration by augmenting the normally feeble regenerative powers of the spinal cord fibers. Other studies also indicated that neurilemmal cells harvested from peripheral nerve promoted regenerative growth when placed into the gap of a transected cord.[89]

To account for the lack of regenerative power of the spinal cord fibers, Kiernan advanced the hypothesis that some protein taken up by the growing tips of the fibers and carried by retrograde transport to the cell body stimulates the synthesis of materials required for regeneration.[90] It is the failure of uptake of such proteins or the relative scarcity of such a factor(s)

---

[82] (Lee, 1929), (Ramón y Cajal, 1968), and (Kiernan, 1979).
[83] (Windle, 1955), (Mark, 1969), (Guth, 1974), (Guth, 1975), and (Puchala and Windle, 1977).
[84] (Guth, Bright, and Donati, 1978) and (Kiernan, 1978).
[85] (Kao, Chang, and Bloodworth, 1977).
[86] (Sugar and Gerard, 1940), and (Ramón y Cajal, 1968).
[87] (Brown and McCouch, 1947), and (Feigin, Geller, and Wolf, 1951).
[88] (Richardson, McGuinness, and Aguayo, 1980).
[89] (Liu et al., 1979).   [90] (Kiernan, 1979).

that accounts for the feeble power of regeneration in the CNS, compared with peripheral nerve.

To restore function, the nerve fibers that regenerate through an implant must be able to engage their original neuronal elements in the local spinal cord mechanisms. If they do not, and it is unlikely that a full reconnection would occur considering the multiplicity of the connections that would have to be restored, they would have to engage some significant portion of the neuronal elements of the local neural machinery of the spinal cord. The number of such reconnections may not have to be too great. It is impressive to observe how much function is retained when most of the tract fibers of the cord are cut leaving only a small number remaining. Also, the regenerating fibers may not have to synapse on exactly those of its original target neurons. It is possible that "foreign" fibers could establish some functional return through other interneurons. In experiments along those lines, the ventral roots of the L7 spinal cord segment of cats were cut on one side and grafted to the cut ends of the L7 dorsal roots of the other side. Regenerating ventral root fibers were found to grow into the spinal cord via the dorsal root.[91] The entry of regenerating ventral root fibers into the dorsal root was shown by injecting [^{3}H]leucine into the L7 ventral horn of the cord on the side supplying the ventral roots and tracing radioactive proteins into the dorsal root.[92] This was also seen by AChE carried into the cord by the regenerating motor fibers.[93] That such entering fibers could make synaptic connections was shown by changes in ventral reflexes recorded when stimulating the new input.[94]

[91] (Barnes and Worrall, 1968).   [92] (Ochs and Barnes, 1969).
[93] (Ranish, Ochs, and Barnes, 1972).   [94] (Barnes and Worrall, 1968).

11

CHARACTERIZATION OF AXOPLASMIC TRANSPORT

The ancient concept of animal spirits moving in hollow nerve fibers to account for sensation and motor control was replaced in the Renaissance with such surrogates as a gas, a thin vapor, a fiery fluid, vibrating particles, and so on, until eventually the nerve impulse was recognized as being electrical in nature. However, this left still unaccounted for the slow onset of Wallerian degeneration appearing a day or so after nerve transection, along with the later slowly developing atrophy of muscles and sensory organs. Some other nerve principle was involved. With the establishment of the neuron doctrine, the question turned on the possibility of the loss of supply of some substance from the nerve cell to its fibers and the tissues innervated, a hormone, an enzyme, or whatever. And, another related question arose, the nature of the mechanism that transports that principle in the fibers.

EARLY HYPOTHESES OF TRANSPORT BASED ON CELL BODY CHANGES
The chromatolysis of cell bodies, loss of Nissl particles staining dark blue with aniline dyes that was seen to follow the transection of its nerve fibers (Chapter 9), drew the attention of Scott to this phenomenon.[1] He found the particles to consist of a "nucleoproteid," later identified as ribonucleic acid (RNA), having a remarkable resemblance to the granular material present in secretory gland cells, such as those in the fundus of the stomach and pancreas.[2] Pointing to that similarity, Scott hypothesized that "the nerve cells are true secreting cells, and act upon other cells by the passage of a chemical substance of the nature of a ferment or proferment from the first cell into the second." His hypothesis was more definitively asserted in his second paper where he stated that; "in the body of nerve cells a substance is formed from the nucleus and Nissl bodies which gradually passes into the nerve fibres; and also that stimulation of other cells by a nerve fibre is brought about by the passage of some of this substance into the cell on

---

[1] (Scott, 1905).   [2] (Scott, 1906).

which the fibre [terminal] acts." The hypothesis of the passage of a substance from neuron terminals to affect the actions of other neurons they synapse on anticipated the process we now recognize as neurotransmission. The establishment of such a downflow in fibers and exactly what substances were moving remained to be determined.

Parker noted that the outflow from the T-shaped cells of the dorsal root ganglia of substances "originating in the region of the nucleus of the cell body, [passes] down its neck to the tract of nervous transmission where they separate into two streams, one flowing peripherally over the neurite [sensory nerve fiber branch] and the other centrally over the central [dorsal root] nerve fiber process."[3] The direction of flow is in opposition to afferent nerve impulses in the sensory branch. Parker thought that what was being transported in the fibers was related to its metabolism, but he was uncertain about the nature of the material(s) transported: "What the metabolic influences are, it is impossible at present to say. It seems hardly reasonable to think of them as streams of material in the nature of a hormone, emanating from the region of the nucleus and percolating throughout the neurone." Torrey, a student of Parker, was not so restrained. He made explicit the hypothesis that the substance supplied by the cell body to its fibers was a "hormone-like" material necessary for the maintenance of the nerve fiber.[4] He further proposed that "the substance involved may be enzymatic in character, produced by the nucleus of the neuron, and transported peripherally by the way of the neurofibrils."

Goldscheider had theorized late in the nineteenth century that a "ferment-like" substance transported from the cell body down into and along the whole course of the axon to their extremity is required for the nutrition of the axons.[5] In a review of nerve and its properties written in 1932, Gerard summarized the then new evidence that nerve impulses depended on the energy supplied by oxidative metabolism. He believed that the enzymes needed to support its metabolism were synthesized in the cell bodies and transported down within the fibers (Gerard, 1932):

> Oxygen and substrate easily reach the axons via its own blood supply, but the catalytic substances necessary to their reaction are not so readily available. If the respiratory ferment or accessories are formed in the cell body (from the nucleus?) and continually travel down the axons where they are slowly destroyed in the course of the oxidizing reactions, much of the behavior of a severed nerve [its Wallerian degeneration] becomes clear.[6]

[3] (Parker, 1929a).   [4] (Torrey, 1934).
[5] (Goldscheider, 1894a), and (Goldscheider, 1894b).
[6] (Gerard, 1932). The discovery of metabolic paths leading from glucose to the production of ATP and its role in the transfer of energy to cellular functions was mainly carried out in the study of muscle in the 1920s. The history of that chapter in biochemistry has been related authoritatively by Dorothy Needham (Needham, 1971). Soon after, the same metabolic components and processes were found present in nerve.

The revolutionary demonstration in the 1950s by Watson and Crick that DNA and RNA control the synthesis of proteins in cell bodies clarified the relation of the neuron cell body to its fiber. The chromatolysis of the cells after transection of their fibers was shown to result from changes associated with an increased production of nucleic acids and the proteins that were presumed to be needed for the regeneration of the transected fibers (Chapter 10). The means by which the proteins produced by the cells are transported down within their nerve fibers posed a pressing challenge.

DAMMING, THE HYPOTHESIS OF A BULK FLOW OF AXOPLASM

The large number of nerve injuries suffered during World War II stimulated a search for methods to enhance nerve regeneration. In animal studies directed to that end, Weiss and Davis used segments of artery to bridge nerve transections, the thought being that the arterial segments would better guide regenerating neurites into the distal amputated stump of nerve.[7] Serendipitously, they found that adrenaline in the circulation caused some contraction of the arterial sleeves to produce a partial compression of the nerves. They turned their attention to a study of the effects of such compression on the fibers. Arterial sleeves were slipped up onto nerves and left for 8 to 13 days. A swelling of as much as three times the normal diameter of the nerve was seen just above the constricted region and a lesser increase of about 50 percent just below the sleeve. They considered that the swelling above the compression resulted from an accumulation of endoneurial fluid between the fibers. But, soon after, swellings and tortuosities were found in the fibers (Figure 11.1A), and the hypothesis advanced was that this was due to a *damming* up of the axoplasm within them (Figure 11.1B).[8] The axoplasm was envisioned as continually being generated in the cell bodies and moved down by the pressure of that ongoing production within the fibers as a semisolid column (much as lead is propelled in a mechanical pencil). On meeting with an obstruction such as that offered by the arterial sleeve, the moving column of axoplasm builds up in the fibers above that site to form the swellings and tortuosities seen in them. These swellings occupy only a few millimeters above the obstruction, and from the increases in fiber size of the enlargements, Weiss calculated from the axoplasm dammed up that the column of axoplasm was moving at the rate of 1 to 3 mm/day. As this rate appeared to correspond to the rate of nerve regeneration,[9] Weiss proposed that the continual movement of axoplasm down within the fibers serves not

---

[7] (Weiss and Davis, 1943).   [8] (Weiss and Hiscoe, 1948).
[9] The rate of regeneration reported in clinical cases, especially those seen during wars (Woodhall and Beebe, 1956) were generally lower than those determined in experimental studies because of complications resulting from extensive injuries. Rates of 1–2 mm/day were most commonly recorded (Sunderland, 1978b), Table 8.1, p. 121.

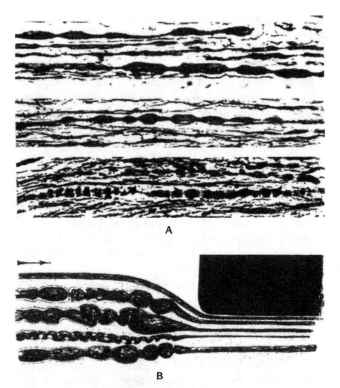

11.1. (A) Swellings and tortuosities of the fibers within a few millimeters above a site of partial constriction pictured by Weiss as being from a "damming" up of the axoplasm. Variations in the form of fiber changes resulting are indicated in the three panels. The upper panel is of fibers closest to the region of partial constriction, the others a little further away more proximally. Note that some of the fibers in the latter sites appear to be beaded. From (Weiss and Hiscoe, 1948), Figure 2. (B) Weiss' schematization of the fiber form changes just above a region of partial compression (black bar). The partial constriction was considered to restrict the downflow of axoplasm (its direction indicated by the arrow), causing the fibers to swell and show tortuous form changes. The smaller amount of axoplasm passing through fibers at the compression site was taken to account for the thinner fibers seen in and below the region of compression. From (Weiss, 1961), Figure 10.

only for the constant renewal of the fiber, but also the supply of materials needed for the regeneration of transected fibers. Weiss strongly emphasized that the swelling of the fibers was limited to only a few millimeters above the compression: This indicated that the axoplasm was a semisolid gel. If

Higher rates of 3.5–4.5 mm/day have been determined in experimental studies in animals in which the conditions of regeneration could be controlled (Chapter 10).

The discrepancy between rates estimated by damming and regeneration is further discussed in this and the next chapter.

the axoplasm were fluid, it would readily spread back up within in the fiber above an obstruction rather than accumulate in so localized a fashion.[10] Another consequence drawn from the hypothesis was that the reduction in the amount of axoplasm moved through the narrowed fibers could account for the small diameter fibers seen in and distal to the constrictions.

Weiss' concept of damming attracted much attention and was for a time widely accepted. Questions, however, were raised about the basic premise, namely that axoplasm is moved out within the fibers in the form of a gel with a high viscosity. This posed the problem of how the gel-like column of axoplasm could divide at fiber branches so that each daughter branch receives its appropriate amount of axoplasm. This problem would be faced by the regenerating fiber, where a large number of neurites arise from its proximal cut end.[11] All of these neurites would have to be supplied by an appropriate division of the semisolid axoplasm in the parent fiber. An even greater impediment to the concept was the finding that the total volume of axoplasm contained by the daughter branches was much greater than that of the parent fiber, by as much as 10-fold.[12]

There was also the question of the reality of the tortuosities of the dammed fibers and the nature of the small fibers distal to the obstruction portrayed by Weiss. A study of the effects of a partial compression of nerve using ligations had earlier been made by Ramón y Cajal.[13] He found a series of alternating swelling and constriction in the fibers just above the region of partial compression, whereas in the nerve below it regenerating neurites were seen (Figure 11.2). These were identified as such by the growth cones seen at their ends (cf. Figure 10.4). Thus, along with the partial obstruction of fibers that Weiss envisioned, transected and regenerating fibers were likely present. Ramón y Cajal conjectured that the beaded fibers he saw above the partial compression was the result not of damming, but of a starting and stopping of fiber regeneration, a hypothesis for which no evidence has been found. The appearance is rather one of beading (Chapter 8).

The form changes seen in fibers just above the site of a partial nerve compression were closely investigated by Spencer.[14] Where Weiss had reported swellings and tortuosities in the fibers, Spencer found normal-appearing myelinated fibers, along with clusters of regenerating sprouts (Figure 11.3). Those normal-appearing fibers of somewhat smaller size could have escaped interruption because of the lesser susceptibility to damage of the smaller fibers to mechanical compression.[15] And some of the fibers would have been spared: as Berthold and Skoglund had shown, a partial constriction

---

[10] (Weiss, 1961).   [11] (Perroncito, 1905).
[12] (Zenker and Hohberg, 1973), and (Friede and Bischausen, 1980).
[13] (Ramón y Cajal, 1968).   [14] (Spencer, 1972).
[15] (Aguayo, Nair, and Midgley, 1971).

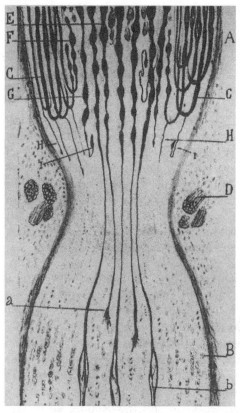

11.2. Partial constriction of a rabbit nerve carried out 13 days beforehand by Ramón y Cajal shows beading in the fibers (**E**, **F**, and **G**) above the constriction (**A**), with fibers turning up (**C**) and regenerating fibers below it (**B**) identified by the growth cones at their terminations (**a**). Fiber reunion (?) shown at (**b**). Remnants of the ligation used to partially constrict the nerve can be seen (**D**). From (Ramón y Cajal, 1968), Figure 100.

of nerves does not produce a uniform pressure on its fibers; those in the outer part of the nerve are more apt to degenerate than those in the interior that may not be affected at all.[16] Weiss had interpreted the small-sized fibers distal to the compressed site to result from a reduced supply of axoplasm to them. On the release of the compression, these fibers he supposed them to be gradually restored in size as the axoplasm in the convoluted and ballooned axons moved into them from above. The fibers above are becoming "straightened and drained of their axonal material."[17] Alternatively, the increase in fiber size more likely represented a maturation of regenerating fibers (Chapter 10).

[16] (Berthold and Skoglund, 1967).   [17] (Weiss, 1961).

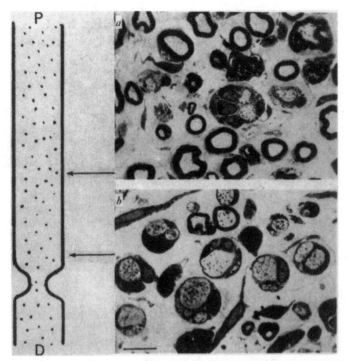

11.3. Cross-section of fibers just above a partial ligation (**b**) shows degeneration and lack of tortuosities and large swellings with evidence of regeneration. Normal-appearing fibers (**a**) are seen at a more proximal site. Proximal (**P**) and distal (**D**) part of nerve. (Spencer, 1972), Figure 1.

THE HYPOTHESIS OF AN OUTFLOW OF FLUID AXOPLASM FROM THE CELL BODY

Another theory of a bulk outflow of axoplasm that gets around some of the difficulties of Weiss' hypothesis of the flow of a gel-like axoplasm was advanced by Young, based on his observations made on the giant nerve fiber of the squid: when the fiber was cut, axoplasm emerged from it.[18] Young theorized that the continual production of axoplasm by the cell body exerts a pressure driving the fluid axoplasm down within the fiber.[19] That pressure is opposed by a lateral pressure exerted by the wall of the fiber so that when the fiber is cut, the unopposed lateral pressure causes the outflow of axoplasm (Figure 11.4). Young extended his model to mammalian nerve fibers. As evidence, he pointed to the beading seen in the amputated

---

[18] (Young, 1936a), and (Young, 1936b). The rediscovery and introduction of the giant fiber into neurobiological research by Young was of major importance in showing the ionic basis of membrane potentials (Chapter 8).

[19] (Young, 1944b).

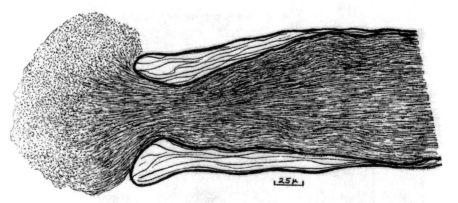

11.4. The outflow of axoplasmic fluid from the cut end of a giant axon. Fibrillary material seen in the axoplasm becomes dispersed when it comes in contact with the medium. The giant fibers, rediscovered by Young, have much larger diameters than that of mammalian myelinated nerve fibers. Note thickening of the fiber sheath and narrowing of the axon at the cut end. From (Young, 1934), Figure 1.

myelinated fibers distal to the transection.[20] Young assumed that this form change was occurred because of the lateral pressure of the myelin sheath, with surface tension causing the beading seen. He based this on the similar appearing beading seen in physical models of surface tension described by Plateau.[21] However, the beading seen in the distal amputated fibers does not occur immediately, but after an interval of a day or so (Chapter 9). This delay would not be expected on Young's model, the change should occur soon after cutting. The organized structure of the myelin sheath (Chapter 8) militates against the simplification that a fluid lipid with surface tension properties is coating the axon.

Young's model seemed to have been given support by Lubińska, who described the outflow of axoplasm from the proximal ends of cut single myelinated nerve fibers that took the form of a sphere (Figure 11.5).[22] The outflow, however, appears to be the result of a local contraction of the fibers at 1, 1.5, 2.5, 4, 5 and 7 hours after sectioning at their cut ends, as can be seen in the figure. A similar contraction near the cut end was seen also in giant nerve fibers, where an outflow of axoplasm can be seen not only at the central end, but also at the distal cut end.[23] Such local constriction at the cut ends of fibers appears to be related to beading, its role being to seal the ends of cut fibers preserving their contents and thereby assisting in its regeneration (Chapter 8).

[20] (Young, 1944a), and (Young, 1945), pp. 56–57.
[21] (Plateau, 1873), and (Thompson, 1942).   [22] (Lubińska, 1956).
[23] (Flaig, 1947).

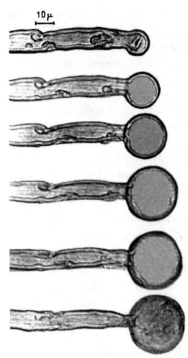

11.5. The time course of the outflow of axoplasm from the end of a cut myelinated fiber at 1, 1.5, 2.5, 4, 5, and 7 hours after sectioning. The form taken by the exudation is spherical, indicating that it is fluid and its surface is coated with myelin lipid. The outflow increases and then levels off within the observation period of thirty-six hours. Note the narrowing of the fiber near the cut end. Modified from (Lubińska, 1956), Figure 5a.

CHARACTERIZATION OF TRANSPORT USING ISOTOPE LABELING

A fundamental challenge to the theory of a simple bulk flow was the biochemical analysis of the axoplasm accumulated in the fibers just above a ligation or a cut showing that some axonal components were present in much greater concentrations than others, indicating a selective transport of substances in the fibers rather than a bulk flow.[24] The analysis of the outflow of different components was furthered by the use of new techniques that became available after World War II. One of the most important was the use of isotopes as tracers of the common elements present in biochemical components. In an early study to show transport in nerves, Samuels et al. injected orthophosphate labeled with the radioactive isotope ^{32}P systemically into guinea pigs.[25] They reasoned that the labeled precursor would be carried by the circulation to the spinal cord where, on entering, it would

[24] (Friede, 1966), pp. 390–396, and (Martinez and Friede, 1970).
[25] (Samuels et al., 1951).

be taken up by the motoneurons and become incorporated into phosphorous compounds that would then be transported down within the motor fibers of the sciatic nerve. Labeled phosphates were indeed later found in the sciatic nerve and their apparent rate of outward movement in the fibers assessed by taking the ratio of the radioactivity in a distal portion of the nerve with respect to a proximal one. The increase in the ratio with time indicated a slow outflow of isotopically labeled phosphorous compounds. However, some features of the technique made analysis difficult. The systemic injection of the labeled precursor gave complicated results because the blood-brain barrier in the spinal cord allows relatively little of the labeled phosphate precursor to enter the cord. This, coupled with the uptake of the tracer from the circulation locally into nerves and its incorporation there made it difficult to differentiate between those labeled compounds carried down in the nerve by transport from those locally synthesized.

To overcome that problem, phosphate labeled with ^{32}P was injected directly into the spinal cord of cats in the vicinity of the pool of motoneuron cell bodies (the lumbar 7th region).[26] A high level of labeled phosphates was then seen to move out in their ventral roots. Taking small sections along the length of the ventral roots, the shift with time of the outflow gave a calculated flow rate of 4.5 mm/day (Figure 11.6). To show that the outflow was not simply the result of a leakage of the labeled precursor from the cord out between the ventral root fibers, the motoneuron cell bodies of the lumbar 7th region were destroyed, either by making the region anoxic by compression or by the injection of cyanide. On injection of the isotope, the outflow of labeled activity in the roots was then almost completely eliminated. That the outflow seen was indeed within the fibers was confirmed later by autoradiography, which showed grains due to radioactivity located over the axons of the fibers.

When tritium (^{3}H) became generally available in the early 1960s, Droz and Leblond injected ^{3}H-labeled amino acid systemically into rats, envisioning, as in the experiments of Samuels et al. that the labeled precursor would be taken up from the circulation, incorporated into proteins by the motoneuron cell bodies, and transported out into the motor fibers.[27] Taking pieces of sciatic nerve at a set distance from the cord from animals at different times after injection of the precursor, thin cross-sections cut from them were coated with photographic emulsion for autoradiography. After allowing adequate time for exposure to the radioactivity emitted by the labeled proteins, the nerve sections were developed. In sections taken early after injection, silver grains were seen only over the outer part of the fibers, in the region of their myelin sheaths, indicating only a local uptake and incorporation of the labeled precursor. Then, in sections taken sixteen

---

[26] (Ochs and Burger, 1958).   [27] (Droz and Leblond, 1962).

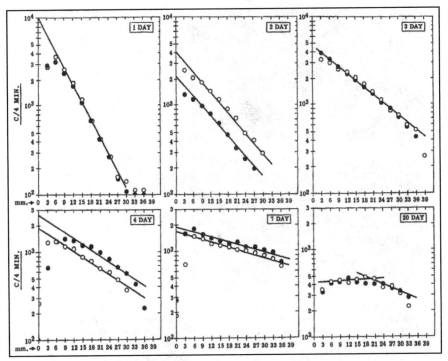

11.6. Slow transport of labeled activity in a cat ventral root after ^{32}P injection into the L7 motoneuron regions of the spinal cord. Open and closed circles represent consecutive small 3-mm segments taken from the ventral roots of the two sides. As the time between the injection and removal of the roots for assaying for radioactivity increased, more labeled activity appeared in the distal, compared with the proximal segments as would be expected of a slow outflow of ^{32}P-labeled compounds. The radioactivity in the segments in counts/minute is given on the ordinate with a logarithmic scale. The abscissa is the distance of the nerve segment in millimeters from the most proximal segment near the cord. From (Ochs, Dalrymple, and Richards, 1962), Figure 2.

days after injection, silver grains appeared over the axons in the center of the fibers showing the transport of labeled proteins within them (Figure 11.7). By finding grains over the axons at increasingly greater distances from the spinal cord, the rate of outflow of labeled proteins was estimated at 1 mm/day. This was consistent with Weiss' estimation of a slow flow of axoplasm.

For further analysis of what protein species were labeled, it was necessary to increase their amount in the fibers, to overcome the blood-brain barrier and limit the local synthesis by the Schwann cells of the nerve. This was accomplished by injecting ^{3}H-labeled amino acid directly into the spinal cord of cats, in the region of the 7th lumbar (L7) or first sacral (S1)

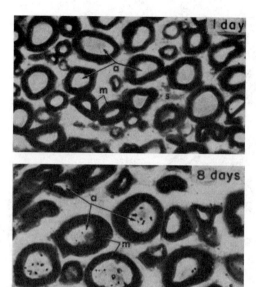

11.7. Evidence for slow transport of labeled proteins. One day after parental injection of [³H]leucine and uptake from the circulation by the motoneurons (m), autoradiography of the sciatic nerve showed no grains due to radioactivity associated with the axonal regions (a) in the upper panel. Then, at eight days, the lower panel shows grains due to radioactivity that have appeared in the axons. From (Droz and Leblond, 1963), Figures 8 and 9.

motoneurons.[28] Then, at different times after injection, small equal pieces of their corresponding ventral roots were taken, along their ventral roots with time indicated a slow rate of the outflow estimated at 4.5 mm/day, but unexpectedly, a break in the outflow curves indicated the presence of an additional faster moving phase of transport (Figure 11.8).

This was further investigated by injection of the labeled ³H-precursor into the L7 dorsal root ganglion to examine fast transport of labeled proteins in the longer length of sensory fibers of the sciatic nerve or injecting it into the motoneuron region of the L7 segment for transport into the motor fibers. The methodology is shown in Figure 11.9, where the small equal pieces (3 or 5 mm) cut from along the nerve are assessed separately for their content of radioactivity using a scintillation counter.[29] As shown in the figure, a very

[28] (Ochs and Johnson, 1969), and (Ochs, Sabri, and Johnson, 1969). [³H]leucine was the more commonly used labeled precursor, in that leucine is present to a greater degree than other amino acids in most proteins. Leucine it also not as readily metabolized as other amino acids. It is not unique. Other labeled amino acids serve as precursors in transport studies and give similar results.

[29] (Ochs, 1977). Reprinted (Ochs, 1982), Figure 2.10, p. 21.

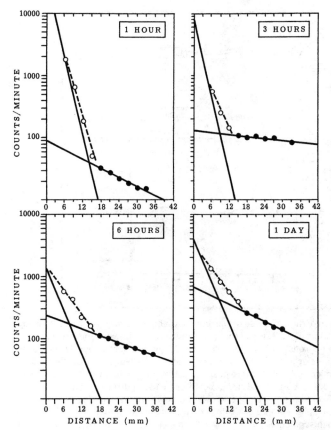

11.8. The precursor [³H]leucine injected directly into the cat spinal cord near their L7 motoneuron cell bodies was followed within hours by a rapid phase of outflow (O) of labeled proteins in the ventral root in addition to the slower phase (●). Radioactivity in the small segments taken from the ventral roots is given in counts/minute on a logarithmic scale on the ordinate. Distances of the segments from the cord given in millimeters are shown on the abscissa. From (Ochs and Johnson, 1969), Figure 3.

large amount of radioactivity remains in the region of the injected dorsal root ganglion dropping down to a plateau that extends distally into the nerve before rising to a crest and falling steeply to baseline levels at the front of the outflow.[30] The front represents those labeled proteins exiting earliest from the cell bodies into the fibers, followed by the crest material and then those contributing to the plateau. In this example, given in Figure 11.9, the

---

[30] To encompass the great range of radioactivity in the outflow, the amounts in the segments were plotted logarithmically on the ordinate against a linear scale on the abscissa when plotting the position along the nerve taking the injected ganglion as zero.

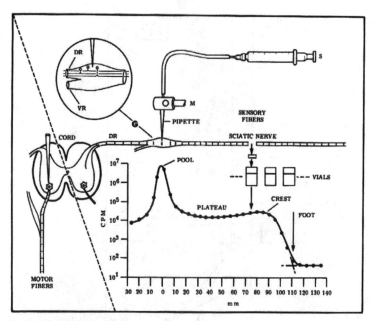

11.9. Schematization of the technique of injection and sampling showing the characteristic pattern of fast transport. Cat L7 dorsal root ganglion cells are shown in the circular insert with their T-shaped neurons, with one branch ascending in the dorsal root (**DR**), and the other descending in the sciatic nerve. A pipette containing [^{3}H]leucine is inserted into the ganglion (**G**) with a micromanipular (**M**) to position it. After injection, the precursor is taken up by the cells, rapidly incorporated into labeled proteins, and these then transported down the sensory fibers of the sciatic nerve. The outflow is determined by removing the sciatic nerve from the animal after a given time of outflow and sectioning it into equal 3- or 5-mm segments. Each segment is placed in a vial and solubilized, scintillation fluid added, and its radioactivity assessed in a scintillation counter. The counts per minute (**CPM**) are displayed on the ordinate on a logarithmic scale, and the distances of the segments in millimeters from the center of the ganglion taken as zero are shown on the abscissa. Radioactivity remains high in the ganglion, falling distally in the nerve to a plateau that rises to a crest before abruptly falling to baseline levels at the front of the outflow. The transport distance is calculated from the foot (arrow) of the front of the outflow to the height of radioactivity in the ganglion. To the left of the dashed line, the injection of the precursor into the L7 motoneuron cell body region in the spinal cord is represented where, following its injection into the cord near the motoneuron cell bodies, a similar outflow pattern in the motor fibers of the ventral root (**VR**) and sciatic nerve is seen. From (Ochs, 1975c), Figure 1. Reprinted (Ochs, 1982), Figure 2.10.

nerve was taken after injecting the L7 dorsal root ganglion injection and 6.5 hours of downflow. Taking that time and the distance of the foot of the front to the center of the injected ganglion gave a rate of fast transport of 17 mm/hr or 410 mm/day. The same outflow pattern of the plateau, crest, and front was found after injecting the precursor into the ventral horn of the L7 or S1 segments.

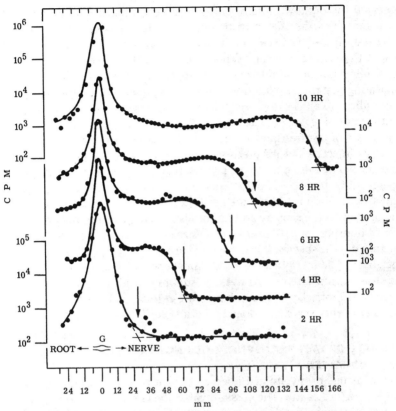

11.10. A characteristic outflow pattern of labeled proteins showing fast transport in the nerves of five cats taken 2–10 hours after injection of [³H]leucine into the L7 ganglia, with the nerves prepared as in Figure 11.9. The nerve taken two hours after injection at the bottom gives little evidence of the outflow pattern; the time allowed was too short. The other nerves show the characteristic plateau, crest, and front advancing distally in the nerve with time. Ordinate scales give counts per minute (**CPM**) at the left for nerves taken at 2 and 10 hours, and partial scales are given at the right for the nerves taken 4, 6, and 8 hours after injection. The abscissa gives distance in millimeters from the center of the ganglia. The arrows at the fronts of outflow show their progressively greater distal displacement with time. From (Ochs, 1972a), Figure 1.

CHARACTERISTICS OF FAST TRANSPORT; PATTERN AND RATE

To better determine the pattern of fast transport as a function of time, the L7 ganglia of a group of five cats were similarly injected and their nerves taken at times from 2 to 10 hours afterward (Figure 11.10).[31] They all showed the same characteristic pattern of outflow. In the nerve taken two hours after injection, the large amount of radioactivity remaining in the ganglion

---

[31] (Ochs, 1972a).

obscured the outflow pattern; but, in the nerves taken after 4, 6, 8, and 10 hours, the same characteristic outflow pattern is seen with the fronts progressively shifted distally in the nerves. The advance of the fronts found from a large number of such experiments was 410 +/− 50 (S.D.) mm/day. No diminution of the rate over the longer transport distances was seen, indicating that the mechanism of transport is present everywhere along the nerve fibers. As will be discussed further, the rate is sensitive to temperature, and the normal rate was determined at a body temperature of 38°C.[32]

The rate of fast transport in motor and sensory fibers was directly compared by injecting labeled precursors into the L7 dorsal root ganglia on the one side of animals and into the L7 motoneuron region in the ventral horn region of the spinal cord on the other side. Taking into account the different antero-posterior positions of their cell bodies on the two sides, the fronts of the outflow in the sensory and motor fibers were shown to have the same rate of transport of 410 mm/day. The size of the animal did not affect the rate nor did the species. It was the same in animals ranging from the rat to the goat.[33] The rate is also relatively independent of age. A slightly lower rate was seen in kittens two weeks postpartum, and the same fast rate of close to 410 mm/day was found in young animals and in the aged, in cats as old as twenty years, and dogs 13.5 years old.[34]

THE RATE OF FAST TRANSPORT DOES NOT VARY WITH FIBER DIAMETER

The diameters of myelinated fibers in the cat and other larger mammals range from 2 to 24 μm. The possibility that the rate of transport might vary with fiber size was examined. This was done by injecting the dorsal root ganglia with ^{3}H-labeled amino acid precursor and taking small pieces of nerve from the front of downflow, cross-sectioning them and coating them for autoradiography.[35] After exposure and development, grains of radioactivity were found located over the axons of myelinated nerve fibers of all diameters at the front (Figure 11.11). This indicated that they all have the same rate of fast transport. If the rate were slower in the small-diameter fibers, no grains would have been present in them at the front. Conversely, if the rate of transport were faster in the small-diameter fibers, only they would be

[32] A body temperature of 38°C was determined in a group of anesthetized cats. Care was taken to make sure that the legs of the animals and the nerves in them were kept at that temperature during the course of transport. More often, a temperature of 37°C is used.

[33] (Ochs, 1972b), (Table 2.1).

[34] (Ochs, 1973). These animal ages are comparable with those of humans aged 75 years or older.

[35] (Ochs, 1972b), Figure 8. The outflow pattern determined from the remaining part of the nerve showed that the segments were in fact taken from the fronts of downflow.

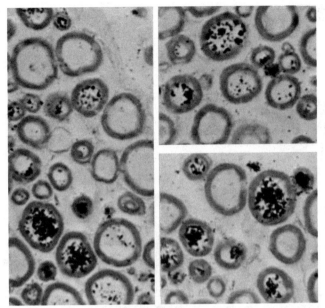

11.11. Labeled proteins are transported in fibers at the same rate regardless of their diameter. Sections were taken from the front of fast transport outflow after injection of the L7 ganglia with [^{3}H]leucine-labeled precursor. The sections were prepared for autoradiography, and after exposure and development, grains from radioactivity were seen present in the fibers of all diameters, from the smallest to the largest – indicating that the rate of fast transport is the same in all of them (see text). From (Ochs, 1972b), Figure 8.

seen to contain grains and not the larger ones. Using the same reasoning, electron micrography of sections taken from the front of downflow showed grains in autoradiographs that were located over axons of the unmyelinated fibers, at the same position as that found in the myelinated fibers – evidence that they have the same fast rate of transport.[36] Byers et al. reported a similar fast rate in unmyelinated fibers using a different method.[37] The same fast rate was also found present in the small neurites of regenerating nerve (Chapter 10).[38]

THE TRANSPORT MECHANISM IS PRESENT ALL ALONG THE LENGTH OF THE NERVE FIBER

A force exerted by the cell body is not responsible for the fast transport; its mechanism is present all along the fiber acting in an "all-or-none" fashion.[39]

[36] (Ochs and Jersild, 1974).   [37] (Byers et al., 1973).
[38] (Griffin, Drachman, and Price, 1976).
[39] (Ochs and Ranish, 1969). The term "all-or-none" is borrowed from its use in describing the nondecrementing behavior of the propagated nerve action potential (Chapter 8).

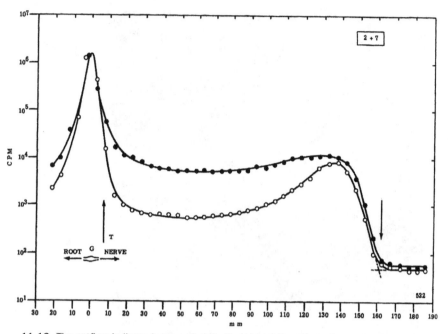

11.12. The outflow in ligated nerve after ligation shows that the transport mechanism is present within the fibers. After injection of cat L7 ganglia (**G**) on each side with a labeled precursor, two hours lapsed for labeled proteins to gain entry into the fibers before the nerve on one side (O) was tied (**T**) just below the ganglion to prevent further outflow into it from the cell bodies. After an additional period of seven hours downflow, the nerve was removed and processed as shown in Figure 11.9. The ligated nerve has a lower plateau height because of the prevention of a later outflow of labeled proteins, compared with the unligated nerve (●), but it has a similar crest height and position of its front to the unligated nerve (down arrow), indicating their similar transport rates. **CPM**, counts per minute. From (Ochs, 1975c), Figure 1.

This was shown by injecting the ganglion, and after allowing several hours or so for labeled proteins produced by the cells to enter the fibers, the nerves were then ligated just below the ganglion to prevent a further outflow from the cells into their fibers. The labeled proteins that had gained entry into the fibers were transported at the same fast rate in them as in the unligated nerve, this shown by the similar positions of their fronts (Figure 11.12). The major difference between them was the lower level of the plateau in the ligated nerve due to the lack of later additions of labeled proteins from the cell bodies: When more time was allowed before making the ligation after injection, the plateaus were higher. The level of the plateau also represents in part some of the labeled components of the crest dropping off as it advances down the nerve, the crest as a consequence becoming smaller with distance. The drop-off from the crest in cat nerve was calculated to amount to

1.5 percent of the transported protein carried in each 5 mm portion of nerve.[40] A similar drop-off with distance was shown in the long olfactory neurons of the garfish (Figure 11.16).[41]

On injecting the dorsal root ganglia, transport into the dorsal root branch of their T-shaped neurons is continued on into the spinal cord where a long branch ascends in the dorsal columns to the brain stem. By taking segments of the dorsal columns along their length after injecting labeled precursor into the L7 dorsal root ganglia of the cat, the outflow pattern in the dorsal root column was seen to be similar to that in the sensory fiber branches of the T-shaped neurons passing down in the sciatic nerve; the same positions of their fronts indicating that they have the same fast transport rates.[42]

TRANSPORT IN DENDRITES

The limited extent of dendrites arising from cells in the central nervous system (CNS) prevents a definite time of transport. Nevertheless, motoneuron cells of the cat spinal cord injected iontophoretically with [³H]glycine showed a rapid outflow of labeled proteins into their dendrites.[43] Portions of the spinal cord were taken at timed intervals after injection of their cell bodies and sections prepared for autoradiography. Grains were found to have spread into the dendrites within a matter of minutes after injection. Injection of certain dyes, such as cobalt blue or Procion Yellow, into cell bodies also showed a fast spread into their dendrites.[44] Most likely, the dyes bind to fast-transported proteins as do other substances such as horseradish peroxidase and other substances carried by the fast transport mechanism and serving as fiber markers.[45]

ROUTING: DIRECTED TRANSPORT OF COMPONENTS INTO FIBER BRANCHES

Some components transported into the two branches of the T-shaped neurons of the dorsal root ganglia must be different. The branch descending in the peripheral nerve carries receptors to its sensory fiber terminals, while the branch entering the dorsal root fibers carries neurotransmitter and transmitter-related substances to their terminals synapsing on neurons in the cord. However, the type and amounts of these components are too small to be detected by gel electrophoresis in the face of the wide range of similar labeled proteins carried into the two branches. The species of labeled proteins transported will be discussed in a later section. What was seen was a large difference in the amounts of labeled protein transported into the two branches. In the mature monkey, the long length of their L7

---

[40] (Muñoz-Martínez, Núñez, and Sanderson, 1981). [41] (Gross and Beidler, 1975).
[42] (Ochs, 1972b), Figure 6. [43] (Schubert, Kreutzberg, and Lux, 1972).
[44] (Kater and Nicholson, 1973) [45] (Mesulam, 1982).

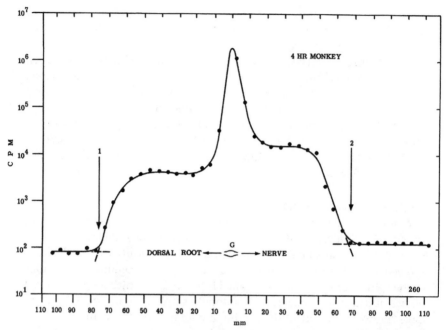

11.13. Asymmetry in the amount of labeled components transported from the dorsal root ganglion (**G**) neurons into their dorsal root and sensory nerve fiber branches shows *routing*. The L7 dorsal root ganglia of large rhesus monkeys have long lengths of 10 cm or more, allowing the full pattern of fast transport to be seen in them. Arrows at the foot of the fronts show the rates to be identical in the root and nerve branches, but with a much larger crest height and plateau in the sensory nerve branch, as much as three to five times that in the dorsal root branch. Distances along the abscissa are given from the peak of activity in the center of the ganglion taken as zero. From (Ochs, 1972b), Figure 4.

dorsal roots allowed the full pattern of the outflow in them to be compared with that in the peripheral nerve branch (Figure 11.13).[46] They had the same characteristic pattern and the same rate as shown by the positions of their fronts. However, the amplitude of the crests in the dorsal root was one-third to one-fifth that in the peripheral nerve. This asymmetry is not caused by differences in the number of fibers or fiber sizes in their populations.[47] The difference lies in the mechanism responsible for transport in the fibers. The labeled proteins exiting from the cell body moves into the initial segment before diverging out into the branches of the T-shaped neurons. Some isolating mechanism in the initial segment common to the branches.

---

[46] (Ochs, 1972b).

[47] (Ochs et al., 1978). The small diameter myelin fibers do show a difference in the diameters of their branches, but not the larger fibers in which the overwhelming bulk of labeled protein is carried.

Otherwise the various components there would mix and become equalized before passing out in the branches. The basis for this isolation is found in the early development of the ganglion neurons. At birth, they are originally bipolar, and early in development the branches come together to form the initial segment of what becomes a T-shaped neuron (see Figure 8.8). As discussed in Chapter 12, the microtubules within the fibers are the "rails" for fast transport, and the inference is that the microtubules originally present in the fibers at each pole of the embryonic neuron preserve their individuality without intermingling.[48] It is this that allows the "routing" of materials over specific subsets of microtubules moving specific components into the axon branches.

TURNAROUND AND RETROGRADE TRANSPORT

The uptake of endogenous nerve growth factor by sensory and sympathetic nerve fiber terminals in the course of their early development[49] is followed by retrograde transport to the cell bodies where it serves to change the program of synthesis in the cells. This uptake mechanism in the terminals opens up a back door to the entry of exogenous substances with pathological consequences. These may be taken up by endocytosis, through special channels, or by carriers and then conveyed to the cell bodies by retrograde transport.[50] Viruses such as herpes simplex, rabies, diphtheria, and tetanus toxins were listed by Gerard in his review of 1932[51] on evidence that was generally circumstantial. An example is that of the polio virus. It was shown to traverse the spinal cord of the monkey in two days, a rate much too fast to be accounted for by diffusion.[52] Rather than a passage intraaxonally, some studies indicated it could occur by an extraaxonal path. Tetanus toxin, when injected into muscles and taken up by the motor nerve terminals, causes motoneuron cell changes and local tetanus to appear within a day after injection.[53] This appeared to occur too soon on the basis of the slow rate of several millimeters/day – then currently taken to account for axonal transport – and an extraneuronal path was proposed. This seemed to be supported when nerve trunks were injected with ethanolamine oleate to cause a sclerosis that supposedly blocked the endoneural space between the fibers with fibrous tissue, while leaving the nerve fibers functional. This blocked passage of tetanus toxin injected below the sclerotic region.[54] Other evidence,

---

[48] (Ha, 1970).
[49] (Hendry et al., 1974), (Levi-Montalcini, 1976), (Mobley et al., 1977), (Schwab, Heumann, and Thoenen, 1982), and (Thoenen, 1991). See Chapter 10.
[50] (Erdmann, Wiegand, and Wellhoner, 1975), (Price et al., 1975), (Schwab et al., 1977), and (Schwab, and Thoenen, 1978).
[51] (Gerard, 1932).   [52] (Fairbrother and Hurst, 1930).
[53] (Brooks, Curtis, and Eccles, 1957).   [54] (Wright, 1955).

however, for an intraaxonal path was given. Freezing nerve briefly with dry ice causing nerve fibers to rapidly degenerate without apparent effect on the endoneural space prevented the passage of tetanus toxin through that region.[55] This evidence was, however, little attended to at the time. Its intraaxonal retrograde transport was later definitively shown by the use of tritium-labeled tetanus toxin. Injected into muscle and taken up by the motor nerve terminals, autoradiography showed the labeled toxin to have ascended within the axons into the ventral roots and then into the motoneuron cell bodies.[56] The rate of its retrograde transport after injection of ^{125}I-labeled tetanus toxin into muscle was given by Sabri et al. as 6.4 mm/hr,[57] a rate fairly close to that generally found for retrograde transport of labeled proteins.

The intraaxonal transport of other toxins, viruses, enzymes, and other neuroactive species has been verified.[58] Cholera toxin, ricin II, wheat germ agglutinin, phytohemagglutinin, and concanavalin A are some of the substances taken up by the nerve and retrogradely transported intraaxonally. The enzyme horseradish peroxidase is of major importance for its use in anatomical studies. Taken up by the terminals and retrogradely transported, it serves as a marker to trace the path of fibers in the CNS.[59] An important technical advance is the use of fragments of tetanus toxin specific for retrograde transport to which various substances bound to it are carried back to the cell body to alter its DNA processing.[60]

A number of components transported down into the fiber terminals undergo a *turnaround* there to be then carried back to the cell bodies by retrograde transport. Such turnarounds also occur at a crush or ligation of a nerve. This was shown by allowing anterograde transport to carry labeled proteins down to a distal ligation, to where they accumulate, and then subsequently making a ligation higher up on the nerve to trap the labeled materials that had turned around and ascended by retrograde transport.[61] Retrograde transport is an important process, whereby endogenous components participating in synaptic transport, such as the membranes of spent neurotransmitter vesicles, are carried back from the nerve terminals to be reutilized by their cell bodies.[62] The transport of neurotransmitters and their associated components will be discussed in a later section.

---

[55] (Roofe, 1947).   [56] (Erdmann et al., 1975), and (Price et al., 1975).
[57] (Sabri et al., 1987), and see (Forman et al., 1977).
[58] (Kristensson, Lycke, and Sjöstrand, 1971), (Kristensson, 1978), and (Brooks 1991).
[59] (Price et al., 1975), (Kristensson, 1978), and (Mesulam, 1982).
[60] (Alkon, Vogl, and Tam, 1991), and see (Forman et al., 1977).
[61] (Lasek, 1967), (Ochs, 1975b), (Bisby and Bulger, 1977), and (Bisby, 1980).
[62] (Brimijoin and Helland, 1976), and (Brimijoin, 1982).

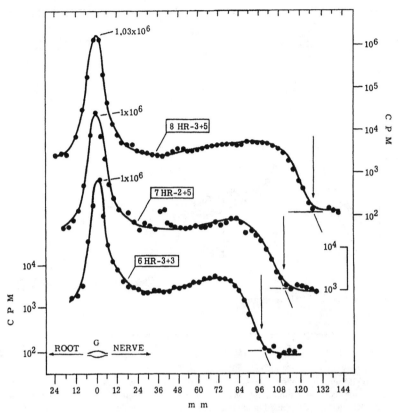

11.14. Characteristic outflow pattern of fast transport in nerves in vitro matches that in vivo. Dorsal root ganglia (**G**) were injected, and a period of time of 2 to 3 hours allowed for labeled proteins to enter the fibers in vivo before the nerves were removed and placed in a chamber for a further three or five hours of transport in vitro. The bottom curve consisted of three hours of in vivo outflow plus three hours in vitro for a total of six hours, the middle curve two hours of in vivo plus five hours of in vitro outflow for a total of seven hours, and the upper curve three hours of in vivo plus five hours of in vitro transport for a total of eight hours. The fronts were all seen to have advanced at the same fast rate of transport and with the same characteristic form as that calculated for a comparable total time of transport in vivo (cf. Figure 11.10). From (Ochs and Smith, 1975), Figure 1.

TRANSPORT IN VITRO

An in vitro preparation was developed to further assess the relationship of oxidative metabolism to transport and of neurotoxins and other agents acting on axoplasmic transport.[63] Dorsal root ganglia were injected with a labeled precursor as usual and then several hours were allowed for labeled

[63] (Ochs and Ranish, 1970).

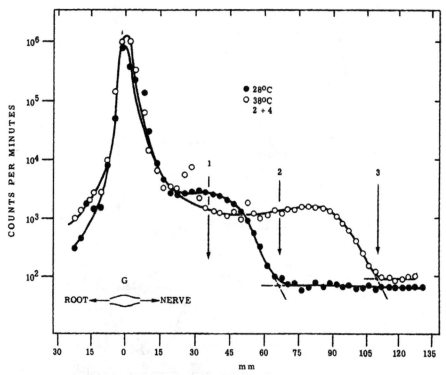

11.15. The dependence of fast transport on temperature shown in an in vitro preparation. Peripheral nerves taken from an animal two hours after injection of the dorsal root ganglia (G) with [³H]leucine were placed in chambers for a further four hours of in vitro downflow. One nerve was kept at 38°C (O), the other at 28°C (●). Arrow **1** indicates the time of transfer to the chambers, arrow **2** at the foot of the downflow in the chamber at 28°C, and arrow **3** at the foot after downflow at 38°C. The rate of transport was approximately halved at the lower temperature. From (Ochs and Smith, 1975), Figure 2.

proteins to pass down partway into the fibers. The nerves were then removed from the animals and placed in a chamber supplied with oxygen at a temperature of 38°C for an additional period of in vitro downflow. The nerves in which both in vivo and in vitro transport had taken place showed the same pattern and fast rate of transport as that usually seen for nerves in vivo (Figure 11.14), indicating the similarity of in vitro to in vivo transport. Using the in vitro preparation, a close dependence of transport on oxidative metabolism was shown. Transport failed within 20 to 30 minutes when the nerve was made anoxic by replacing oxygen with nitrogen, indicating that the transport mechanism is an energy-driven system.[64]

[64] (Ochs et al., 1970), and (Ochs and Hollingsworth, 1971).

The effects of temperature on transport made using the in vitro preparation further showed the dependence of transport on metabolism. When the temperature was reduced to 28°C, a marked decline in the transport rate was seen, to 212 mm/day, a rate about half that seen at the normal body temperature of 38°C (Figure 11.15). At successively lower temperatures, the rate was decreased with this same ratio, the behavior expressed by the Q_{10}, the ratio of the rate change seen with a 10°C difference in temperature. In the cat sciatic nerve the Q_{10} found was 2.0 to 2.5[65] – a value that indicates the action of a chemical (metabolic) process rather than a physical one, where a Q_{10} in the range of 1.1 to 1.3 would be expected. At successively lower temperatures, the rate of transport decreased until, at a temperature of 11°C, it was completely arrested – the phenomenon termed *cold block*. This was reversible with transport resuming on rewarming the nerve. Brimijoin and colleagues found a similar cold block close to this temperature with rabbit and bullfrog nerves, the bullfrog nerves blocking at a somewhat lower temperature.[66]

FAST TRANSPORT IN THE NERVES OF NONMAMMALIAN SPECIES

Outflow characteristics similar to those seen in mammal nerves were found in a variety of nonmammalian nerves, many of them poikilothermic. Abe et al. obtained an in vitro outflow pattern similar to that seen in mammalian nerves.[67] The experiments were carried out at a lower temperature. When scaled to the body temperature of mammals of 38°C using a Q_{10} determined to be 2.6, the rate of transport found was 405 mm/day, one similar to that of the mammal. Edström and Hanson applied labeled precursors directly to the ganglia of frog nerves placed in a chamber for in vitro transport.[68] Peaks of radioactivity were found to move distally in the nerve at a fast rate, which on taking the Q_{10} into account and scaled to a temperature of 37°C, was calculated to be 400 mm/day. Transport in the long lengths of olfactory nerves in the garfish containing small unmyelinated fibers was studied by Gross and Beidler.[69] ^{3}[H]Leucine was applied to the olfactory mucosa from which it was taken up by the cell bodies and after its incorporation labeled proteins then transported in the long fibers with a characteristic pattern similar to that of the mammal (Figure 11.16). The rate determined by taking the Q_{10} into account and scaling it to a temperature of 37°C was 403 mm/day, again a rate similar to that of the mammal.

Some species showed different rates of transport and temperature characteristics. The CNS of the swan mussel, *Anodonta cygnea*, consists of three pairs of ganglia joined by long connectives. On injecting ^{3}H-labeled leucine,

[65] (Ochs and Smith, 1975).   [66] (Brimijoin, Olsen, and Rosenson, 1979).
[67] (Abe, Haga, and Kurokawa, 1973).   [68] (Edström and Hanson, 1973).
[69] (Gross and Beidler, 1973).

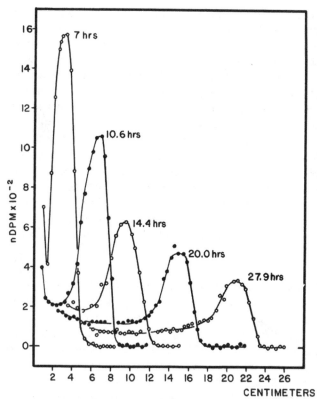

11.16. Transport of labeled proteins in the long lengths of five garfish olfactory nerves after exposure of the olfactory fiber cells in the nose to the labeled precursor. The labeled proteins that have gained entry to the fibers show the characteristic pattern of fast transport, a plateau followed by a rise to a pronounced crest with a sharp drop to background levels at the foot of their fronts of outflow. The crest heights are seen to fall off with distance. Note the linear scale on the ordinate exaggerates the height of the crests in comparison to the use of a logarithmic scale (cf. to Figures 11.10, 11.12). From (Gross and Beidler, 1975), Figure 3.

valine, or tryptophane into the cerebral ganglia for uptake by its cells, Heslop and Howes found the rate of transport of labeled proteins to decrease down to a temperature of 4°C, and then to level off without showing a definite cold block.[70] Apparently, a special adaptation in the nerves of this poikilotherm enables it to function at the lower environmental temperature. Using the Q_{10} of 2 found for this animal, a rate of 240 mm/day at 37°C was ascertained, one substantially lower than that found for garfish olfactory

[70] (Heslop and Howes, 1972).

and frog nerves. Crayfish nerves were also reported to have a slower fast rate of about 270 mm/day.[71]

MEASUREMENTS OF TRANSPORT IN INTACT NERVE

Rather than using successive pieces of nerve to assess transport, Takenaka and his colleagues[72] devised a miniature detector enabling the measurement of radioactivity from intact nerves in vivo using isotopes with higher particle energy, such as ^{32}P and ^{35}S. This technique was used to take the outflow patterns of ^{32}P-labeled ATP repeatedly over a period of 10 days in the same nerve.[73] Smith developed a chamber consisting of a series of sensitive counters for radioactivity by which transport was continuously measured in vitro in the intact nerve.[74] The ganglia of frogs were injected with amino acids containing the isotope ^{35}S, which has a higher particle energy than tritium. The movement of labeled proteins could be followed down to its distal cut end and then, after a turnaround, its retrograde transport. The transport rates at the lower temperature, when scaled to mammalian body temperature with the Q_{10} taken into account, were also similar to those of mammalian nerves.

TRANSPORT IN THE VISUAL SYSTEM

Transport in the visual system was examined by injecting labeled amino acid precursor into the posterior chamber of the eye. Taken up by the ganglion cells of the retina, it is incorporated into labeled proteins and polypeptides and these transported in the optic nerve fibers to their termination in the optic tectum in fish, or in mammals, the lateral ganglion. This approach was first used by Weiss and Holland, who injected [^3H]leucine into goldfish eyes and then used autoradiographs to assess transport into the tectum.[75] The results were inconclusive until Grafstein and colleagues used scintillation counting to assess the radiolabeled proteins accumulated in the tectum of fish or the lateral geniculate in the mammal.[76] In the fish, the optic nerves are crossed (Figure 11.17), and as a result the labeled proteins are transported into the tectum of the opposite side. There, the bulk of the labeled components accumulates but for a small amount lost from the terminals.[77] This was shown to be of importance in that the leaked labeled proteins taken up by the terminals of second-order neurons in the lateral geniculate of cats and macaques is carried to the primary visual area of the cortex to terminate there in alternate layers interdigitating with the fibers entering

[71] (Fernandez, Huneeus, and Davison, 1970).
[72] (Takenaka, Horie, and Sugita, 1978).   [73] (Takenaka and Ochs, 1980).
[74] (Snyder, 1986)   [75] (Weiss and Holland, 1967).
[76] (Grafstein and Forman, 1980).   [77] (Grafstein, 1971).

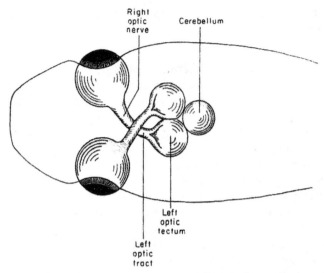

11.17. The visual system of the goldfish. The retinal fibers of each eye crosses in the optic tracts to terminate in the contralateral optic tectum. Injection of an eye with a labeled precursor is followed by transport of labeled proteins to the tectum on the opposite side. From (McEwen, Forman, and Grafstein, 1971), Figure 1.

from the opposite eye. This laminated pattern is referred to as the *ocular dominance bands*, which is held to be responsible for binocular vision.[78]

Although the technique of injecting the eye to study transport is relatively simple, an important consideration had to be taken into account. Circulation within the retina is extensive, and a considerable amount of the labeled precursor [^{3}H]leucine injected into the eye enters the circulation and from it taken up by cells in the tecta of fish or the lateral geniculate ganglia in the mammal. To determine the portion of the radioactivity axonally transported from that locally incorporated, the radioactivity on the side injected, which represents the locally incorporated precursor, is subtracted from the radioactivity on the crossed side that includes both transported and locally incorporated labeled proteins. As much as half the radioactivity seen on the crossed side represents locally incorporated proteins.[79] Much less local synthesis occurs when using [^{3}H]proline or [^{3}H]asparagine as precursors, as these amino acids do not readily pass the blood-brain barrier of the tectum or lateral geniculate.[80]

The rate of transport in the retinal ganglion fibers was determined by the lapse of time after injection of the precursor into the eye to its earliest accumulation in the tectum or lateral geniculate, and the distance of the tectum

---

[78] (Wiesel, Hubel, and Lam, 1974), and (LeVay, Stryker, and Shatz, 1978).
[79] (McEwen and Grafstein, 1968).   [80] (Grafstein, 1977).

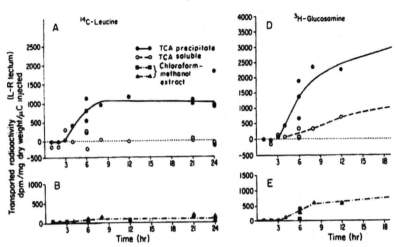

Transported Radioactivity

11.18. Labeled protein [as tricarboxylic acid **(TCA)** precipitate] accumulated in the optic tectum. After injection of labeled precursor [^{3}H]leucine in one eye, labeled proteins accumulate in the tectum on the opposite side as shown at left in **(A)**. The use of [^{3}H]glucosamine gives a similar pattern of accumulation as shown at right in **(D)**. The time between the injection and the sharp rise of radioactivity accumulated in the tecta is a measure of the rate of fast transport in the optic tract fibers. The curves of transport were derived after subtracting the radioactivity in the tecta on the same side, which represents leakage of the labeled precursor into the circulation with direct uptake by the tecta: left minus right (L-R). **(B, E)** represent radioactivity of non-TCA precipitates remaining in the supernatant. From (McEwen et al., 1971), Figure 2.

or lateral geniculate from the retina (Figure 11.18). The rate of fast transport when scaled to a temperature of 37°C using a Q_{10} of 3, was determined to be 190 mm/day.[81] A later wave appears that represents a much slower transport (Figure 11.19).[82] These observations were taken by many to show that only two separate transport mechanisms are operating: one fast, the other slow. However, similar studies of accumulation in other animal species indicated the presence of a range of different rates. Karlsson and Sjostrand described two fast waves of accumulation and two slow waves.[83] Others have reported intermediate rates. Schonbach and Cuenod found a fast rate in the avian retino-tectal path of 100 to 500 mm/day, and a later appearing slower rate of 20 to 60 mm/day.[84] They also found two other waves with a rate of 40 mm/day and a slow rate of 1 mm/day.[85] Other evidence for multiple rates was found when using separation methods to analyze the outflow and rate of specific proteins, as discussed in the following section. These observations

[81] (Grafstein, 1967), and (McEwen, Forman, and Grafstein, 1971).
[82] (Grafstein, 1977).   [83] (Karlsson and Sjöstrand, 1971b).
[84] (Schonbach and Cuenod, 1971).   [85] (Cuenod and Schonbach, 1971).

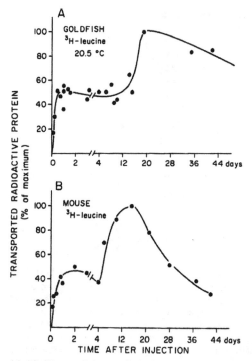

11.19. Waves of slow outflows found when tecta or geniculate were taken at longer times than those of Figure 11.18. The curves of accumulation show a first fast wave and then a later slow wave similarly seen in the goldfish tectum (**A**) and mouse geniculate (**B**). Note the time scale on the abscissa in days versus hours in Figure 11.18. From (Grafstein, 1977), Figure 2.

of multiple rates of transport are important when assessing the nature of the transport mechanism. The problem to be faced is whether different transport mechanisms are needed to account for the different rates of transport or whether they can be accounted for by a single mechanism as will be addressed in the next chapter.

CLASSIFICATION OF TRANSPORTED PROTEINS

A wide range of labeled proteins transported in nerves was shown by homogenizing, centrifuging them, and then isolating proteins from the soluble fraction.[86] Upon passing the proteins through Sephadex columns, proteins with molecular weights ranging from above 450,000 to less than 50,000 were identified. Isoelectric focusing was also used to separate a wide range of alkaline and acidic proteins.[87] Differences in the amounts transported at

[86] The method used is the precipitation of the soluble proteins in the supernatant fraction with trichloracetic acid.
[87] (James and Austin, 1969), and (Kidwai and Ochs, 1969).

different rates were shown; the bulk of the labeled proteins transported was found to be carried at the slow rate.[88]

A powerful technique used to determine the rate at which various labeled protein species are transported is that of sodium dodecyl sulfate-polyacrilamide gel electrophoresis (SDS-PAGE): The proteins in each segment are solubilized with the detergent SDS, and the monomer subunits separated on the basis of their molecular weights by use of PAGE. After injection of the labeled amino acid precursor [^{35}S]methionine into the spinal cord of rats, sciatic nerves were taken at different times afterward and cut into segments, the proteins in each determined using SDS-PAGE. Hoffman and Lasek found a wave of labeled proteins in the sciatic nerve they estimated to move at a rate of 1 to 2 mm/day, labeling it *slow component a* (SCa) (Figure 11.20).[89] The major portion of the wave, 70 to 85 percent, consists of cytoskeleton proteins, of the microtubules and neurofilaments (Chapter 8). A smaller and somewhat faster wave estimated to move at the rate of 3–5 mm/day was termed *slow component b* (SCb). It contained smaller amounts of the cytoskeleton proteins, including actin, the protein constituting the microfilaments, and a variety of other proteins.

They theorized that the SCa wave consists of microtubule and neurofilament polymers that have been assembled in the cell bodies and the organelles, then moved down within the axons as part of an interconnected matrix.[90] However, in using SDS-PAGE, the polymers are dissagregated into their individual monomers. The form in which these cytoskeletal proteins are transported, whether as organelles, oligomers, or their monomeric protein subunits, cannot be determined by this technique alone. Other evidence is required to judge the validity of their hypothesis, as will be discussed in Chapter 12.

The SDS-PAGE procedure was used by Willard et al. in a study of the rate of transport of labeled proteins in the visual system.[91] Taking successive small pieces of the visual system – the optic nerve, optic tract, lateral geniculate, and superior colliculus of rabbits at different times after injecting [^{32}S]methionine into their eyes – they found the rates of outflow of labeled proteins to fall into four groups: group I, the fastest with a rate greater than 200 mm/day; group II, 34–68 mm/day; group III, 4 to 8 mm/day; and group IV, 2 to 4 mm/day. In addition, a very slow, group V, later found in the visual system of the guinea pig, had a rate of 0.7 to 1.1 mm/day.[92] These multiple rates will be discussed in Chapter 12 in relation to the mechanism advanced to account for their transport.

---

[88] (Ochs, Johnson, and Ng, 1967), (McEwen et al., 1968), and (Kidwai and Ochs, 1969).
[89] (Hoffman and Lasek, 1975). [90] Ibid., and (Black and Lasek, 1980).
[91] (Willard, Cowan, and Vagelos, 1974).
[92] (Willard and Hulebak, 1977), and (Levine and Willard, 1980).

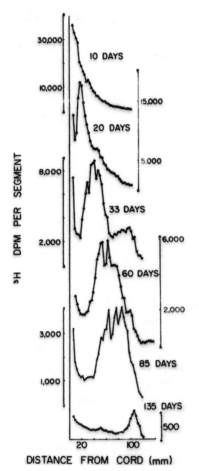

11.20. After injection of ^{3}H-labeled amino acid into the rat motoneuron region of the spinal cord, waves of radioactivity are found in the sciatic nerve. The displacements of the peaks of the waves indicated a slow outward flow at the rate of 1.0 to 1.2 mm/day (labeled slow component a). Radioactivity is shown on the ordinate with a linear scale. Their distance from the cord is given on the abscissa in millimeters. From (Hoffman and Lasek, 1975), Figure 1.

A further advance in the analysis of the various labeled proteins transported was the two-dimensional technique in which both SDS-PAGE and isoelectric focusing are used to resolve as many as several thousand proteins in small pieces of nerve.[93] These proteins arise from many sources: membranes, endoplasmic reticulum, mitochondria, and structural components present in the nerves, the cytoskeletal proteins in the axoplasm, the membrane skeleton and its constituents, ion channels and ion pumps, specific enzymes, neurotransmitters, and transmitter-related components. The transport of some of these specific components have been assessed by their special functional properties as described in the following sections.

---

[93] (O'Farrell, 1975), and (Wilson and Stone, 1979).

TRANSPORT OF NEUROTRANSMITTERS AND TRANSMITTER-RELATED COMPONENTS

a. The cholinergic system

In Chapter 7, the transmitter acetylcholine (ACh) was discussed with respect to its release from the motor nerve terminals to activate skeletal muscle. Using bioassay methods, ACh was found not only at the neuromuscular junction, autonomic ganglia, and in certain CNS neurons,[94] but also
all along their fibers, indicating a transport within them to the terminals.
Along with the transmitter, its hydrolyzing enzyme acetylcholinesterase
(AChE) and its synthesizing enzyme choline acetyltransferase (ChAc) are
present in the synaptic vesicles. An early study that showed a declining
proximo-distal gradient of AChE in the sciatic nerve suggested to Lubińska
and colleagues that it is being transported.[95] A similar decreasing gradient
found for ChAc by Hebb and Silver was also considered evidence for its
transport.[96]

To better establish the transport of AChE, Koenig and Koelle reasoned –
on the basis of the assumption of its slow rate of transport of several mil-
limeters/day, the rate given by Weiss at the time – that by treating animals
with difluorophosphate, an agent that irreversibly blocks AChE, the supply of newly synthesized AChE should appear in the nerves over a period
of weeks with the proximo-distal gradient of a slow transport.[97] Instead,
the enzyme appeared uniformly all along the whole length of the nerve
within a few days. They interpreted this result to indicate that AChE is locally synthesized in the nerve fibers rather than in the cell bodies. However,
AChE was subsequently found by Lubińska and colleagues to collect above
an interruption at a fast rate (Figure 11.21A).[98] A fast retrograde transport
was also indicated by its rapid accumulation below the crush. On making
two crushes to isolate a segment of nerve, they found that with time the
accumulation of the enzyme above the distal crush within the segment and
just below the proximal crush (lower graph, Figure 11.21A). The accummulation leveled off at a time that depended on the length of the nerve
segment isolated by the two crushes. From a series of experiments made by
varying the length of segments, Lubińska and Nemierko calculated a fast
rate for the anterograde transport of AChE of 220 mm/day, the retrograde
rate of 100 mm/day.[99] The total amount of AChE within the nerve segment
remained unchanged, it just becomes redistributed within the segment.

---

[94] (Eccles, 1964), and (Phillis, 1970).
[95] (Lubińska, Niemierko, and Oberfield, 1961), and (Lubińska et al., 1962).
[96] (Hebb and Silver, 1961).   [97] (Koenig and Koelle, 1961).
[98] (Lubińska, 1964), and (Lubińska and Niemierko, 1971).
[99] (Lubińska and Niemierko, 1971).

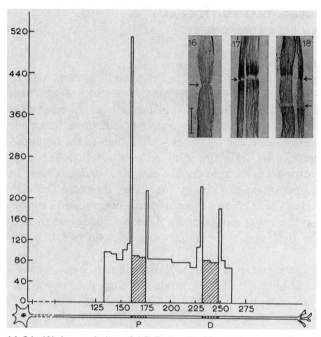

11.21. (A) Accumulation of AChE at a double crush preparation. (On the nerve shown below, proximal at (P) distally at (D).) An increase of the enzyme occurs just above the crush because of the anterograde transport, and an increase below the crush is indicative of a retrograde transport. The segment isolated by the crushes is indicated by hatching. Both anterograde and retrograde transports within the isolated segment are shown in the graphical presentation by the accumulation just above the lower crush and just below the upper crush, respectively. The figure on the right shows the darkened staining of the nerves for AChE at the crushes where the enzyme has accumulated (cf. Figure 11.23). From (Lubinska, 1964), Figures 17 and 23.

The amount of the enzyme in the middle of the segment did not fall to zero. This is because most of the enzyme is bound, with only a relatively small portion, 10 to 15 percent, that of the isoform characterized as 16 S, free to move.[100] Although the results obtained by Lubińska and Nemierko placed the enzyme in the class of rapidly transported proteins, the rate fell significantly short of the 410 mm/day rate found for labeled proteins. Did this mean that a different mechanism exists for the transport of AChE, compared with that found for labeled proteins?

---

[100] (Brimijoin, 1979), (Fernandez, Duell, and Festoff, 1979), and (Younkin and Younkin, 1988). S refers to Svedberg units of molecular weight determined by ultracentrifugation.

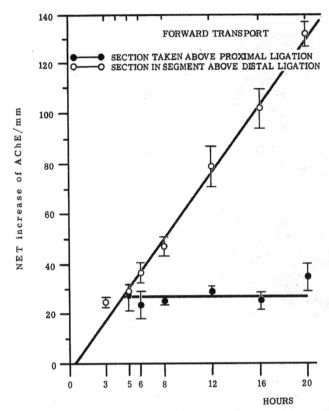

11.21. (*cont.*) (**B**) The "departure" method to determine the rate of AChE in segments of nerves isolated by a double ligation. Accumulation above the upper ligation continues uniformly for twenty hours or more (O). That above the lower ligation within the segment (●) shows a similar increase at first, but then an abrupt departure from that curve occurs, at a time that depends on the distance between the ligations. The departure represents the clearance of the AChE free to move within the ligated segment of nerve. From the time of departure and the length of the isolated segment, the rate of its fast transport was determined to be 431 mm/day, a rate statistically similar to the rate of 410 mm/day found for labeled proteins measured by the fronts of labeled protein outflow (cf. Figure 10). From (Ranish and Ochs, 1972), Figure 2.

Rather than varying the distances between crushes to isolate a segment of nerve, a "departure" technique was used to determine the rate of AChE.[101] Two ligations were made far apart at the same distance in a group of cat sciatic nerves. The enzyme accumulated above the distant ligation within the isolated segment at a linear rate, as it did above the upper ligation, until the accumulation within the segment abruptly leveled off (Figure

[101] (Ranish and Ochs, 1972).

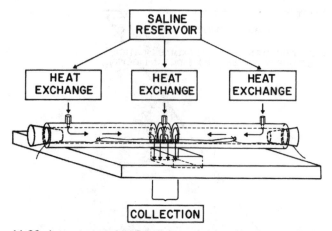

11.22. Apparatus used by Brimijoin to characterize transport by his "stop-flow" procedure. A nerve is placed within a tube-shaped chamber in which temperature is kept at body temperature by flow of saline in heat exchangers with a narrow chamber in the middle controlled by a separate heat exchanger set to a low temperature of 2°C to cause a cold block. AChE transported in the nerve accumulates above the low temperature region. After several hours of accumulation, the cold blocks were relieved and the subsequent waves of outflow followed for different time periods give a measure of the rate of fast transport (cf. Figure 11.25). From (Brimijoin, 1975), Figures 1 and 6.

11.21B). The departure occurred when the segment was cleared of the AChE free to move. Calculation of the rate of transport of the enzyme by taking the length of the nerve segment between the two ligations and the time of departure gave a rate of fast transport of the enzyme of 431 mm/day, one statistically similar to that of labeled proteins.[102] The retrograde rate of ACh similarly determined by its departure from an accumulation below the upper ligation of the doubly ligated segment was 220 mm/day.

A similar fast rate of anterograde transport of AChE was found by Brimijoin and colleagues using the "stop-flow" technique that he developed (Figure 11.22).[103] A nerve, the sciatic nerve of a rabbit, is placed in a chamber with a narrow region in the middle that can be brought down to a temperature of 2°C to cause a cold-block damming of transport, the rest of the nerve kept at 37°C. After a suitable period of time of damming (e.g., three hours), the cold part is quickly warmed to 37°C, allowing the dammed up components to move out into the distal portion of nerve. Its analysis showed the outflow to contain AChE moving as a wave. By taking the position of the fronts of a series of such waves taken at different times after rewarming, a fast transport rate of 400 mm/day was determined for it.

---

[102] (Khan, Ranish, and Ochs, 1971), and (Ranish et al., 1972).
[103] (Brimijoin, 1979), and (Brimijoin and Wiermaa, 1978).

Using the technique of doubly crushing nerves, the fast transport of the neurotransmitter ACh was shown (Dahlström et al., 1974). The technique of chilling the nerve before assaying for ACh prevented local changes due to the action of AChE to hydrolyze or choline acetylase (CAT) to augment the level of the ACh (Hewall, 1978). In his review, Hewall gives evidence based on their fast rates that ACh and AChE are associated with the same transport vesicle, and likely also CAT.

The presence of ACh synthesizing and hydrolyzing enzymes carried along with ACh requires that the neurotransmitter be compartmentalized or else it would be hydrolyzed by the AChE.[104] Electron microscopic studies showed the presence of vesicles in the fibers and a much larger number in the fiber terminations at the motor end-plates of muscle.[105] The vesicles were found to contain ACh by harvesting the vesicles using differential ultracentrifugation.[106] Studies of the neuromuscular junction had shown that several hundred synaptic vesicles release their content of ACh when an impulse sweeps over the motor nerve terminals to bring about a muscle contraction.[107] At rest, there is a constant random discharge of a few such vesicles that have the same characteristic form of the end-plate potential, except for their much smaller amplitude. These *miniature end-plate potentials* (MEPPs) are discharged in too few numbers to cause a muscle contraction. Miledi and Slater made use of their presence to assess their rate of transport in the nerve fibers.[108] On cutting the nerve close to the sartorius muscle of frogs, the MEPPs fell off in number sooner than when cutting nerve further from the muscle. Taking difference in the time of the fall off in the discharge of MEPPs in the two cases, and the distance between the cuts, a rate of 360 mm/day was determined. Although a loss of vesicles carried down within the fibers to their terminals could account for the findings, the loss of some fast-transported component controlling the release of ACh from terminal vesicles may also account for the findings.

b. The adrenergic system

The compound noradrenaline (NA) had been known since 1904 to be chemically related to adrenaline, the amine identified as having sympathetic nervous system activity. NA received little attention by neurochemists until its presence in nerve was discovered by von Euler.[109] This raised the possibility that NA is carried down to the sympathetic nerve terminals from which it is released as a neurotransmitter. However, even as late as 1956, von Euler

---

[104] (Feldberg and Vogt, 1948).   [105] (van Breemen, 1958).
[106] (De Robertis, 1958). The rich concentration of vesicles in electric organs was the source used for their isolation.
[107] (Katz, 1966).   [108] (Miledi and Slater, 1970).
[109] (von Euler, 1956). Adrenaline is the methylated form of noradrenaline.

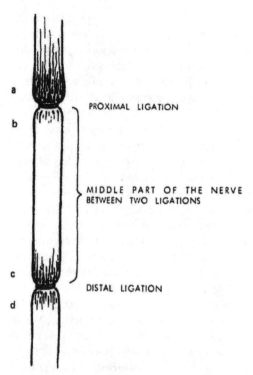

PROXIMAL LIGATION

MIDDLE PART OF THE NERVE
BETWEEN TWO LIGATIONS

DISTAL LIGATION

11.23. Transport of NA shown by making double ligations to isolate a segment of nerve. Accumulation of NA was seen as an amine fluorescence just above the segment (**a**) and above the distal ligation (**c**) within the segment. Retrograde transport is indicated by accumulation just below the isolated segment (**d**) and below the upper ligation within the isolated segment (**b**). The latter shows little flourescence (cf. that for AChE in isolated nerve segment shown in Figure 11.21A). From (Dahlström and Haggendal, 1966), Figures 1 and 2.

could write that there was "no evidence that the NA present in the nerve trunk is released under physiological conditions." The problem was that von Euler found no reduction of NA after a period of intense stimulation of the sympathetic nerves, though he did consider the possibility that this could be the result of a rapid resynthesis of NA in the terminals. The situation changed when Dahlström and colleagues gave evidence for the transport of the amine within sympathetic nerve fibers.[110] To visualize the relatively small amount of NA present in the fibers, they made use of the fluorescent method of Falck et al.[111] In doubly ligating nerves to produce an isolated segment, Dahlström and colleagues found the amine to accumulate in the segment just above the upper ligation and above the distal ligation at a rate

[110] (Dahlström and Fuxe, 1964), and (Dahlström, 1971).   [111] (Falck, 1962).

consistent with its fast transport (Figure 11.23). Only a small amount of fluorescence was seen just below the ligations, indicating a smaller amount of the amine. Using a bioassay for NA, Livett et al. gave its anterograde rate as 450 mm/day,[112] a value reasonably close to the fast transport rate of 410 mm/day found for labeled proteins.

In the accumulations, dense core vesicles (DCVs) were seen in electron micrographs. Two groups of vesicles were present, one 400 to 600 Å in size, with a larger one of 750 to 1200 Å.[113] When isolated by differential centrifugation, the smaller vesicles were found to contain NA, the larger ones dopamine (DA) in addition. DA, the precursor of NA, is converted to it by the enzyme dopamine-β-hydroxylase (DBH),[114] which is also carried in the larger vesicles.[115] With the conversion of DA to NA, the vesicles are reduced in size. The small accumulation of NA carried by retrograde transport and the absence of DA in them indicates that the DCVs are depleted of nearly all their content of amines before the spent vesicles are carried back to the cell bodies by retrograde transport.[116]

A fast rate of transport of NA was determined by the departure method that had been used to show AChE transport.[117] In the segment isolated by doubly ligating nerves, the amine was found to accumulate just above the distant ligation at a linear rate until its departure when no further accumulation was seen (Figure 11.24). From the time of departure and the length of the nerve segment, the rate of anterograde transport was determined to be 392 mm/day. This rate was somewhat lower than the rate of 450 mm/day given by Livett et al., and less than the rate of 410 mm/day found for labeled proteins. Furthermore, unlike the maintained level of AChE, which accumulated at the distant ligation after departure (Figure 11.21B), NA at the lower ligation underwent a slow decline. This was accounted for by diffusion of the amine from its vesicles and its degradation by the monoamine oxidase (MAO) present in nearby mitochondria. The anterograde rate of DA found using the departure method was still lower, 347 mm/day, and its level also declined at the lower ligation. The decline with time was ascribed to its conversion to NA in the vesicles. That this was the case was indicated by treating animals with disulfiram, an agent that blocks DBH. This raised the level of DA accumulated in the double-ligated nerve segments and its calculated fast rate of transport to 400 mm/day, one closer to that of labeled proteins. A faster rate of transport of NA was also seen when using the stop-flow method.[118] When the cold block was relieved, the outflowing waves

---

[112] (Livett, Geffen, and Austin, 1968).
[113] (Stjärne, 1966), (Hokfelt, 1969), and (Banks and Mayor, 1972).
[114] (Dahlström, 1999). [115] (Livett, Geffen, and Rush, 1971).
[116] (Geffen and Livett, 1971). [117] (Ben-Jonathan, Maxson, and Ochs, 1978).
[118] (Brimijoin and Wiermaa, 1977a).

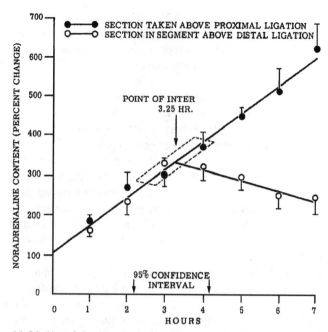

11.24. Use of the departure method to assess the transport rate of NA. As in the determination of the rate of AChE (cf. Figure 11.21), an accumulation of NA above the upper ligation of the segment continues (O), whereas that at the lower ligation within the segment of the doubly ligated nerves (●) shows a departure, its abrupt cessation of accumulation. The time of departure depends of the length of the isolated nerve segment from which the amine is cleared, giving a rate of 392 mm/day for it. Unlike the case for AChE, where the accumulated enzyme at the lower ligation remained at a fixed level, NA showed a slow decline with time. This was accounted for by its diffusion from its transport vesicles and degradation by MAO (see text). From (Ben-Jonathan, Maxson, and Ochs, 1978), Figure 2.

showed a rate of NA transport of 430 mm/day (Figure 11.25). The cold-block technique obviates the defects seen when the vesicles remain concentrated at the ligations leading to loss by conversion and degradation of the amine to give a lower rate and the decline seen in Figure 11.24.

Another component of the adrenergic system, the enzyme tyrosine hydroxylase (TH), which converts the precursor tyrosine to *l*-DOPA, is transported in the nerve as a soluble enzyme rather than within vesicles. Earlier estimates indicated that it had a slow rate. But, as Dahlström pointed out,[119] those determinations did not take into account the fact that only a relatively small portion of the enzyme is free to move, with most of the enzyme present in the fibers in a relatively immobile state. On reevaluating TH transport using the stop-flow technique, Brimijoin and Wiermaa found

---

[119] (Dahlström, 1973).

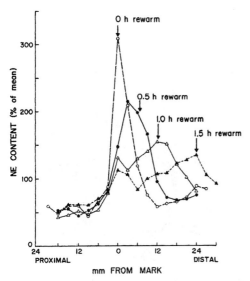

11.25. The time of transport of NA using the stop-flow procedure. Waves of accumulated NA are seen that have moved down into the distal nerve after rewarming the cooled region of a stop-flow preparation (cf. Figure 11.22). The times of rewarming indicated by the arrows gave the rate of transport of 430 mm/day for the advancing fronts of the waves of NA. (**NE**, norepinephrine). From (Brimijoin and Wiermaa, 1977a), Figure 3.

the enzyme to be transported at both a slow and a fast rate.[120] The significance of two different rates of the enzyme with respect to the mechanism of its transport will be discussed in Chapter 12.

c. Serotonin transport

The similarity of the outflow patterns and rates seen in nerves of various types containing a range of fibers of different sizes was accounted for by the similarity of the transport mechanism in their individual fibers. This was shown to be the case by the investigations of Goldberg and Schwartz, who examined transport of the neurotransmitter serotonin in the single nerve fibers of *Aplysia*.[121] After labeling, the serotonin contained in vesicles was seen to be fast transported with a pattern of outflow closely similar to that of labeled proteins in the multifibered nerves (Figure 11.26). Their rate was close to 400 mm/day, when the Q_{10} was taken into account and the temperature scaled to 37°C. Of further interest is the phenomenon of down-regulation and routing shown in this nerve. It divides into two branches: one supplying the buccal mass and the other the lip musculature

---

[120] (Brimijoin and Wiermaa, 1977b).
[121] (Schwartz et al., 1975), (Goldman, Kim, and Schwartz, 1976), and (Goldberg, Schwartz, and Sherbany, 1978).

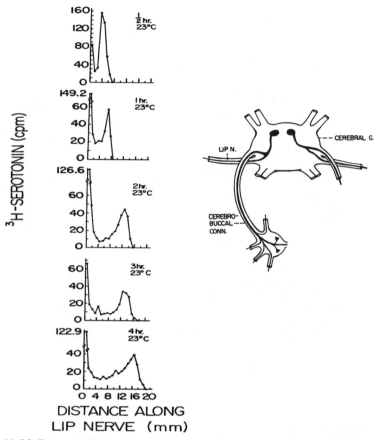

**DISTANCE ALONG
LIP NERVE (mm)**

11.26. Transport of serotonin in single fibers of *Aplysia*. After injection of the cell body with labeled serotonin, the outflow in its fiber shown in the column on the left has the same characteristic pattern as that of fast transport in multifibered nerves – a plateau that rises to a crest and then falls abruptly at the front. The rate determined from the advance of the fronts with time scaled to a temperature of 37°C was close to 400 mm/day. On the right, the neuron is shown with its fiber branches, one branch innervating the lip of the animal (**LIP N**), the other the CEREBRO-BUCCAL CONNECTIVE (CONN). The serotonin transported into a branch can be altered by cutting the other one (see text). **G**, ganglia. Modified from (Goldberg, Harris, and Schwartz, 1982), Figures 1 and 2. (Schwartz, et al., 1975) and (Goldberg, et al., 1978).

(Figure 11.26). On transecting the branch innervating the buccal mass, a down-regulation was seen by the decrease of serotonin transported into that branch and an up-regulation with an increase of vesicles passing into the branch innervating the lip musculature.[122] A signal from the cut branch carried to the cell body is the likely cause of the channeling of granules

---

[122] (Aletta and Goldberg, 1984).

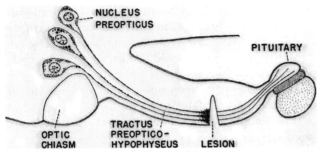

11.27. Transport in the hypothalamo-neurohypophysial system. Cell bodies in the nucleus preopticus in the hypothalamus are shown with their fibers passing in the stalk leading to the posterior lobe of the pituitary gland. On cutting the stalk, neurosecretory material staining dark was seen accumulated above the transection and depleted below it and from the pituitary gland. This figure is based on the experimental studies of Hild published by (Bargmann and Scharrer, 1951), Figure 2.

to the appropriate branch, an example of routing described in a previous section.

d. Transport of vasopressin and oxytocin hormones in the stalk of the hypothalamo-neurohypophysial system

Historically, studies of the transport of hormones in the hypothalamo-neurohypophysial system (HNS) had been carried out with little reference to those made in the peripheral nervous system and vice-versa.[123] The study of transport in the HNS began in the 1940s when the Scharrers noted that the large cell bodies in the hypothalamus were similar in appearance to secretory cells.[124] These large cells are located in the supraoptic nucleus and in the paraventricular nuclei of the hypothalmus with their fibers passing down via the pituitary stalk to terminate in the posterior pituitary gland. The fibers transport the hormone *vasopressin*, an antidiuretic hormone that also acts to increase blood pressure, and *oxytocin*, the milk-releasing factor that also acts to control uterine contractions in parturition. Excitation of these cells causes action potentials to propagate down their stalk and release the hormones from the posterior pituitary gland into the bloodstream, the term *neurosecretion* coined to refer to this process.[125] On transecting the pituitary stalk and using Bargmann's chrome-alum-hematoxylin to stain for these hormones, Hild found them to accumulate just above the transections and to become depleted from the fibers below the transections and from the posterior pituitary gland (Figure 11.27).[126] The pituitary hormones are transported in the axons of the stalk within electron-dense vesicles that range from 1,100 to 1,900 Å in diameter along with *neurophysins*, proteins

---

[123] (Ochs, 1977). [124] (Scharrer and Scharrer, 1940).
[125] (Bargmann and Scharrer, 1951). [126] (Hild, 1951).

to which the hormones in the vesicles are bound.[127] The rate of transport of the hormones was determined using the labeled precursors [^{35}S]cysteine and [^{35}S]methionine.[128] Injected into the hypothalamus near the large cell bodies, they are taken up by them, incorporated into their hormones, packaged into vesicles and these transported down within the stalk fibers.[129] Although earlier studies had suggested a slow rate of their transport, Gainer and colleagues found a faster rate of 190 mm/day for the hormones and a similar fast rate for the neurophysins. After the release of the contents of the vesicles from their terminals in the posterior lobe of the hypophysis, the spent vesicle membranes are returned to the cell body by retrograde transport for reuse of their components.[130]

VISUALIZATION OF PARTICLE TRANSPORT IN LIVING FIBERS AND AXOPLASM WITH NOMARSKI OPTICS AND AVEC-DIC MICROSCOPY

Although the neurotransmitter and neurotransmitter-related components transported in vesicles require the use of an electron microscope to visualize them, the larger mitochondria could be seen to move in fresh nerve fibers using dark-field and Nomarski optic microscopy.[131] They move both in the anterograde and retrograde directions at a relatively rapid rate approaching that of fast transport, but do so for only short periods of time so that their net rate of transport is slow, of the order of a few millimeters per day. This slow rate is matched by biochemical studies carried out with markers for mitochondria, such as its content of MAO.[132]

When computer techniques were used in conjunction with Nomarski optic and differential interference microscopy, smaller particles of the size of vesicles could be visualized moving in the axoplasm of living nerves. This remarkable advance in methodology was achieved by Robert Allen and colleagues and is termed *Allen video-enhanced contrast differential interference contrast* (AVEC-DIC) microscopy.[133] Vesicles and other small particles 40 to 50 nm in size are at the wavelength of light and cannot be seen directly as such, but AVEC-DIC microscopy gives a close representation of them. Using this technique, particle movement in freshly extruded axoplasm of giant axons was seen to occur along tracks identified as microtubules, at rates close to 400 mm/day. The particles moved in both the anterograde and retrograde directions, even on the same microtubule, passing one

---

[127] (Sachs, 1969).
[128] Labeled leucine cannot be used as the precursor because this amino acid is not present in the small polypeptides of the pituitary hormones.
[129] (Gainer, Sarne, and Brownstein, 1977).
[130] (Fawcett, Powell, and Sachs, 1968).   [131] (Cooper and Smith, 1974).
[132] (Zelena, 1968), and (Khan and Ochs, 1975).
[133] (Allen, Allen, and Travis, 1981), (Allen et al., 1982), (Allen et al., 1985), (Allen, 1987), (Vale, 1987), and (Sheetz, Steuer, and Schroer, 1989).

another without collision. This was traced to their movements along different protofilaments forming the wall of the microtubules (Chapter 8). Using particle movement as an assay, two proteins essential for transport were isolated from the axoplasm of giant fibers: *kinesin* subserving anterograde transport and *cytoplasmic dynein*, in brief *dynein*, retrograde transport. The micromechanical details by which these "motor" proteins effect vesicular transport are given in Chapter 12, where the mechanism of such movements is described.

AXOPLASMIC TRANSPORT: AN ENERGY-DRIVEN MECHANISM

In the early part of the twentieth century, conduction of the action potential in nerve was considered to be a physical process with no evidence for its being dependent on a metabolic source of energy. In large part, this was because nerve studies were carried out on frog nerves that have a low metabolism. This began to change in the earlier decades of the twentieth century when Tashiro, using a sensitive method he developed to measure small concentrations of carbon dioxide, found it to increase when nerve was activated.[134] Then, the metabolic processes that were discovered in muscle were also found present in nerve.[135] While metabolic processes were shown to be needed to maintain resting and active nerve potentials,[136] in the middle of the twentieth century the main theories proposed to account for transport in nerve, those of Weiss and of Young, were of a physical, non-energy requiring process.

The situation changed when it was found that transport depended on oxidative metabolism to provide the energy required for transport. Transport was shown to fail rapidly, and stop completely within 20 to 30 minutes when nerves were made anoxic by rapidly bleeding out the animal, or when nerves in vitro were made anoxic by replacing oxygen in the chamber with nitrogen.[137] After a lag period, transport underwent a decline and then ceased in twenty to thirty (Figure 11.28).[138] As in muscles and other cells, oxidative metabolism in nerve generates ATP, which drives chemical and micromechanical processes by means of its high-energy phosphate bond ($\sim$P).[139] To relate $\sim$P to transport, it was necessary to take into account not only the level of ATP, but also creatine phosphate (CrP) because it contains $\sim$P that can be rapidly transfered to ATP. The sum of ATP and CrP therefore represents the amount of $\sim$P available to supply energy to

[134] (Tashiro, 1917).
[135] (Downing, Gerard, and Hill, 1926), and (Hill and Howarth, 1958).
[136] (Gerard, 1932), Chapter 7.
[137] (Ochs et al., 1970), (Ochs and Hollingsworth, 1971), (Leone and Ochs, 1978), and (Ochs, 1982).
[138] (Ochs, 1972a).   [139] (Lehninger, Nelson, and Cox, 2000).

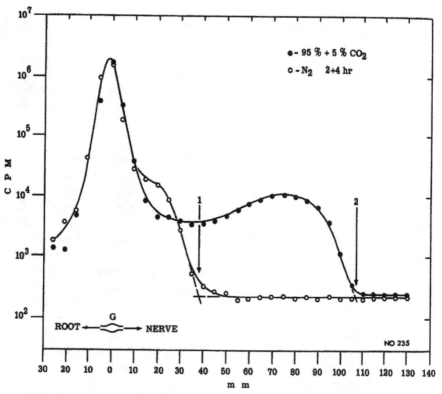

11.28. Block of transport by anoxia. After injection of a cat's ganglia (**G**) with labeled precursor, two hours were allowed for labeled proteins to gain entry into the fibers before the sciatic nerves were removed for a further period of in vitro transport. One nerve (●) was oxygenated in the chamber and showed the usual fast rate of outflow indicated by arrow **2** at its foot. The other nerve (○) exposed to nitrogen in the chamber to make it anoxic shows a failure of transport soon afterward – the outflow advancing no more than that seen after 20 to 30 minutes of in vivo downflow. **CPM**, counts per minute. From (Ochs, 1972a), Figure 7.

the transport mechanism. When nerves are made anoxic and transport was blocked, the level of $\sim$P fell to a critical level, to approximately half of its control level of 1.0–1.4 μM/g.[140] A similar fall of $\sim$P was seen when agents such as azide or sodium cyanide were used to block the terminal stage of oxidation, the respiratory chain of oxidative metabolism in the mitochondria, where production of $\sim$P takes place as ADP is converted to ATP (Figure 11.29). The agent 2,4-dinitrophenol acts to uncouple the phosphorylation

[140] (Sabri and Ochs, 1972), and (Ochs, 1974).

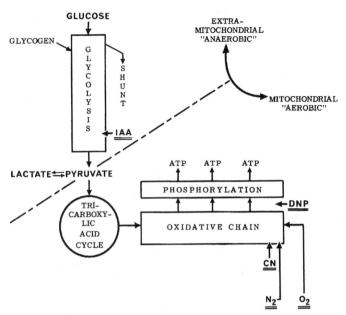

11.29. Metabolic pathways in the production of ATP. Glucose passing from the circulation into the fibers enters the glycolysis chain and its product pyruvate entering as acetyl-CoA into the tricarboxylic acid cycle. Reducing equivalents from it enters the mitochondrion, where transfer in the oxidative chain forms ATP. Agents shown acting at various points along the metabolic pathways block ATP production (see text). **IAA**, iodoacetic acid, **DNP**, dinitrophenol, **CN**, cyanide, N_2, nitrogen anoxia. From (Ochs, 1974), Figure 4.

of ADP to ATP leading to the same loss of ~P, and it similarly rapidly blocks transport.[141]

In the first stage of metabolism, glucose enters the fibers all along their length[142] to be metabolized by the glycolytic chain forming pyruvate and from it acetyl-coenzymeA that enters the tricarboxylic acid cycle (Figure 11.29).[143] Its step-wise degradation in the cycle gives rise to reducing equivalents that enter the mitochondria to become oxidized by a series of electron transfers leading eventually to oxygen. Agents that interfere with glycolysis block transport, though after a longer delay than that seen following anoxia. The block of glycolysis with iodoacetic acid, which does so by its inhibition of glyceraldehyde phosphate dehydrogenase,[144] took 1.5 to two

[141] The electron transfer in the mitochondria continues when nerve is treated with dinitrophenol, although it is futile because ~P production is blocked. This finding was the basis of its use to reduce body weight without the inconvenience of dieting. Its popularity soon waned when it was found to produce blindness.

[142] (Gaziri, 1984), and (Gaziri and Ochs, 1987).   [143] (Lehninger et al., 2000).

[144] (Sabri and Ochs, 1971).

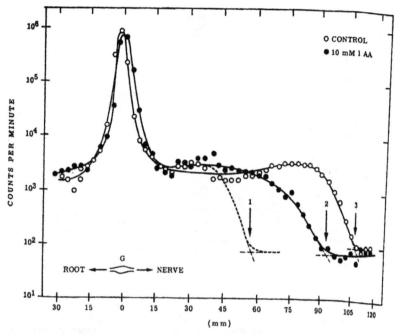

11.30. The agent iodoacetic acid (**IAA**) acts to block the enzyme glyceraldehyde-3-phosphate dehydrogenase in the glycolytic path (cf. Figure 11.29). Arrow **1** indicates the downflow of labeled proteins in the animal before removing the nerve for a subsequent period of in vitro downflow. Transport in the nerve exposed to IAA (●) continues for 1.5 to 2 hours before it is completely blocked as shown by arrow **2**. This is a much longer time than the block seen in 20 to 30 minutes with anoxia (cf. Figure 12.28). The control nerve (○) shows the expected transport to the distance indicated by arrow **3** at the front of its outflow. From (Ochs and Smith, 1971), Figures 1 and 2.

hours to block transport (Figure 11.30).[145] It took an hour or so for transport to fail when the tricarboxylic acid cycle was blocked by fluoracetate.[146] These longer survival times are considered to be due to a storage of some metabolite(s) able to generate ATP for a time. Another possibility to account for the longer survival time is that ATP is produced, at least for a short time, by some metabolite other than glucose. This could be lipid, as suggested by the prolongation of transport found with γ-hydroxybutyric acid after blocking glycolysis.[147] The ATP produced by metabolism supplies the energy needed for transport, the molecular mechanisms proposed for transport described in the following chapter.

---

[145] (Ochs and Smith, 1971), (Sabri and Ochs, 1971), and (Edström and Mattsson, 1976).
[146] (Ochs, 1974).
[147] (Gaziri and Ochs, 1983), (Gaziri, 1984), and (Gaziri and Ochs, 1987).

12

MOLECULAR MODELS OF TRANSPORT

The characteristics of transport discussed in the preceding chapter were shown to depend on the energy supplied by oxidative metabolism. In this chapter, models advanced to account for how that energy is utilized for the movement of proteins, and the vesicles and other particles visualized by means of allen video-enhanced contrast differential interference contrast (AVEC-DIC) microscopy, are described. The view that has emerged is that all these materials are moved out along the microtubules by specific "motors." The development of this model of fast transport is described in this chapter.

Slow transport on the other hand has remained a matter of contention. The old view that axoplasm moves down in bulk (Chapter 11) was replaced by the hypothesis that only the microtubules and neurofilaments are moving down at the slow rate. An opposing theory holds that these cytoskeletal organelles are stationary in the fibers with their protein subunits moving in the fluid axoplasm. The question raised is whether this requires the presence of a different mechanism of transport other than that serving for fast transport with the further complication that, in addition to fast and slow transport, a number of intermediary transport rates have been found. The hypothesis that a single mechanism termed the *unitary hypothesis*, can account for all the different transport rates will be taken up at the end of the chapter.

EARLY VIEWS ON FIBRILLARY MATERIAL IN THE FIBERS IN RELATION TO TRANSPORT

When, in the latter part of the nineteenth century, Schultze substantiated Remak's finding that nerve fibers contain a fibrillary component, he hypothesized that their role is to conduct the action potential.[1] This concept was abandoned when the membrane was shown by the end of the

---

[1] (Schultze, 1870).

nineteenth and the early part of the twentieth century to be the seat of the nerve action potential (Chapter 7).[2] This opened up the possibility that the neurofibrils might serve for the transport of materials, a hypothesis advanced by Parker[3] in 1929 on the basis of the T-shape of the dorsal root ganglion neurons:

> In the ordinary sensory neurons, nerve impulses originate at the peripheral end, make their way centrally over the neurite, and without entering the body of the cell, pass on to discharge at the central end of the neuron. The metabolic influences on the other hand originate in the region of the nucleus of the cell body, pass down its neck to the tract of nervous transmission where they separate into two streams, one flowing peripherally over the neurite and the other centrally over the central nerve fiber process. And thus, the course of the neurofibrils does not follow that of the nerve impulses but does duplicate exactly that of the metabolic influences. I conclude therefore that the neurofibrillary system in the neuron is concerned specifically with the distribution of the metabolic influences and not with the conduction of nerve impulses. These influences start in the region of the neuronic nucleus and spread over the lines of neurofibrils throughout the whole neuron.[4]

Parker, speculating on whether the influence "is of the nature of a hormone emanating from the region of the nucleus and percolating throughout the neurone," concluded that this was unlikely. Torrey, a student of Parker's, was not so circumspect. He was favorably inclined to the hypothesis that a "hormone-like material necessary for the maintenance of the nerve fiber [one which] may be enzymatic in character, [is] produced by the nucleus of the neuron, and transported peripherally by the way of the neurofibrils."[5]

MODELS OF FAST TRANSPORT
With the recognition that the neurofibrils could be differentiated into defined classes – neurofilaments, microtubules, and microfilaments (Chapter 8) – attention was directed to which one, or perhaps all of these cytoskeletal organelles, is essential for transport. Porter, in summing up what was known of microtubules in a review published in 1966, considered as a working hypothesis that they not only determine the form of the fiber, but also that they "may define not only the channels for streaming but also channels which orient and distribute other fibrous or filamentous components."[6] He also thought it possible that the presence of ATPase on the surfaces of the microtubules could be of significance in playing such a role, a prescient view.

---

[2] The notion that neurofilaments could serve as the conduction of the action potential was still entertained by Maximow and Bloom in their well-known textbook of histology in an edition published as recently as 1930 (Maximow and Bloom, 1930).
[3] (Parker, 1929b).   [4] (Parker, 1929a).   [5] (Torrey, 1934).   [6] (Porter, 1966).

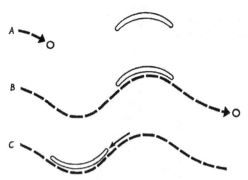

12.1. Observations in living nerve of *Xenopus*, made using dark-field illumination, shows a relatively long structure, most likely a mitochondrion, moving along the sinuous course of microtubules. A particle (**A**) was seen to travel along the path to the right (**B**), whereas the mitochondrion was moving to the left over that same path (**C**). From (Cooper and Smith, 1974), Figure 10.

A major finding attesting to the essential role played by the microtubules in transport was that colchicine, vinca alkaloids, and other tubulin-binding agents that dissagregated microtubules or otherwise interfered with their action (Chapter 13), also blocked transport. The close relationship of vesicles to microtubules often seen in electron micrographs[7] was suggestive of a transport role for microtubules, the concept given support by the studies made by Cooper and Smith using dark-field microscopy.[8] A structure in the fiber that could only be that of mitochondria was seen to bend so as to conform to the curvature of the microtubules as it moved down along it (Figure 12.1).

Schmitt proposed a model in which he pictured vesicles rolling down along the microtubules with the making and breaking of specific projections on their surfaces as the means of transport (Figure 12.2). An ATPase or a GTPase was indicated to be present to utilize ATP and provide the energy required.[9] Some considerations argue against Schmitt's model. The vesicles by temporarily attaching at only a single point as it rolls forward to the next point of attachment would not have much force to move against the viscosity of the axoplasm. Although axoplasmic viscosity is relatively low in mammalian nerve (wherein a value of 5 cps was measured using electron spin resonance),[10] it would be enough to hinder the forward movement of such a relatively large structure, particularly so when meeting with the greatly increased resistance expected when the cytoskeleton is compacted in the

[7] (Smith, Jarlfors, and Cameron, 1975).   [8] (Cooper and Smith, 1974).
[9] (Schmitt and Samson, 1968).
[10] (Haak, Kleinhans, and Ochs, 1976). The measure is given in centipoise (cps) units, with water having a cps of 1.

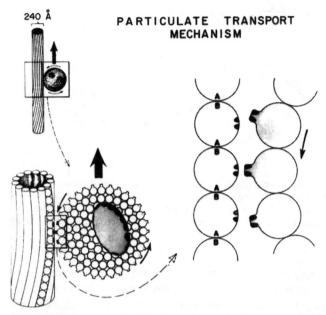

12.2. A model proposed by Schmitt to account for fast transport. A large globule is pictured with sites on its surface binding to those on the microtubules, the making and breaking of the bonds utilizing ∼P to roll the globule down along the microtubule. From (Schmitt and Samson, 1968), Figure 6.

constrictions of beaded fibers (Chapter 8) leaving less room between the cytoskeletal organelles than the size of the vesicle. A similar obstruction would be met with by the greatly increased content of neurofilaments produced in nerve fibers exposed to aluminum and other neurotoxins (Chapter 13). Despite such an impediment, fast transport still has its usual characteristic shape and rate in those fibers. The model also does not take into account the evidence that a wide range of soluble proteins of different molecular weights is moved down the fiber at fast rates. These could conceivably be carried inside these vesicles, but the model would not account for the movement of the specific vesicles carrying neurotransmitter and neurotransmitter components, or hormones in the case of the hypothalamo-neurohypophysial system (Chapter 11). Neither would the model account for the transport of large oblong or sausage-shaped structures, such as the mitochondria and segments of the endoplasmic reticulum.

To account for the transport of the wide range of components, from soluble proteins to large structures, such as vesicles, segments of the endoplasmic reticulum, or mitochondria, a *transport filament* model was advanced on analogy to the sliding filament theory of skeletal muscle contraction

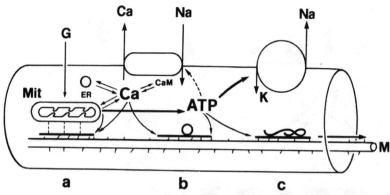

12.3. The transport filament model. Glucose (**G**) enters the fiber and, after glycolysis, its products pass to the tricarboxylic cycle and then the mitochondrion (**Mit**), where oxidative phosphorylation gives rise to ATP. The $\sim$P of ATP supplies energy to the sodium pump controlling the levels of Na^+ and K^+ in the fiber and to the side-arms on the microtubules to move transport filaments along them. The transport filaments are shown as linear structures to which the various components transported are bound and thus carried down along the microtubules. The components transported include: the mitochondria (**Mit**) attaching temporarily, as indicated by dashed lines, to the transport filament (**a**), vesicles (**b**), and soluble proteins (**c**) shown as a folded globular protein. Small polypeptides not figured are also transported, along with a wide range of labeled and endogenous components at the same fast rate bound to the carriers. Calcium (**Ca**) is required for transport, its level regulated in the axoplasm to an optimal level sub-μM level by calcium-binding protein, the endoplasmic reticulum (**ER**), and a Ca–Na exchange (or Ca pump requiring ATP). Calcium controls the function of calmodulin (**CaM**), which in turn, regulates a wide range of enzymes. Among them is a Ca-Mg ATPase associated with the side-arms proposed to utilize ATP to move the transport filaments along the microtubules (**M**). From (Ochs, 1982), Figure 10.24.

(Figure 12.3).[11] The various components transported were considered to bind to transport filaments that are moved down along the microtubules by their side arms utilizing the $\sim$P of ATP to provide the needed energy. The various materials bound to the carriers were envisioned as having varying degrees of binding. Some components, such as the mitochondria, by binding for only brief periods of time would thus appear to have a slow net rate of transport, whereas those binding for longer times a fast rate of transport. Neurotransmitter vesicles remaining bound to the carriers down to the terminals thus would have the fastest rate. In the model, the side arms of the microtubules contain, or have associated with them, a Ca-Mg ATPase that utilizes the $\sim$P of ATP for movement. The participation of Ca^{2+} was indicated by the block of transport seen when Ca^{2+} was depleted from the nerves or when present at too high a level altering the optimal concentration

[11] (Ochs, 1971), (Ochs, 1972a), and (Ochs, 1982).

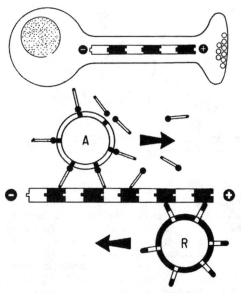

12.4. Model of transport in which the transport motors kinesin and dynein subserve anterograde and retrograde transport, respectively, along the microtubules. The motor shown attached to vesicle (**A**) and to the microtubules is kinesin, which is responsible for movement in the anterograde-positive (+) direction (from the cell body to the terminal). The motor for movement of vesicle (**R**) in the retrograde, negative (−), direction (from the terminal toward the cell body) is dynein. The alternating black and white regions of the microtubules represent the tubulin dimers aligned longitudinally in the protofilaments forming the microtubules along which the motors move. From (Schnapp et al., 1986).

of Ca^{2+} in the axon of 10^{-8}–10^{-7} M at which it binds to calmodulin, which in turn controls the activity of Ca-Mg ATP and a number of other enzymes (Chapter 13).[12]

However, transport filaments were not found. More recently, studies using AVEC-DIC microscopy have indicated that the specific "motor molecules" could serve the function that had been proposed for transport filaments. Instead of microtubule side-arms propelling transport filaments, the motor molecules have ATPase activity when their heads are in contact with the microtubules along which they move, the energy for the movement, provided by ATP. Vesicles and other cargo tethered at the ends of the elongated chains of the motor molecules are pulled down along the microtubules as the heads advance.

---

[12] (Ochs and Iqbal, 1983), (Ochs, Gaziri, and Jersild, 1986), and (Iqbal, 1986).

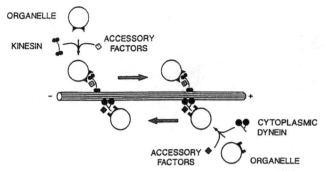

12.5. Diagram indicating that accessory factors help determine the direction in which organelles are moved by the kinesin and dynein motors; in the anterograde (+) and retrograde (−) directions, respectively. From (Sheetz, Steuer, and Schroer, 1989), Figure 2.

a. Models of fast transport based on motor molecules

Using AVEC-DIC microscopy of particle movement as an assay method, proteins essential for transport were found. These are the motor molecule *kinesin*, which moves vesicles or other cargo in the (+) direction (away from the cell body) at the rate of fast transport,[13] whereas another motor molecule *cytoplasmic dynein*, sustains movement in the retrograde (−) direction (back toward the cell body) (Figure 12.4).[14] The direction of movement of the two motor proteins depends on "accessory factors," which determine the different binding properties of the motors (Figure 12.5).[15] Dynein itself is transported in the anterograde direction in an inactive form and then, at fiber terminals, or at an interruption of the nerve fiber,[16] it is transformed to an active form enabling it to move in the retrograde direction. A third motor molecule, myosin, transports particles along the actin microfilaments, as will be discussed later.

i. Kinesin: the motor for long-haul transport in the anterograde direction.
As knowledge of the molecular structure of kinesin evolved over the last decade or so, the means by which it moves along the microtubule has become clarified. The kinesin molecule consists of two globular heads with heavy chains in a coiled-coil arrangement forming an elongated rod with light chains at its tail. ATPase activity is conferred on the heads when they come into contact with the microtubules permitting them to utilize the energy of the ∼P of ATP (Chapter 11) to move down along the microtubules, with vesicles and other cargo bound to their tails pulled along.[17] Their

[13] (Vale et al., 1985), and (Manning, Erichsen, and Evinger, 1990).
[14] (Schnapp et al., 1986), and (Vallee, Shpetner, and Paschal, 1989).
[15] (Sheetz, Steuer, and Schroer, 1989).   [16] (Li et al., 2000).   [17] (Vale, 1987).

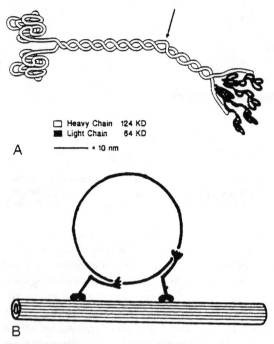

12.6. Micromechanical model of vesicle transport based on the cyclic bending and straightening of the kinesin chain. **(A)** Kinesin consisting of heavy (coiled-coil) and light chains (shown black at the end of the heavy chain). Arrow points to the hinge region in the heavy chains of the molecule. **(B)** The light chains bind to a vesicle being transported, whereas the head region of the heavy chains attaches to a microtubule. A cyclic straightening and folding of the hinge region was proposed to lift up and thereby move the vesicle forward. From (Hirokawa et al., 1989), Figure 9.

movement is blocked when inhibitors of kinesin-activated ATPase – such as vanadate, 5'-adenylylimidodiphosphate or ATP in the absence of magnesium – are present.[18]

An early model proposed by which the micromechanical action of kinesin effects transport involves the *hinge* region in the long heavy coiled-coil chain of kinesin that allows the chain to bend. With the head attached to a microtubule and the tail to a cargo, such as a vesicle, the straightening of the hinge was pictured propelling the vesicle forward (Figure 12.6).[19] Later studies showed kinesin to have a more complex structure, with several regions where the chain can fold (Figure 12.7). With the heads folded back onto a region of the chain behind it, kinesin is in a soluble inactive state

[18] (Cohn, Ingold, and Scholey, 1987).
[19] (Hirokawa, 1989), and see (Shpetner, Paschal, and Vallee, 1988), (Vallee and Shpetner, 1990), and (Hirokawa et al., 1991).

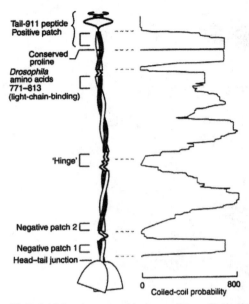

12.7. Additional structural features of the coiled-coil kinesin chain. The heads of the heavy chain are seen at the bottom, the tail containing light chains at the top. In addition to the hinge region approximately in the middle of the heavy chain, it can fold at several other sites (cf. Figure 12.8). On the right, the probability of finding a coiled-coil (nonbending) and bending regions, where the chain can fold, are shown. From (Cross and Scholey, 1999), Figure 1.

without ATPase activity (**III** in Figure 12.8). When the tail attaches to vesicles or other cargo, the rod straightens out (**I** in Figure 12.8) and the kinesin heads bind to microtubules, with ATPase activity conferred on them.[20] The heads are then able to utilize ATP, and move down along the microtubules: At the ends of the long chain, cargo bound to their tails is pulled along as the heads advance.[21] When the vesicle cargo is released the kinesin becomes folded (**II** in Figure 12.8) and then released from the microtubule.

The heads bind specifically to the β-tubulin of the dimers that line up longitudinally to form the protofilaments that then join laterally to constitute the wall of the microtubule (Chapter 8). The details by which the heads of kinesin move along the protofilament tracks of the microtubules, has recently emerged.[22] In the free state, each head binds ADP. When one of the heads attaches to the β-tubulin of a protofilament, its ADP is released and replaced by ATP. The ATP is hydrolyzed, its energy allows the head behind it to swing forward and attach to the next β-tubulin further along the protofilament, with the process repeated over and over in cyclic fashion. This manner

---

[20] (Cross and Scholey, 1999).   [21] (Vale and Milligan, 2000).
[22] (Mandelkow and Johnson, 1998).

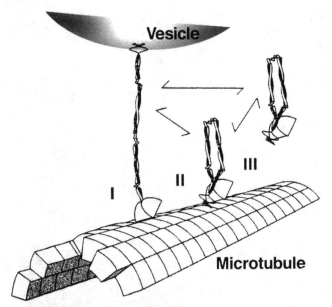

12.8. Model showing a shift of kinesin from a folded, soluble inactive state to an extended form to which cargo is bound at its tail for its transport. The kinesin heads bind cyclically to a microtubule, along which it progresses pulling the cargo along (I). In the folded state, (II) the molecule does not have ATPase activity and is detached and soluble (III). Only when the heads bind to the microtubules does it have ATPase activity enabling it to transport cargo. From (Cross and Schokley, 1999), Figure 2.

of procession is like the alternate swinging of feet when walking from one stepping stone to another, or the hand-over-hand movement when climbing a rope. In each cycle of movement, the kinesin heads move ~80Å forward, the distance between the β-tubulins along the protofilament. Such a processive movement of single kinesin molecules along the microtubule has been directly observed.[23]

ii. Dynein: the motor for long-haul transport in the retrograde direction. The protein cytoplasmic dynein, the protein motor subserving retrograde transport, consists of two large subunit heads with long polypeptide chains to the ends of which cargo can attach and be hauled along as the heads move down along the microtubules. As with kinesin, ATP is the energy source used by the heads to effect their movement. Unlike kinesin, dynein motor movement is brought about by contact of a single head to the microtubules, its movement effected by the repeated folding, extension

---

[23] (Vale et al., 1996).

of the chain behind the head and its release and reattachment further along the microtubule.

iii. Myosin: the motor for short-haul transport along actin microfilaments. In addition to the long-haul transport system subserved by microtubules, a local supply of ion channels and ion pumps to the membranes is subserved by a short-haul system utilizing actin microfilaments as the rails. The microfilaments, unlike the linearly organized microtubules, are irregularly oriented in the axons, coursing laterally to cross the microtubules at oblique angles and present with a greater concentration at or near the axolemma. Particles moving down along microtubules were seen with AVEC-DIC to cross over onto actin microfilaments and then move along them.[24] The motor for such actin-based transport is myosin V, a member of a family of myosins having properties similar to those of muscle myosin. Myosin differs in its manner of its progression along actin from that of kinesin along the microtubules.[25] It consists of a dimer with two identical heads with a coiled-coil rod behind them. The two heads act independently, and both are not necessary for its micromechanical action. A head containing ADP-Pi attaches to the actin causing the release of phosphate from the active site with the lever-like arm of the molecule swinging back from its neck, much as a ratchet action or an oar stroke moving a boat forward. After this stroke phase in which ADP is still bound, ATP binds to the active site, the energy released by its hydrolysis allows the neck of the lever arm to recoil back to its extended form ready to bind to the actin at a site further along, the action repeated cyclically. By this means, the heads are moved forward at incremental distances of ~360 Å. The carrying over of organelles from the microtubules – where kinesin is the motor, to the actin microfilaments[26] where myosin is the motor[27] – appears to be brought about by a special system.[28]

b. Models of fast transport not based on motor molecules
Before strong evidence had been found for the role of motor molecules moving along microtubules as the mechanism of fast transport had emerged, a variety of other models of transport had been proposed.[29] Although they cannot account as well as motor molecules for long distance transport, some may be involved in local processes. Endoplasmic reticulum (ER) was pictured as a continuous saccular network passing down from the cell bodies to the nerve fiber terminals, serving as the route for transport.[30] This hypothesis was based on the finding of transported labeled proteins inside the

---

[24] (Kuznetsov et al., 1994).   [25] (Vale and Milligan, 2000).
[26] (Kuznetsov, Langford, and Weiss, 1992).   [27] (Goldstein and Philip, 1999).
[28] (Brown, 1999).   [29] (Ochs, 1982).   [30] (Droz, Rambourg, and Koenig, 1975).

ER. However, the continuity of the ER was not upheld by studies of pike olfactory fibers, it was shown to be discontinuous.[31] The labeled proteins seen
in the ER appear to have been translocated into them locally in the fiber.[32]
This was made evident by the uptake of horseradish peroxidase (HRP) in
nerve terminals, where it enters vesicles that coalesce to form short segments of ER that are then retrogradely transported.[33]

Channels of low viscosity within a gel-like axoplasm and polyelectrolyte
contraction were advanced as a means of transport.[34] These could be the
basis for some local transport, but cannot account for the characteristic
form of fast transport, its maintained rate, and the wide range of proteins
and the larger particulate components transported at the same fast rate in
an "all-or-none" manner (Chapter 11).

A hypothesis for transport that had attracted considerable attention for a
time was based on the extensive cross-linking of all axonal structures, into
a *microtrabecular network*, that was seen in electron micrographs when using
the technique of *critical-point drying* for fixation.[35] Transport was proposed
to be carried out by the making and breaking of the cross-bridge connections
of the network. However, such hyperconnectivity was shown by Ris to be
an artifact brought about by remaining traces of water in the course of this
method of preparation.[36]

MODELS OF SLOW TRANSPORT

a. Bulk flow hypotheses of transport

Proposals for slow transport began with the theory of a bulk flow of axoplasm advanced by Weiss on the basis of evidence for a damming of axoplasm just above a region of partial constriction (Chapter 11). In addition
to evidence against the hypothesis given in that chapter, Young pointed out
that it would take an enormous force exerted by the cell body to move axoplasm down lengths of nerve that could be as long as several feet in the larger
animals.[37] In response, Weiss proposed a local process for slow transport, a
microperistalsis of the fibers. He based this hypothesis on observations made
on cultured myelinated nerve fibers using time-lapse cinematography. Indentations were seen moving along the surface of the fibers that he took
to be peristaltic-like waves moving at an average rate of 2.14 mm per hour.

---

[31] (Kreutzberg and Gross, 1977).   [32] (Byers, 1974).
[33] (Lavail and Lavail, 1974), and (Lavail and Lavail, 1975).
[34] (Gross and Weiss, 1982), and (Weiss and Gross, 1982).
[35] (Ellisman, 1982).   [36] (Ris, 1985).
[37] The theory of Young, in which axoplasm is moved down the fibers by the pressure
of its production in the cell bodies (Chapter 11), suffers from the same objection
in that the force needed for such movement of axoplasm over a long length of
nerve would be too great.

Weiss concluded that the waves were produced by constrictions of their myelin sheaths. He stated that the axis cylinder within the sheath of the fiber "keeps growing forth throughout life from its base in the cell body, its macromolecular substance being... conveyed as a cohesive semisolid mass toward the distal ending... at a standard rate of the order of 1 mm/day, driven by a microperistaltic wave generated at the axonal surface."[38]

However, the indentations described by Weiss were asymmetrical. If they were truly due to microperistalsis, they should appear as annular constrictions of the fibers.[39] The distinguished neuroanatomist Szentagothai found that the indentations pictured by Weiss were more likely those changes known to occur in the early stages of Wallerian degeneration and Weiss himself later referred to the indentations as "incipient Wallerian degeneration." A further telling argument against the model was given by Samson, who pointed out that the postulated constrictions of the myelin could not account for transport in unmyelinated fibers.[40]

A somewhat similar proposal advanced to account for slow transport was advanced on the basis of a declining exponential curve of labeled outflow seen in early isotope studies of downflow (Chapter 11). This was taken to be brought about by a random series of fiber constrictions and relaxations all along the fibers to produce in effect a facilitated diffusion.[41] Using a transillumination technique by which microscopic observations of nerve fibers of the mouse peroneal nerve could be made in situ repeatedly over a period of days in the same animal, Williams and Hall did not see such a microperistalsis.[42] There was "no change in either external or internal myelin sheath contour: the majority of internodes appeared to be smooth-walled and cylindrical." And, in cultured nerve fibers that had recently become myelinated, no indentations indicative of a microperistalsis were observed.[43]

b. Slow transport of cytoskeletal organelles: the "structural hypothesis"

A hypothesis that has some similarity to the bulk flow model of Weiss was proposed by Hoffman and Lasek.[44] Rather than all the axoplasm continuously moving out within the fibers, they envisioned that only the linearly organized microtubules and neurofilaments were moving at a constant slow rate down within the fibers. The hypothesis was based on the slow waves of labeled proteins they reported moving down the nerve fibers (Figure 11.20). The major outflow was the slow component a (SCa) wave they found to move at a rate of about 1 mm/day (Chapter 11). When analyzed

---

[38] (Weiss, 1972a), (Weiss, 1972b), and (Biondi, Levy, and Weiss, 1972).
[39] (Heslop, 1975).   [40] (Samson, 1976).   [41] (Ochs, 1965a), and (Ochs, 1966).
[42] (Williams and Hall, 1970), and (Williams and Hall, 1971).
[43] (Murray and Herrmann, 1968).   [44] (Hoffman and Lasek, 1975).

by sodium dodecyl sulfate-polyacrylamide gel electrophoresis (SDS-PAGE), the wave was found to consist predominantly of microtubule and neurofilament proteins (Chapter 11). The hypothesis advanced was that these polymeric proteins are assembled in the cell bodies before being moved out in the fibers, "irresistibly" in a "coherent" fashion as an interconnected matrix, in what they termed the "structural hypothesis."[45] The motor for the movement was left somewhat tentative with the actin filaments associated with the axonal membrane suggested as being involved in some fashion. Because the hypothesis was inferred from the data given using SDS-PAGE, which only shows the presence of individual labeled proteins, the supposition that cytoskeletal organelles are transported in polymeric form requires substantiation.

Some features of the theory are difficult to support. One such is the requirement that the microtubule and neurofilament polymers need to be degraded in the nerve terminals, the degradation proposed to be due to the presence of a Ca^{2+}-activated protease. However, the Ca^{2+} level measured in nerve endings is in submicromolar concentrations, too low to activate Ca^{2+} protease.[46] Microtubules have also been shown to be present in nerve terminals, terminating either in the inner membrane of the terminals[47] or in the form of loops,[48] evidence that protease activity is not sufficient to dissemble them there. A further difficulty with the theory is that the cytoskeleton proteins must endure for months or a year or more in long nerve fibers – a requirement at variance with evidence that neuronal proteins generally have short half-lives of days or weeks.[49]

Another difficulty with the theory is that the rate of 1 mm/day given for the SCa movement of neurofilaments and microtubules falls short of the rate of regeneration of nerve fibers that has been experimentally determined to be 3.5–4.5 mm/day (Chapter 10).[50] This rate of fiber growth would soon outstrip the availability of microtubules and neurofilaments to supply the regenerating nerve fibers. This problem could be resolved if there were an increase in the rate of transport of the cytoskeleton during regeneration, a possibility Hoffman and Lasek suggested.[51] However, although there was an increase in the amount of tubulin produced during regeneration, there was no change in the rate of SCa in the regenerating neurites,[52] and the slow component b (SCb) rate even fell slightly.[53]

Another serious difficulty was that the velocity of the SCa wave was found to decrease with distance, falling to as little as half its original rate at the

[45] (Lasek and Hoffman, 1976) and (Lasek and Black, 1988).
[46] (Baker, Knight, and Whitaker, 1980).
[47] (Gray, 1978).   [48] (Chan and Bunt, 1978).   [49] (Lajtha, 1964).
[50] (Gutmann et al., 1942), and (Griffin, Drachman, and Price, 1976).
[51] (Hoffman and Lasek, 1980).   [52] (McQuarrie, Brady, and Lasek, 1980).
[53] (Hoffman, Griffin, and Price, 1981).

farther reaches of the nerve.[54] Such slowing of the rate would lead to a pile-up of organelles behind the advancing polymers, much as a slowing of cars at some point on a busy freeway would cause congestion of the traffic behind it. Nixon found marked variations in the content of neurofilaments along the length of the mammalian fibers, a finding incompatible with the concept of their uniform continual outflow in the fibers. After pulse labeling, a 4-fold greater amount of the labeled material was found to remain in the fibers even four months after the front of the slow-moving wave had passed by.[55] The label left behind the advancing wave indicated that neurofilament subunits had become incorporated into the stationary organelle. Similar patterns seen in rat peripheral nerves, guinea pig optic nerves, and mouse optic nerves have given further support to the concept that the neurofilaments are stationary in the fibers.[56]

c. Slow transport of cytoskeleton subunits turning over in a stationary cytoskeleton

An alternate to the structural hypothesis holds that the slow-moving waves of labeled cytoskeletal proteins represent their subunits moving down within the soluble axoplasm to turn over in their stationary organelles.[57] Bamburg showed that the assembly of microtubules occurs in the growth cones of the neuron-like PC12 cells and cultured dorsal root ganglia.[58] This was demonstrated by applying colcemid, an agent that interferes with microtubule assembly, alternatively to the cell body and to the growth cone. At the cell body, the agent was ineffective, whereas growth was readily blocked when applied to the growth cone. Examination of the growth cones with AVEC-DIC microscopy showed a complex shortening and elongating of microtubules in a dynamic process of assembly and disassembly going on in them.[59] Growth entails a dominance of assembly over disassembly. As the microtubules are assembled at the growth cones, their extension in the neurite growing behind them allows more tubulin to be brought forward over them to the growth cones for further assembly into microtubules, and continued growth of the neurite.

Whereas the cytoskeleton laid down behind the growth cones remains stationary in the fiber, it is not static. As is the case for other proteins in the body, the microtubule and neurofilament proteins have a short half-life,[60] and they require new subunit proteins to turn over in them. This was

---

[54] (Watson et al., 1989).   [55] (Nixon, 1998).   [56] (Nixon and Logvinenko, 1985).
[57] (Ochs, 1975a), (Ochs, 1982), and (Ochs, 1999). See Heidemann for a recent review of the controversy (Heidemann, 1996).
[58] (Bamburg, Bray, and Chapman, 1986), and (Bamburg, 1988).
[59] (Sammak, Gorbsky, and Borisy, 1987), and (Sammak and Borisy, 1988).
[60] Even hours in neurites (Heidemann, 1996).

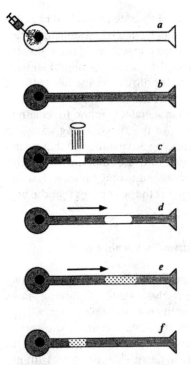

12.9. Schematic representation of fluorescent labeling to determine if microtubules move slowly down within the fibers at the SCa rate, or are stationary. (**a**) Fluorescent-labeled tubulin is injected into the cell body of a neuron in tissue culture. (**b**) After an interval, labeled tubulin is incorporated into the microtubules throughout the length of the neurite. (**c**) A narrow region of the neurite is then bleached using a laser beam, and the fate of the bleached area is followed over a period of hours. Three possible outcomes are shown. (**d**) The microtubules are translocated distally in the axon (arrow pointing in the direction to the right) shown by the movement of the bleached region distally. (**e**) The microtubules are both translocated and exchanges its bleached tubulin subunits for fluorescent tubulin present in the axoplasm, the spot moving down the neurite and gradually recovering its fluorescence (shown by stippling). (**f**) The microtubules remain stationary in the neurite, with the bleached spot gradually recovering its fluorescence as the bleached tubulin sub-units are exchanged for fluorescent ones. The last case represents what is experimentally found. From (Hollenbeck, 1990), Figure 1.

shown in cultured neurites. After injection of their cell bodies with tubulin tagged with a fluorescent dye, the tagged tubulin was transported out into the neurite and incorporated into the microtubules. Then, low-level laser illumination of a narrow region of the neurite was use to bleach the fluorescently tagged tubulin that had been taken up by the microtubules.[61] If the microtubules were moving, the bleached region would be seen to

[61] (Lim, Sammak, and Borisy, 1989), (Lim et al., 1990), (Okabe and Hirokawa, 1990), and (Hirokawa and Okabe, 1992).

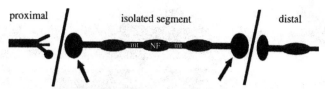

12.10. Diagram showing slow retrograde transport in the nerve fibers of a genetically altered mouse, C57BL/6/Ola. Sciatic nerves were transected (oblique lines) at two levels to create an isolated segment of nerve. After seven days, giant swellings containing neurofilaments were found at the proximal and distal ends of the transected segment (arrows). In the midportion of the isolated segment, bead-like swellings contained concentrations of neurofilaments (**NF**) alternating with narrow regions containing high concentrations of microtubules (**mt**). Large swellings were also present in the fibers below the distal stump and above the isolated region. From (Glass and Griffin, 1991), Figure 1.

move down the neurite at the SCa rate (Figure 12.9)[62]. The result was that the bleached region did not move at all. With time, the bleached region regained its fluorescence as soluble fluorescent-tagged tubulin in the axoplasm replaced the bleached tubulins in the microtubules. The bleaching was carried out at a low strength so as not to damage the microtubules. That they remained functional was shown by the transport of subunits through the bleached region to the growth cones, where they were assembled into microtubules allowing for a continued growth of the neurites.

Similar bleaching experiments were carried out with fluorescent-labeled neurofilament protein forming the neurofilaments.[63] After injection of the cell body with fluorescent-labeled protein – and these carried down into the neurite to be incorporated in the neurofilaments – a narrow region of the neurite was bleached. Again, as with the microtubules, the bleached region did not move, showing that neurofilaments also remain stationary in the fibers.

A key tenet of the structural hypothesis, namely that only an outward movement of the cytoskeleton can take place, was negated by the evidence given by Glass and Griffin showing transport of cytoskeleton proteins in both the anterograde and retrograde directions.[64] This was seen in the nerves of the C57BL/6/Ola mouse (Ola), a genetic strain in which Wallerian degeneration is much delayed,[65] thus allowing for the observation of slow transport over a period of weeks instead of a few days after nerve transection (Chapter 9). The nerves in these animals were transected at two sites to create a "double axotomy." Within the isolated segment, swellings were found to develop just above the distal transection and additionally just below the proximal that were filled with neurofilaments (Figure 12.10).

---

[62] (Hollenbeck, 1989), and (Hollenbeck, 1990).   [63] (Loomis and Goldman, 1999).
[64] (Glass and Griffin, 1991), and (Glass and Griffin, 1994).
[65] (Lunn et al., 1989), and (Brown et al., 1991).

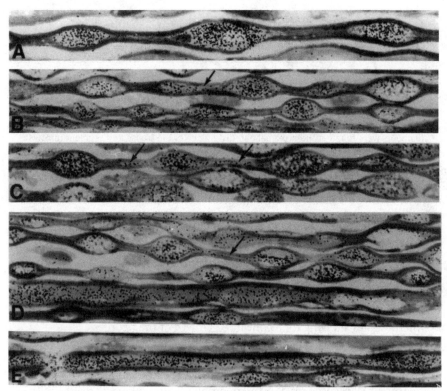

12.11. Partitioning the cytoskeletal organelles in the axons from the soluble components shows that the cytoskeleton is stationary in the fibers, whereas its subunits are slowly transported. Cat L7 ganglia were injected with the labeled precursor [^{3}H]leucine and dorsal roots taken after times of outflow at 7, 14, 21 and 25 days to encompass the slow downflow rates of SCa and SCb. The roots were then stretched to bead them, quick-frozen and freeze-substituted, and sections taken from them prepared for autoradiography. On photographic development, the grains indicating labeled proteins were found located over the expansions of the beaded fibers (**A-C**) and not over the constrictions (arrows) where the cytoskeleton is compacted, as was seen in nerves taken after seven days of downflow. The constricted regions would be heavily labeled if the cytoskeleton were moving down within the fibers at the SCa or SCb slow rates. The same location of grains was seen in the expansions rather than in the constrictions in fibers after allowing 7 days of downflow (a similar result was seen after 14, 21 and 25 days of downflow). After 25 days, a few of the fibers did show a heavier concentration of grains over the constrictions, indicating that some turnover had occurred in them (see text). In unbeaded fibers, grain distribution was uniform (**D, E**). From (Ochs, Jersild, and Li, 1989), Figure 3.

The swellings began three to four days after transection and continued to enlarge until the fourteenth day. These swellings were not simply the result of a translocation of formed neurofilaments from the midportion of the nerve segment. The midportions contained high concentrations of microtubules in the expanded regions and an increased content of neurofilaments in the constrictions. After injection of labeled precursor into the cell bodies and allowing for an outflow of labeled proteins, an isolated segment was created and an increase of labeled cytoskeletal proteins was later found to have accumulated at both its ends.[66] It is likely that cytoskeletal subunits moved into the ends could assemble there into polymeric forms. AVEC-DIC microscopy showed granules shuttling in both in the anterograde and retrograde directions along microtubules.[67] Such movement was seen eight days after isolation of the nerve segment, indicating that the microtubules were functional and that they could carry out the movement of subunits to the ends of the segment.

The question remaining is whether turnover in stationary cytoskeletal organelles occurs in the nerves of mature mammals as it does in cultured neurites and the nerve fibers of the Ola genetic strain of mice. Evidence for movement of labeled subunits of cytoskeletal was gained in nerve fibers of the adult cat by the use of beading and autoradiography.[68] The precursor [^{3}H]leucine was injected into their L7 dorsal root ganglia and sufficient time allowed for the labeled proteins to move out into the dorsal roots at the SCa and SCb rates of transport. After one to four weeks, the outflow labeled proteins had passed fully into the 3–4 cm lengths of the dorsal roots. The roots were removed, stretched to bead their fibers, quick-frozen, and freeze-substituted. Longitudinal sections prepared for autoradiography showed, after exposure and development, a great concentration of silver grains representing radioactivity over the expanded regions of the beads with very few grains over the constrictions where the cytoskeletal was densely compacted (Figure 12.11). If the labeled proteins had been carried down from the cell bodies incorporated into the microtubules and neurofilaments as conceived of by the structural hypothesis, the constrictions would have been heavily labeled. The heavy labeling seen in the expanded regions is rather what would be expected of the cytoskeletal proteins being present in soluble form and able to move with fluid and other soluble components when expressed from the constrictions into the expansions of the beaded fibers (Chapter 8). The same picture was seen in dorsal root fibers that had been taken 7 through 25 days after injection, indicating that few fibers had incorporated the labeled proteins, except for a small proportion of the fibers that did show an increased density of silver grains over the constrictions after 25 days.[69] The turnover rate is thus apparently very much slower in the nerve fibers of the adult mammal than in neurites in tissue culture.

---

[66] (Glass and Griffin, 1994).
[67] (Smith and Bisby, 1993).   [68] (Ochs and Jersild, and Li, 1989).
[69] Ibid., Figure 5D.

THE UNITARY HYPOTHESIS: A SINGLE MECHANISM OF TRANSPORT FOR SLOW, INTERMEDIATE, AND FAST RATES

In discussions on transport mechanisms, usually only two outflow rates were taken into account: fast and slow. As noted in Chapter 11, intermediate rates have also been found. A theory of transport should account for all these various rates of transport. So far, only the mechanism for fast transport has won general acceptance. Slow transport and intermediate rates seemed to suggest the need for other mechanisms. A model that could account for all the rates seen on the basis of one mechanism had earlier been advanced, the "unitary hypothesis." In it, the different species transported – soluble proteins, vesicles, and larger formed organelles like the mitochondria – were visualized as binding to "transport filaments," which are moved down along the microtubules (Figure 12.3).[70] The different rates of transport were ascribed to differing kinetics of drop-off from the carriers. Those components that drop off quickly, remain inert for a time, reattach for short periods and do this repeatedly, such as the mitochondria, tubulin and neurofilament subunits, appear to be slow-transported, whereas those that remain tightly bound, such as neurotransmitter–containing vesicles, are carried down to the terminals show the fastest rate of transport. Those with intermediate degrees of binding and drop-off give rise to the various intermediate rates of transport that have been reported.

With the demonstration of kinesin and dynein motors for fast transport, the unitary hypothesis was modified. The various components transported are now considered to be carried as cargo by motor molecules, attaching to their tails and with different degrees of binding and kinetics of drop-off to give rise to the different rates of transport seen. Kinesin has been indicated to have raft-like regions at the tail, and a small species such as tubulin or neurofilament subunits or oligomeres could be bound to it directly or indirectly through an intermediate binding component, along with the binding of vesicles or larger particles.[71] Superfamilies of kinesin with shorter and longer chains have been found in different cells, and they could have different degrees of binding of cargo to their tails and differing kinetics of drop-off to account for the different rates of transport.[72] Dynein also belongs to a superfamily whose members could carry a variety of cargoes having different drop-off kinetics to account for different rates of transport in the retrograde direction. The slow transport of mitochondria serves as an instructive example of how drop-off can account for the slow rate of transport. Using AVEC-DIC microscopy, mitochondria was seen to move at a fast rate in both the anterograde and in the retrograde direction, though only for short times. Most of the time, they are motionless so that their

[70] (Ochs, 1971), (Ochs, 1972a), and (Ochs, 1982). See also Chapter 11.
[71] (Goldstein and Philip, 1999).   [72] (Hirokawa, 1998).

overall net rate is that of slow transport. Such an attachment and drop-off of tagged small segments of neurofilament was also seen using AVEC-DIC microscopy.[73] The segments moved at a fast rate, though for only a short time before they became detached to remain motionless, giving a slow net rate of transport.

Further insight into the mechanisms of binding and kinetics of drop-off will no doubt lead to a better understanding of the alterations of transport seen in neuropathies and neurotoxicities. What is known of the mechanism of axoplasmic transport gives a rationale for the action as described in the next chapter.

---

[73] (Koehnle and Brown, 1999).

13

ACTIONS OF NEUROTOXINS AND NEUROPATHIC CHANGES RELATED TO TRANSPORT

A number of agents used in the analysis of axoplasmic transport were noted in the preceding chapters. In this chapter, they are set out in systematic fashion on the basis of what is known of its mechanism. In addition, some neuropathies that appear to be accounted for on the basis of an altered transport will be noted. More extensive accounts of the agents used and neuropathies referred to may be found in general works on these subjects and special volumes.[1]

BLOCK OF SYNTHESIS IN CELL BODIES CAUSING FAILURE OF TRANSPORT

The paradigm of an interference of transport leading to pathological changes in the fiber is the Wallerian degeneration seen in the distal amputated nerve after transection (Chapter 9). It results from the loss of substances needed by the fiber that are continually being supplied to it by axonal transport. An interference with synthesis by the cell bodies similarly results in Wallerian degeneration. This is seen when protein synthesis is blocked by *puromycin* or *cycloheximide*.[2] When either of these agents were injected into the dorsal root ganglia shortly before that of a labeled amino acid precursor, or even at the same time, the outflow of labeled proteins was almost completely blocked (Figure 13.1).[3] Puromycin acts by blocking the translation of messenger RNA (mRNA) to protein. It does so because of the similarity of its molecular structure to transfer RNA. Acting on the ribosomes forming the protein chains it causes the premature release of incomplete chains which then are not carried down into the fibers. Cycloheximide

---

[1] (Sunderland, 1978), (Dyck et al., 1993), (Siegel et al., 1999), (Spencer, Schaumburg, and Ludolph, 2000), and (Herkin and Hucho, 1992).

[2] (Ochs, Johnson, and Ng, 1967), and (Nichols, Smith, and Snyder, 1982).

[3] (Ochs, Sabri, and Ranish, 1970).

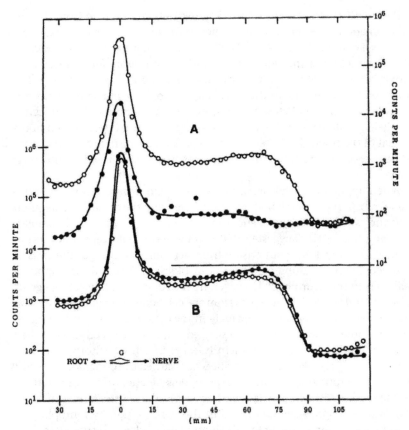

13.1. Block of protein synthesis by cycloheximide. **(A)** Cycloheximide was injected into the left L7 dorsal root ganglion (●) and Ringer's solution as a control into the right (O). Twenty minutes later, the precursor [^{3}H]leucine was injected into each ganglion **(G)**. Nerve taken four hours after [^{3}H]leucine injection shows a block of protein synthesis and outflow produced by cyclohexamide. **(B)** Cycloheximide was injected into the right L7 ganglion (O) and Ringer's solution into the left (●) thirty minutes after [^{3}H]leucine was injected into each ganglion. Little effect of the later injection of cycloheximide on synthesis and outflow was seen indicating that the rapid synthesis of the precursor was over within thirty minutes or less. From (Ochs, Sabri, and Ranish, 1970), Figure 5.

blocks the peptide formation that is directed by mRNA on the ribosomes.[4] The rapidity of their actions indicates that synthesis of labeled amino acids takes place within a short period of time. When they are injected only ten to twenty minutes after injecting a labeled amino acid precursor, the ouflow of labeled proteins was normal, showing that protein synthesis has been completed by that time.

[4] (Alberts et al., 1994).

Doxorubicin (adriomycin) injected systemically is taken up from the circulation and passes into the cell bodies of the dorsal root ganglion, where it forms a complex with DNA.[5] The block of DNA it produces causes complete destruction of the cells.[6] Degenerative changes are found first in the proximal regions of the nerve fibers and then more distally, a "dying forward" pattern of action. In the early stages of toxicity, transport might still have its normal rate of close to 410 mm/day.[7] An increased delay of egress of labeled proteins from the cell bodies, from a normal time of 0.08 hours to that of 0.4 hours, indicates an action on a mechanism proposed to control the egress of synthesized components from the cell bodies into their fibers.[8]

When applied to distal nerve fibers, the toxin *ricin* is carried by retrograde transport to the cell bodies, where it destroys the dorsal root ganglion cells and causes Wallerian degeneration – a process termed "suicide axoplasmic transport."[9] This suggested that it could have a clinical application to control the intractable pain arising from neuromas growing at the ends of transected nerve in amputated limbs – by blocking the continual supply of materials to the neuroma.[10] However, ricin and the similarly acting agents *modeccin* and *volkensin*[11] may pass from the cell bodies into axons synapsing on them and then spread more widely in the central nervous system (CNS) to afflict other neurons. A more restricted action on the cells may be had by use of other agents. Doxorubicin injected into the territory of sensory nerve terminals is taken up by the fibers and carried by retrograde transport to the dorsal root ganglia, where it causes degeneration of the cell bodies[12] ; some success with its use has been reported in the control of pain.[13] *Vinca* alkaloids and *formyl-leurosin* carried in through the skin by iontophoresis are taken up by sensory fiber terminals and retrogradely transported to their cell bodies also causes them to degenerate.[14]

Herpes simplex virus is carried by retrograde transport to the CNS cell bodies (Chapter 8), where it acts in the nucleus.[15] It remains there and when replicated produces the disease. However, a fragment of the virus has been isolated, which retains its ability to be carried by retrograde transport

[5] (Mendell and Sahenk, 1980).   [6] (Cho, Schaumburg, and Spencer, 1977).
[7] (Griffin et al., 1977).
[8] Evidence for a gating mechanism in the cell body had been given by Hammerschlag and colleagues (Hammerschlag and Stone, 1986), and Hammerschlag and Brady, 1989).
[9] (Wiley, Blessing, and Reis, 1982), and (Yamamoto, Iwasaki, and Konno, 1984).
[10] (Nennesmo and Kristensson, 1986), and (Brandner, Buncke, and Campagna-Pinto, 1989).
[11] (Wiley and Stirpe, 1988).   [12] (Bigotte and Olsson, 1987), and (Kato et al., 1988).
[13] (Smith et al., 1989).   [14] (Hermier, Forgez, and Chapman, 1985).
[15] (Kristensson, Lycke, and Sjöstrand, 1971), and see (Goodpasture, 1925).

but without its virility. It has been used to carry components bound to it into the cells, acting as a vector, a "delivery system."[16] Agents bound to the vector can by this means be carried into cell bodies to effect genetic changes. This technique is an important tool for the analysis of the transport mechanism and other cell functions, as well as serving for the selective delivery of therapeutic agents.

RELATION OF MEMBRANE ACTIVITY TO TRANSPORT

The possibility that membrane activity could have an effect on axonal transport was assessed in nerves in which transport of labeled proteins was examined in vitro. Depolarization brought about by K^+ had little effect on transport, and when action potentials are completely blocked with *tetrodotoxin*, which acts by obstructing the Na^+ channels; no effect at all was seen on either anterograde axoplasmic transport[17] or on retrograde transport.[18] When excitability was blocked by the local anesthetic *procaine*, to the point where excitation of action potentials failed, fast axoplasmic transport also continued on as usual.[19] However, at higher levels, it blocked transport, apparently by an action on the transport mechanism.[20]

Maximal action potentials delivered at the maximum repetition rates that would be met with physiologically had no effect on transport. Even stimulating at the very high rate of 350 pps carried out for four hours resulted in only a 15 percent decrease in the rate of fast axoplasmic transport.[21] The decrease was likely the result of augmented entry of Na^+ into the axons. This was indicated by the effect of the neurotoxin *batrachotoxin* (BTX), which blocks excitability by keeping the Na^+ channels in a constantly open state.[22] The hypothesis that the blocking effect can be ascribed to Na^+ seemed to be obviated by the finding that transport in sheathed nerves exposed in vitro to BTX was blocked even when Na^+ was deleted from the medium.[23] This effect, however, was traced to the presence of Na^+ that is normally present within the endoneural space of sheathed nerves, the toxin allowing Na^+ to enter the fibers from the endoneural compartment. In the desheathed nerve preparation, BTX in a Na^+-free medium had no effect on transport, but it was blocked with Na^+ present in the medium along with BTX.[24] A block by BTX of particulate movement in neuroblastoma cells had also been seen when Na^+ was present in the medium, but not with BTX in a Na^+-free medium.[25]

---

[16] (Geller et al., 1990).
[17] (Ochs and Hollingsworth, 1971), and (Lavoie, Collier, and Tenenhouse, 1976).
[18] (Boegman and Riopelle, 1980). [19] (Ochs et al., 1971). [20] (Byers et al., 1973).
[21] (Worth and Ochs, 1976). [22] (Narahashi, Albuquerque, and Deguchi, 1971).
[23] (Ochs and Worth, 1975). [24] (Worth and Ochs, 1982).
[25] (Forman and Shain, 1981).

CALCIUM AND CALMODULIN-BLOCKING AGENTS

A block of transport by an increase of Na^+ within the axon could come about by the release of Ca^{2+} from its sequestration in the mitochondria and other storage sites within the axoplasm. Exposure of isolated mitochondria to levels of Na^+ as low as 10 mM caused a release of Ca^{2+} from them.[26] However, transport in nerve in vitro appeared to be little affected by either high elevations of Ca^{2+} in the medium or deleted by placing nerves in a Ca^{2+}-free medium. This was traced to the low permeability of the perineural sheaths to Ca^{2+}. After slitting the sheath, or removing it altogether to produce a desheathed preparation, transport was found to require Ca^{2+}. With Ca^{2+} removed from the medium, transport declined and was completely blocked within about two hours (Figure 13.2),[27] whereas with Ca^{2+} present transport showed its characteristic pattern (Figure 13.3).[28] The dependence of transport on Ca^{2+} was further shown using the Ca^{2+} ionophore *A23187* to permeabilize the nerve sheath and the axolemma of the fibers to Ca^{2+}. In nerves placed in a Ca^{2+}-free medium with the ionophore present, transport was blocked.[29] So was transport in nerves in vitro treated with *Triton X-100*, which makes the perineural sheath and axolemma permeable to Ca^{2+}, when placed in a Ca^{2+}-free medium, and restored when placed back into a Ca^{2+}-containing medium.[30] In desheathed nerves from which Ca^{2+} was depleted, strontium – which has chemical properties close to calcium – served to restore transport.[31] In some preparations, it appears that Mg^{2+} – which is also somewhat close to it in the periodic table – can replace Ca^{2+}. However, LaVoie et al.[32] and Kanje et al.[33] have shown Ca^{2+} to be as essential for transport in frog nerve as it is for mammalian nerve.

The relative insensitivity shown by desheathed nerve to high levels of Ca^{2+} is accounted for by the low permeability of the axonal membrane to Ca^{2+}. Only with very high concentrations of 50–60 mM present in the medium did it produce a complete block of transport in desheathed nerves in vitro, and then only after 2.5 hours (Figure 13.4).[34] A similar block was seen with *A23187*, which is effective at lower external concentrations of Ca^2 because it permeabilizes the axon to Ca^{2+}.[35] In addition to the low permeability of the axon membrane to Ca^{2+}, additional protection is effected by the sequestering and binding of an excess of the ion entering the axon. This was seen in the axons of nerve fibers that had been exposed to such high concentrations of Ca^{2+} using *pyroantimonate* to reveal the ion in electron

---

[26] (Carafoli and Crompton, 1976).   [27] (Ochs, Worth, and Chan, 1977).
[28] (Ochs et al., 1977), and (Chan, Ochs, and Worth, 1980a).
[29] (Lavoie, Bolen, and Hammerschlag, 1979).
[30] (Kanje, Edström, and Ekström, 1982).   [31] (Ochs et al., 1986).
[32] (Lavoie et al., 1979).   [33] (Kanje, Edström, and Hanson, 1981).
[34] (Ochs et al., 1977), (Chan et al., 1980a), and (Chan and Iqbal, 1986).
[35] (Esquerro, Garcia, and Sanchez-Garcia, 1980), and (Lees, 1985).

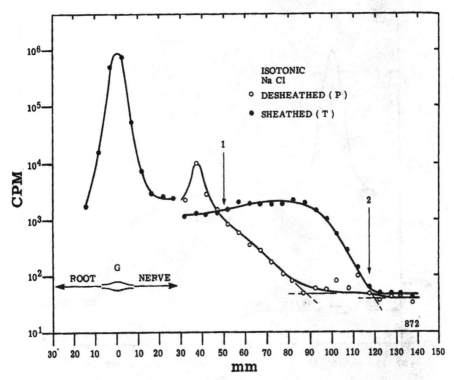

13.2. Block of transport in a Ca^{2+}-free medium shown in the desheathed nerve preparation. After injection of the L7 dorsal root ganglion (**G**) with [^{3}H]leucine and allowing downflow for two hours in the animal for labeled proteins to enter the fibers, the sciatic nerve was removed, the peroneal branch was desheathed, and the tibial branch was left sheathed. The preparation was placed in an in vitro medium containing isotonic NaCl for a further period of transport. Above the site where the peroneal nerve was desheathed (arrow **1**), an adventitious peak of dammed activity is seen. The essential finding was that transport shows a decline toward baseline levels in the Ca^{2+}-free medium with a complete block within three hours. The sheathed tibial nerve shows a typical pattern of fast axoplasmic transport with its front moving to the expected distance (arrow **2**) because of the Ca^{2+} retained within the endoneural compartment enclosed by the sheath of the nerve. **CPM**, counts per minutes; **P**, peroneal; **T**, tibial. From (Ochs, Worth, and Chan, 1977), Figure 2.

micrographs. It was found located at the nodes where Ca^{2+}-binding sites are present, in the endoplasmic reticulum, within the mitochondria, and as well in particulate Ca^{2+}-binding sites scattered throughout the axoplasm.[36] These sequestering sites act to keep the level of Ca^{2+} in the fibers at sub-μM levels in the face of increases in concentration of the ion until, with time,

---

[36] (Chan, Ochs, and Jersild, 1979), (Jersild, and Ochs, 1982), (Chan, Ochs, and Jersild, 1984), and (Ochs and Jersild, 1984).

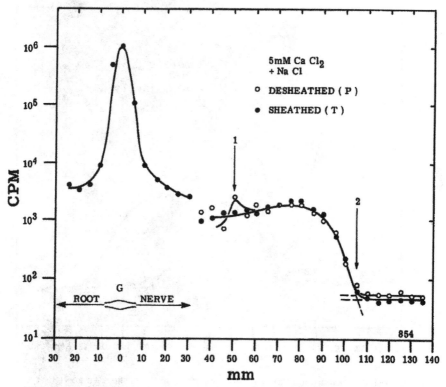

13.3. Transport in the NaCl medium used in Figure 13.2, with the addition of Ca^{2+} shows normal transport. It has the same pattern as that seen in the sheathed tibial nerve (arrow 2). A small degree of adventitious damming is seen at the site upper site, where desheathing of the nerve was initiated (arrow 1). **CPM**, counts per minute; **P**, peroneal; **T**, tibial. From (Ochs et al., 1977), Figure 1.

these regulatory mechanisms are overcome and a block of transport ensues. When nerves treated with such high levels of Ca^{2+} to the point where block would occur were transferred back into a normal Ringer's solution, transport recovered in the course of several hours, during which time the ion was slowly washed from its various storage and binding sites in the axon (Figure 13.5). But, if exposure to such high concentrations of Ca^{2+} was too prolonged, the control mechanisms regulating Ca^{2+} in the fibers are overcome and a Ca^{2+}-dependent proteolytic enzyme is activated that disrupts the microtubules, the axons then undergo degeneration.[37]

The regulatory mechanisms in the axon act to keep Ca^{2+} at a sub-μM level optimal for binding to *calmodulin*, which in turn regulates a number of key

---

[37] (Schlaepfer, 1974).

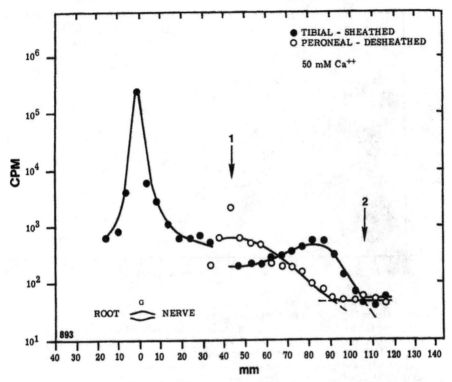

13.4. High levels of Ca^{2+} causes a block transport. Using the same preparation as that in Figure 13.3, transport of labeled protein in the desheathed peroneal nerve (O) exposed to the high concentration of the ion of 50 mM (arrow 1) declined and then was blocked after 2.5 hours. Little effect on transport in the sheathed tibial nerve (●) was seen with the front of outflow (arrow 2) moved to the distance expected of fast transport. The lack of effect of the high level of the ion is due to the relatively low permeability of the sheath to Ca^{2+}. CPM, counts per minute. From (Chan, Ochs, and Worth, 1980), Figure 13.

enzymes, including some critically affecting transport.[38] The intermediating role of calmodulin in transport was indicated by the block of transport seen with *trifluoperazine* (TFP), an anticalmodulin agent that binds to calmodulin preventing its action.[39] It readily blocks fast anterograde transport (Figure 13.6) and retrograde transport.[40] *Amitriptyline, desipramine,* and *imipramine* are equipotent inhibitors of calmodulin effective in a concentration of 0.2 mM.[41] The higher level of 1–3 mM required for TFP to fully block transport in cat nerves in vitro was traced by Iqbal to its lower axon permeability. The anticalmodulins are relatively specific in their action. They

[38] (Iqbal and Ochs, 1980), (Ochs and Iqbal, 1983), (Iqbal, 1986), and (Ekström, 1987).
[39] (Ekström, Kanje, and McLean, 1987), (Lavoie and Tiberi, 1986), and (Iqbal, 1986).
[40] (Tiberi and Lavoie, 1985).
[41] (Lavoie and Tiberi, 1986), (Iqbal, 1986), and (Ekström, 1987).

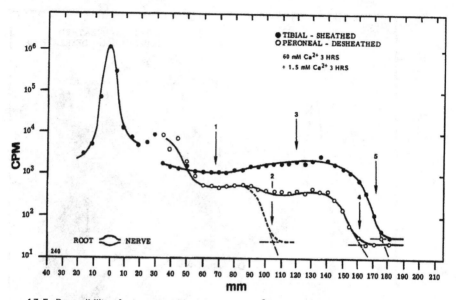

13.5. Reversibility of a transport block by a high-Ca^{2+} medium. In the nerve preparation used in Figure 13.4, the desheathed nerve was exposed to 60 mM Ca^{2+} at arrow **1** and transport (dashed line) would have been carried to the point indicated by arrow **2**. At arrow **3**, the preparation was placed back into a 1.5 mM Ca^{2+} medium, and recovery was indicated by the further transport in the desheathed nerve to the distance shown by arrow **4**. Arrow **5** shows the expected distance of a normal transport seen in the sheathed tibial nerve. **CPM**, counts per minute. From (Chan et al., 1980a), Figure 14.

do not block transport by disassembling microtubules,[42] nor by inhibiting metabolic processes leading to the production of ATP.

AGENTS BLOCKING METABOLISM SUPPLYING ENERGY FOR TRANSPORT

The dependence of transport on the supply of ATP was shown by its failure when oxidative metabolism was blocked (Chapter 11). Block occurs quickly, within 20–30 minutes, when nerves are made anoxic with nitrogen or treated with *cyanide, azide,* or *2,4-dinitrophenol* – agents blocking the production of ATP. When glycolysis was blocked by iodoacetic acid, which inhibits glyceraldehyde phosphate dehydrogenase (GPDH), the block of transport takes longer, 1.5–2 hours. Block of the tricarboxylic cycle, by fluoroacetate causes transport to fail in about an hour. These longer survival times suggested the possibility that the utilization of lipids can continue to

[42] (Lavoie et al., 1986).

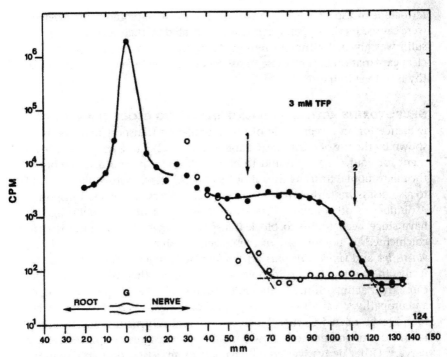

13.6. The effect of the calmodulin-blocking agent trifluoperazine (**TFP**) to interrupt transport. Labeled precursor was injected into the L7 ganglia (G). The nerve in which labeled proteins was initiated was removed from the animal, the peroneal branch (O) was desheathed, and the preparation was placed in a medium containing 3 mM TFP for a further period of in vitro transport. The desheathed branch shows a block of transport (arrow **1**), with little effect on the sheathed tibial branch (●), where the front advanced to the expected distance (arrow **2**), the sheath having a low permeability to the agent. **CPM**, counts per minute. From (Ochs, 1982), Figure 12.11.

supply ATP, at least for a short time, before production fails and ATP falls below the critical level needed to drive the transport mechanism.[43]

One of the distressful consequences seen in human diabetics is the development of a peripheral neuropathy. To examine the possibility that elevated blood glucose levels seen in diabetes could affect transport and cause diabetic neuropathy, *streptozotocin* (STZ) was injected systemically into rats to produce a diabetic model – these animals showing a persistently high level of blood glucose (of 400 mg/dl or more). However, this had little effect on fast anterograde transport.[44] On the other hand, slow transport was decreased in the STZ-treated rats,[45] a defect of turnaround[46] and a decreased

---

[43] (Gaziri and Ochs, 1983).   [44] (Sidenius and Jakobsen, 1987).
[45] (Sidenius, 1982), (Sidenius et al., 1987), and (Takenaka, Inomata, and Horie, 1982).
[46] (Schmidt, Grabau, and Yip, 1986).

retrograde transport of glycoprotein was found.[47] That some of these defects were caused by the hyperglycemia was indicated by their reversal when insulin was given,[48] although retrograde transport remained impaired.[49] A clear explanation of how these transport defects relate to diabetic neuropathy is not yet in hand.

NEUROTOXINS ACTING ON MICROTUBULES TO BLOCK TRANSPORT

Evidence for an integral role of microtubules in transport had first been shown by the use of *colchicine* (Chapters 11 and 12). Long known as a treatment for gout,[50] it was found to block cell division by a disassembly of the microtubules forming the mitotic spindles.[51] It was subsequently found to also disassemble the microtubules in nerve fibers and block transport.[52] A number of other such *mitotic blocking* agents or *tubulin binding* agents, have since been found to block transport at lower concentrations than colchicine.[53] A potent class of these agents is that of the vinca alkaloids, *vincristine* and *vinblastine* plus other vinca derivatives.[54] The vincas are effective in the treatment of neoplastic diseases, but their use is limited because their action on nerve fiber microtubules blocks transport to produce a neuropathy.[55] Variations in the actions of derivatives of the vincas were found to be related to their block of transport. Vincristine, the most potent in this action on nerve,[56] was paralleled by its greater uptake into the nerve.[57] Other derivatives could have less of an action on nerve than on neoplastic cells.

Mitotic blocking agents act by binding to tubulin in equilibrium with the microtubules, causing a shift of tubulin from the microtubules.[58] Whereas the block of axoplasmic transport by the tubulin-binding agents was generally thought to be related to their effect on the microtubules,[59] some contrary reports had appeared. One indicated a block of transport without an apparent decrease of microtubules,[60] another a continuation of transport with few microtubules seen remaining.[61] The latter study was carried out

[47] (Karlsson and Sjöstrand, 1971).   [48] (Jakobsen et al., 1981).
[49] (Schmidt et al., 1986).
[50] Introduced into medicine by Alexander of Tralles in the 6th century (Garrison, 1929), p. 124.
[51] (Dustin, 1984).   [52] (Dahlström, 1968), and (Kreutzberg, 1969).
[53] (Hanson and Edström, 1978), and (Dustin, 1984).   [54] (Gerzon, 1980).
[55] (Weiss, Walker, and Wiernik, 1974).   [56] (Chan, Worth, and Ochs, 1980b).
[57] (Marotta, 1000).
[58] (Inoue and Sato, 1967), (Wilson and Meza, 1973), (Borisy et al., 1975), and (Paulson and McClure, 1975).
[59] (Banks and Till, 1975), and (Dustin, 1984).   [60] (Byers et al., 1973).
[61] (Brady et al., 1980).

on unmyelinated fibers. Their axons have a higher density of microtubules in them and are much more susceptible to the action of tubulin-binding agents than are the larger myelinated fibers in which the great bulk of labeled proteins is transported. To relate a block of transport to their effect on microtubules, it was necessary to examine their effect on the large myelinated fibers.[62] In such studies using the tubulin-binding agents *maytansine*[63] and vinca alkaloid[64] a complete block of transport was seen when the microtubular density was reduced to approximately half its normal value.

The presence of microtubules still remaining when transport was fully blocked was ascribed to the tubulin-binding agents producing a defect of microtubule segments by acting at their ends.[65] The microtubules do not run the length of the nerve fibers. Serial sectioning had shown them to be relatively short, of the order of 1 μm to several hundred microns. In *Caenorhabditis elegans* axons, they are short (1.2–10.7 μm)[66]; in rat dorsal root neurons in tissue culture, they average 108 μm in length[67]; and they have longer lengths in mouse saphenous nerves (306–706 μm).[68] Cross-sections of fibers to determine the loss of microtubular numbers in the treated fibers, compared with controls would thus include those portions of the microtubule segments remaining in the fibers acted on by these agents while disassembly occurs at the ends of their segments, sufficiently so as to cause a block of transport. That such reductions do occur at the ends of the microtubule segments[69] was indicated by observations of particle movement along the microtubules of fibers in vitro using dark-field microscopy.[70] A reduction in the range of particle movement appeared 1–4 hours after exposure to colchicine or vinblastine at intervals, the particles remaining stationary more and more frequently and for progressively longer periods of time until they stopped completely. Observation of particle movement in the neurites of cultured dorsal root ganglia neurons using Nomarski optics microscopy showed a similar process after exposure to colchicine or vinblastine.[71] These agents produced regions of swelling in which transport of particles were at first impeded in their movement, faltered, and then stopped to accumulate at the blocked regions.[72]

---

[62] (Ochs, 1972b). Autoradiography shows greater accummulation of grains over the large fibres (Figure 8).
[63] (Ghetti and Ochs, 1978).   [64] (Chan et al., 1980b), and (Ghetti et al., 1982).
[65] (Paulson and McClure 1975), and (Ochs, 1982).
[66] (Chalfie and Thomson, 1979).   [67] (Bray and Bunge, 1981).
[68] (Tsukita and Ishikawa, 1981).   [69] (Burton, 1987).
[70] (Hammond and Smith, 1977).   [71] (Takenaka, 1986).
[72] (Horie, Takenaka, and Inomata, 1981).

Taxol, a compound extracted from the yew tree (*Taxus brevifolia*), was found to have an opposite action – promoting the assembly of micro-tubules,[73] thus bringing about an increased density of microtubules.[74] Taxol can counter the disassembly of microtubules produced by colchicine, preventing it from blocking transport.[75] The overproduction of microtubules produced by taxol interferes with transport to cause a neuropathy.[76] This stabilizing action of taxol on the microtubules is shared by other agents. *Dimethylsulfoxide*, a polar solvent, also increases the resistance of micro-tubules to disassembly by colchicine,[77] while at higher concentrations it causes a block of transport.[78]

Agents that oxidize the sulfhydryl groups of tubulin reduce their ability to polymerize and form microtubules[79] and block transport.[80] The metallic ions Zn^{2+}, Cd^{2+}, Hg^{2+}, and Cu^{2+} have this action.[81] However, at low concentrations of Cd^{2+} and Zn^{2+}, the amount of labeled materials fast-transported was increased.[82] This could come about by a reduction in the drop-off of fast-transported proteins from the crest, which normally takes place to a small extent all along the length of the fibers (Chapters 11 and 12).[83] The organo-metal compound of mercury, *methyl-Hg*, also acts on sulfhydryls and blocks transport,[84] as do the sulfhydryl blocking compounds *N-ethylmaleimide* (NEM) and *parachloromercuribenzenesulfonic acid* (PCMBS).[85] Lower concentrations of NEM and PCMBS were also seen to increase the amount of labeled protein fast-transported, and, as in the case of Cd^{2+} and Zn^{2+}, this may come about by a decreased drop-off from the crest as it moves down along the length of the fibers.

NEUROTOXIC AGENTS ACTING ON NEUROFILAMENTS TO IMPEDE TRANSPORT

After systemic administration of β,β'-*iminodiproprionitrile* (IDPN) into rats, Chou and Harman found balloon-like swellings in the intraspinal part of motoneuron axons, the swellings containing a greatly increased content of neurofilaments.[86] They interpreted their results on the basis of Weiss' damming hypothesis, namely that all the axonal contents of the fiber flows down within them – the neurotoxin by impeding the flow producing an "axostasis" with the piling up of neurofilaments and the swellings seen.

---

[73] (Schiff, Fant, and Horwitz, 1979), and (Horwitz et al., 1986).
[74] (DeBrabander et al., 1982).   [75] (Horie et al., 1987).
[76] (Röyttä and Raine, 1985).   [77] (Dulak and Crist, 1974).
[78] (Donoso, Illanes, and Samson, 1977).
[79] (Mellon, Rhebun, and Rosenbaum, 1976).   [80] (Edström and Mattsson, 1976).
[81] (Mildvan, 1970).   [82] (Edström and Mattsson, 1975).
[83] (Gross and Beidler, 1975), and (Muñoz-Martínez, Núñez, and Sanderson, 1981).
[84] (Abe, Haga, and Kurokawa, 1975).   [85] (Edström and Mattsson 1976).
[86] (Chou and Hartmann, 1965).

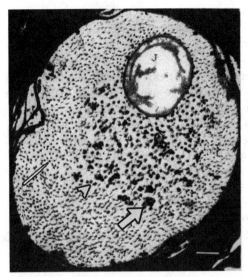

13.7. Cross-section of the axon of a nerve fiber taken from an IDPN-treated rat. Microtubules (arrowhead) and silver grains (arrow) associated with them are seen in the central region of the axon with neurofilaments (pointer) in the outer part of the axon ringing them. The large circular structure (J) is likely a disrupted mitochondrion. Modified from (Papasozomenos et al., 1982), Figure 1.

However, the increased content of neurofilaments in the ballooned region was not accompanied by a pile up of microtubules as would be expected of a stasis of all the axonal contents. Fast transport through the region of increased neurofilaments was not much impeded. A rate of 392 mm/day determined,[87] one comparable with the normal rate, shows that the microtubules passing through the piled up neurofilaments remain functional. In autoradiographs grains, of radioactivity of fast-transported labeled proteins are seen associated with the centrally located microtubules (Figure 13.7).[88] The distal part of the fibers beyond the swellings appears shrunken, with a much reduced component of neurofilaments – these displaced to the periphery of the fibers with the microtubules gathered into the central region.

These findings were taken to indicate a selective effect of IDPN to block the movement of the neurofilaments on the basis of the structural hypothesis (Chapter 12). However, as discussed in that chapter, later evidence indicates that the neurofilaments and microtubules are stationary in the axons with the slow transport of labeled cytoskeletal proteins as soluble subunits. The action of IDPN can best be accounted for on the basis of the unitary

---

[87] (Griffin et al., 1978).
[88] (Griffin et al., 1981), (Papasozomenos, Autilio-Gambetti, and Gambetti, 1981), and (Papasozomenos et al., 1982).

hypothesis (Chapter 12). In this model, slow transport results from an early drop-off of neurofilament subunits from the motor molecules. An increase in the degree of drop-off of the neurofilament subunits would result in a higher concentration of subunits proximally in the motor fibers, where they can assemble to form the greater number of neurofilaments seen. The fewer neurofilament subunits left to be transported distally leads to the decrease of neurofilaments there. That IDPN can have a local action on transport in the fibers was indicated by the studies of Griffin and colleagues. They injected IDPN into the endoneurial compartment of nerve and found neurofilament to increase at the site in the fibers as early as a day later instead of the months required when administered as usual in the animal's drinking water.[89]

IDPN can have a more drastic action when nerves are exposed to the agent in vitro. Not only neurofilaments, but microtubules and actin filaments also become dissaggregated and replaced by an amorphous-appearing material.[90] This finding offered an opportunity to determine if the beading mechanism is dependent on some structural change of the cytoskeleton. In such IDPN-treated fibers, beading was not only elicited, but was even augmented. Ovoids were seen in the absence of cytoskeletal beading constrictions; the membranes were no longer prevented from merging to form them (Chapter 8). This picture was similar to the events seen in Wallerian degeneration (Chapter 9) where, in the first stage, beading appears while the cytoskeleton is still intact; and then when the cytoskeleton becomes degraded, ovoids and spheres are produced. This sequence of events appears in a matter of hours with IDPN rather than days in the case of Wallerian degeneration.

Aluminum is another agent causing a proliferation of neurofilaments in the fibers. This was first seen in cells in the brain and spinal cord after its injection as a phosphate, or applied to the brain surface in a slurry.[91] When systemically injected, it causes an overproduction of neurofilaments in peripheral nerve fibers. This had little effect on fast transport. It retained its characteristic pattern of outflow with a rate of 440 mm/day, one comparable to that of labeled proteins in normal nerve fibers.[92] After its injection into the region of the hypoglossal cell bodies, SDS-PAGE analysis of their nerves showed a great increase of neurofilament proteins present in proximal regions with a diminished amount in the distal portions.[93] This is a distribution similar to that seen with IDPN, and a similar interpretation can be given. Aluminum, by causing a greater drop-off of neurofilament proteins proximally in the fibers, leads to an increased assembly of neurofilaments there, whereas the diminished transport of neurofilament proteins distally

[89] (Griffin et al., 1983).   [90] (Ochs, Pourmand, and Jersild, 1996).
[91] (Klatzo, Wisniewski, and Streicher, 1965).   [92] (Liwnicz et al., 1974).
[93] (Bizzi et al., 1984).

in the nerve with a lesser drop-off of neurofilament proteins there leaves less to assemble into neurofilaments.

AGENTS PRODUCING THE PROGRESSIVE EFFECTS SEEN AS "DYING BACK" AND "DYING FORWARD"

A spatial-temporal development of nerve defects with distinctive morphological features has been seen with a number of agents where defects appear first distally in the nerve and then ascend with time: Or, are first seen proximally and then descend. Accounts given for these spatio-temporal actions are of some metabolic defect or the development of structural changes causing a mechanical impediment to transport. Systemic injection of *acrylamide* produces an accumulation of neurofilaments and smooth vesicular membrane structures that appear early on in the distal regions of nerve fibers and then later found at successively more proximal sites,[94] the pattern characteristic of a "dying-back neuropathy."[95] Little change in the rate of anterograde fast transport was seen in those nerves,[96] while the rate of slow transport was reduced.[97] Retrograde transport was blocked for a while,[98] then returning to normal within 48 hours following after a single injection.[99] A decreased capacity for uptake by the terminals and retrograde transport was seen using ^{125}I-labeled tetanus toxin.[100] The hypothesis given to account for the dying-back action was that it is caused by a decreased glycolysis with a diminished energy supply in the fibers. This proposed action of acrylamide was based on studies of acrylamide on the sulfhydryls of GPDH in in vitro studies.[101] However, in vivo studies showed the action on nerve to be too weak to affect transport: the levels of ATP and creatine phosphate were insufficiently reduced to the point where transport would be blocked.[102] On the other hand, pyruvate was able to reverse the neurotoxic effect of acrylamide,[103] indicating that some metabolic impairment is involved in its toxicity. A defective supply of some needed component(s) to the distal portions of the fibers or an effect on the transport mechanism was indicated by the greater fall off in the height of the crest with distance of labeled glycoproteins seen with acrylamide.[104]

The hexacarbons – *normal hexane* (*n*-hexane), *methyl n-butyl ketone* (MBK), and *2,5-hexanedione* (2,5-HD) – all produce focal swellings in the fibers and an excess of neurofilaments. These are seen in the larger myelinated fibers, at first distally then spreading proximally, the picture of a dying-back

---

[94] (Asbury and Brown, 1980).   [95] (Cavanagh, 1964).
[96] (Sidenius and Jakobsen, 1983).   [97] (Gold, Griffin, and Price, 1985).
[98] (Miller and Spencer, 1984).
[99] (Papasozomenos, Autilio-Gambetti, and Gambetti, 1982).
[100] (Price et al., 1975).   [101] (Spencer et al., 1979), and (Moretto and Sabri, 1988).
[102] (Brimijoin and Hammond, 1985).   [103] (Sabri et al., 1989).
[104] (Harry et al., 1989).

neuropathy.[105] The swellings, often located at the nodes of Ranvier, contain whorls of neurofilaments surrounding centrally located microtubules[106] Three weeks after an injection of 2,5-HD, fast transport of labeled proteins in motor fibers was impaired with an increased amount of the labeled protein seen in the focal swellings.[107] This was viewed as a hindrance of rapidly transported organelles, such as vesicles, through the neurofilament-filled regions. On the other hand, the slow rate of neurofilament proteins was increased,[108] a similar picture also seen with *carbon disulfide*. The explanation given was that these agents freed the neurofilaments from a neurofilament-microtubule meshwork, thereby allowing the neurofilaments to move faster.[109] This account was based on the concept of a slow movement of neurofilaments in polymeric form. Considering that the neurofilaments are stationary and that the movement of slow component a (SCa) consists of neurofilament subunits rather than as polymers (Chapter 12), an alternate explanation is that these agents decrease the drop-off of the neurofilament subunits proximally, allowing more subunits to be carried distally, giving rise to the apparently faster rate observed.[110] An opposite effect was produced by of an analog of 2,5-HD, 3,4-*dimethyl-2,5-hexanedione*. It impairs the slow transport of neurofilament proteins by as much as 75–90 percent, this causing an enhanced early drop-off of neurofilament subunits.

The hexacarbon MBK produces focal swellings in the fibers with an increased amount of neurofilaments in them.[111] The rate of fast transport was reduced to 283 mm/day, compared with the rate in controls of 417 mm/day. The hypothesis advanced to account for the local swellings and the lower rate seen was that the increased presence of neurofilaments produced by MBK mechanically obstructs fast transport.[112] In support of that hypothesis, nerves were partially ligated and a slowing of the rate of fast transport was indeed found. However, an increase in neurofilaments does not necessarily cause an obstruction. As noted previously, IDPN and aluminum greatly increased the content of neurofilaments in the proximal portions of the fibers, but did not decrease the rate of fast transport. Nor did the marked compaction of microtubules and neurofilaments in the constrictions of beaded fibers[113] change the rate or the pattern of fast-transported labeled proteins passing through them.[114]

A single oral dose of *p-bromophenylacetylurea* (BPAU) given to rats produced the effects characteristic of a dying-back neuropathy. After a delay of

---

[105] (Spencer, Couri, and Schaumburg, 1980).   [106] (Cavanagh, 1985).
[107] (Griffin, Price, and Spencer, 1977).   [108] (Monaco et al., 1989).
[109] (Gambetti et al., 1986).   [110] (Ochs, 1992).   [111] (Mendell et al., 1977).
[112] (Sahenk and Mendell, 1979b).
[113] (Ochs and Jersild, 1987), (Ochs et al., 1994), and see Chapter 8.
[114] (Ochs and Jersild, 1985).

eight days or so, a loss of position sense and hindlimb weakness appeared, which then progressed upward to cause a total paralysis within the next two weeks.[115] Their nerve fibers showed an increased content of tubulo-vesicular membranes before going on to degenerate. As early as two days after administration of the toxicant, before the first signs of weakness appeared, the turnaround of transported material in the terminals was delayed and reduced.[116] This observation suggested a failure of retrograde transport causing the accumulation of the tubulo-vesicular membranes.[117] This may act as a signal causing an altered pattern of synthesis in the cell bodies that then initiates the further progression of the neuropathy. The involvement of cell bodies in the neurotoxicity produced by BPAU was indicated by an increase in the rate of egress of labeled proteins from them.[118]

Zinc pyridinethione (ZPT) gives rise to an accumulation of tubulo-vesicular membranes in the motor nerve terminals, followed by their degeneration of the fibers in the pattern of a dying-back neuropathy.[119] ZPT inhibits the turnaround in the terminals, a process correlated with the presence of swellings and tubulo-vesicular profiles. These changes were proposed to come about by the binding of zinc to the sulfhydryl groups of thiol proteases that were deemed to be necessary for turnaround.[120] The evidence given for this hypothesis was that the application of the protease inhibitors leupeptin and E-64 to the central ends of cut peripheral nerves caused a decrease in the amount of labeled proteins carried by retrograde transport to the cell bodies, with the axons at their cut ends distended with membranous tubules. According to the hypothesis given, the membranous tubules are transient intermediates involved in the process of turnaround, their increased presence indicating a failure of their conversion into the organelles that are normally retrogradely transported. Whether the membranous structures are necessary intermediates for the turnaround as hypothesized, or represent an unrelated change generated by the neurotoxin, cannot be judged simply on the basis of their accumulation. The protease inhibitors could interfere with a number of processes in the terminals. An indication of the complexity of the effects that a protease may produce is the augmentation of neurofilaments seen in the terminals after exposure to leupeptin.[121]

The organophosphate di-isopropylfluorophosphate, when injected systemically in cats, gives rise to a dying-back neuropathy.[122] Varicosities appear in the fibers, which then evolve into focal degenerations creating a "chemical transection."[123] Before that stage occurs, anterograde transport appears to

---

[115] (Cavanagh et al., 1968).   [116] (Jakobsen, Brimijoin, and Sidenius, 1983).
[117] (Oka and Brimijoin, 1992).   [118] (Oka and Brimijoin, 1990).
[119] (Sahenk and Mendell, 1979a).   [120] (Sahenk and Lasek, 1988).
[121] (Roots, 1983).   [122] (Bouldin and Cavanagh, 1979a).
[123] (Bouldin and Cavanagh, 1979b).

be little affected, indicating that a defect of transport is not the primary cause of the structural changes.[124]

The effects of an agent on the nerve may not be direct. The organophosphate *di-n-butyl dichlorvas* injected systemically into hens causes a defect of retrograde transport, as indicated by injecting ^{125}I-labeled tetanus toxin into the gastrocnemius muscles for uptake by motor nerve fiber endings.[125] The defect produced by the organophosphate reached a maximum in seven days, the slow temporal evolution of neuropathy was termed *organophosphate-induced delayed polyneuropathy.* The delayed onset was ascribed to the phosphorylation of a *neurotoxic enzyme* that was considered the agent responsible for the neuropathic changes.[126]

DIFFERENTIAL EFFECT OF A NEUROTOXIN ON SENSORY AND MOTOR FIBERS

A neuropathy in humans and experimental animals resulting from the ingestion of the *tulidora toxin* present in the endocarp of the fruit of *Karowinskia humboldtidiana* appears as a flaccid quadraparesis beginning distally in the hindlimbs before progressively ascending to involve the more proximal musculature.[127] Nerve biopsies taken at an early stage of the neuropathy show a segmental demyelination of the fibers with, subsequently, focal axonal swellings and later a Wallerian degeneration. In tissue-cultured neurons, the toxin was seen to cause stacks of smooth endoplasmic reticula displaced to a marginal position in the fibers with the central region largely occupied by neurofilaments.[128] After a single oral dose of the toxicant given to cats, axoplasmic transport was found to be impaired during the flaccid stage, the motor fibers more so than sensory fibers.[129] Differing sensitivities of motor and sensory neurons to the toxin was shown by injecting labeled precursor into the L7 dorsal root ganglion on one side and into the L7 ventral horn on the other before administering the toxin. On comparing the patterns of fast transport in the two nerves, a depression of transport in the motor fibers was seen at a time when transport remained relatively normal in the sensory fibers. This may be because of a greater sensitivity of the motoneuron cell bodies to the toxicant with a decreased synthesis of some needed component(s) moving into the distal reaches of the long nerve fibers. Alternatively, the transport mechanism in motor fibers may be more susceptible to a direct action of the toxicant on them than the sensory fibers.

[124] (Pleasure, Mishler, and Engel, 1969), and (Bradley and Williams, 1973).
[125] (Moretto et al., 1987).   [126] (Johnson, 1975).
[127] (Weller, Mitchell, and Daves, 1980).   [128] (Heath et al., 1982).
[129] (Muñoz-Martínez, Massieu, and Ochs, 1984).

TRANSPORT DEFECTS IN HUMAN NEUROPATHIES

Transport in the CNS has the same character as that found in peripheral nerve.[130] A defect of transport could be implicated in some diseases of the CNS. The striking increase in the amount of the unusual neurofibrillary tangles in neurons seen in Alzheimer's disease led Gajdusek to ascribe an interference with slow transport in them.[131] The hypothesis was based on Weiss' theory of bulk flow. An alternative account holds that, in this disease, an augmented drop-off of altered neurofilament subunits is followed by their local assembly into the abnormal fibrils characteristic of the disease.

Studies of an altered transport in peripheral neuropathies in the human were initiated by Brimijoin taking biopsies from the sural nerve[132] and using his stop-flow technique to show the movement of the endogenous enzymes DBH and AChE (Chapter 11).[133] Transport differences with respect to the distribution, content, and storage of these enzymes were seen. Another method used to examine transport in biopsied nerve is to visualize particle movement in the fibers with AVEC-DIC.[134] In amytotrophic lateral sclerosis, kinetic abnormalities of particles were seen in their nerve fibers: an increased speed of anterograde transport and a decrease in the amount of particles transported. These are pioneering studies. Further studies using these techniques will no doubt help reveal changes in axonal transport that underlie the appearance of peripheral neuropathies.

Our present knowledge of transport in nerve function indicates that toxins and neuropathies can have an action at a number of points to give rise to changed form and function. Variation in the kinetics of the drop-off of components carried by the transport motors has been given as the basis of changes in the amount and location of various organelles in the fibers. How the drop-off from transport filaments is regulated requires more understanding, as well as how components are switched from microtubules onto actin microfilaments for local transport, to be steered to their proper sites in the fiber and inserted into their targets. Those targets include not only structural organelles within the axon, but also components associated with the axolemma, the membrane skeleton, that appears to be responsible for changes in the form of the fiber seen in a number of neuropathies, including the beading seen in the first stages of Wallerian degeneration (Chapter 8). The failure of supply of some component carried by transport changes appears to be responsible for the characteristic beaded form changes brought about. Just as important as the proper insertion of components into their

---

[130] (Cowan and Cuenod, 1975).   [131] (Gajdusek, 1985).
[132] (Dyck, Gianni, and Lais, 1993).
[133] Section by Brimijoin in (Ochs and Brimijoin, 1993), pp. 346–347.
[134] (Breuer et al., 1987).

targets is the process by which components are removed from the target sites. Some of those carried back by retrograde transport are reutilized by the cell bodies. Neurotoxins and neuropathies could have their basis of action at any of these points in the process of transport. In assessing the role of transport, it is necessary to determine what is primarily affected: A defect in what is produced by the cell body causing a lack of supply of a needed component or an altered substance could be the primary cause. As more is known of the transport mechanism, a larger group of neuropathies will no doubt be better understood.

14

PURPOSEFUL REFLEXES AND INSTINCTIVE BEHAVIOR

The ancient view that the spinal cord is a nerve-like prolongation of the brain received experimental support from Galen, who showed that by cutting the cord transversely, sensation from the body below the level of the cut was lost, as was motor power – effects mimicking those seen when cutting a peripheral nerve (Chapter 2). Even as late as the seventeenth century, Descartes looked on the cord as only a conduit for nerve tubules passing sensation to the brain, where reflexes are controlled (Figure 5.5). But, study of the decapitated animal indicated that reflexes remained present in the spinalized animal, with its purposive-like behavior leading to the hypothesis that some mind-like principle was present in the cord. As the anatomy and physiology of the nervous system became better understood, aided in large part by the discovery of the Bell-Magendie law in the early part of the nineteenth century, the question was then asked whether mind-like behavior could be accounted for by the complex interconnectivity of neurons in the spinal cord. This question was also raised with respect to the instinctive behavior seen in lower forms and the emergence of higher functions in the course of evolution. The development of the brain with centers for higher functions of learning and memory; in man ideation; caused the lower centers of the spinal cord to become more machine-like in its reflex behavior.

REFLEXES AND THE SPINAL SOUL

As had been noted in (Chapter 1), Aristotle cut insects and found that each part could move and was sensible. On that basis, he concluded that the soul is divisible and present in each cut portion of the animal in that movement remained in each small section, and movement was equated with the action of the soul (Chapter 3). In Descartes' model, reflexes and other movements of the body are due to animal spirits, the corporal soul, but with the human soul responsible for reason ruling over the animal spirits presiding in the pineal gland (Chapter 5). Schuyl, in his edition of Descartes' book,

differentiated between the soul of animals and the higher soul of man that
enables reasoning, writing that:

> the divisibility [of animals] all the Philosophers say in common [in] accord with
> a property of matter... [that proves] that the soul of animals is corporal, since
> serpents, lizards, and an infinity of insects that are divided do not lose all marks
> of life... one should not imagine that insects have a soul more vile than other
> animals. St. Augustine had pointed out that we admire with astonishment the
> agility of the fly, the industry of bees, the gait of a mare.... The separated members
> by their agility is witness that they are activated by soul. The cut parts of a centipede
> can walk as well as an intact one. Touching a separated part with a needle, it
> wrinkles and folds at the place touched. The centipede can be cut in many parts
> and each shows this behavior. This cannot mean that the insect is provided with
> by many souls before cutting it, but that the soul is divisible... this follows from
> the material soul of beasts which is extended in space, unlike that of the soul of
> man.[1]

Schuyl points out that this does not apply only to insects. The soul of a dog
is divisible, "each little part recognizes only itself when touched."[2] It must
have been common knowledge that chickens with their heads cut off can
run around in a coordinated manner for a time. This is seen not only in
animals, movement can occur in the limbs of men and in the head of one
decapitated for a time. This is vividly recorded in the poem of Lucretius:

> Since therefore we feel sure that mind and soul
> Pervade the entire body, that the whole
> Is animated by their force, perhaps a blow
> Is struck, so strong it cuts the body in two –
> Undoubtedly, the spirit is also halved,
> Is shaken, split, from the bisected torso.
> But something split, divided into parts,
> Surely denies its nature is immortal.
> It is said scythe-bearing battle-chariots,
> Red-steaming from their killing course, can cut
> Limbs off so quickly you can see them tremble
> Or quiver on the ground... a comrade
> Lifts his right arm to scale a wall, and sees
> His right arm isn't there, or attempts to rise
> While his leg is kicking at him from the ground,
> Even a severed head can lie in dust
> With an alert expression, open-eyed,
> Until the spirit is entirely lost... a snake... hacked

[1] Schuyl in his preface to the 1644 French edition of Descartes' *De Homine* (Descartes, 1664).
[2] Ibid., p. 441.

To little bits, each one of which keeps writhing...
Are we going to say each of these little segments
Contains a complete spirit? That would seem
To make it follow that a single being
Has a whole host of spirits in his body.
Let's be more sensible: that which was split
At the same time as body, and divided
Into as many parts, we'd have to say
Was every bit as mortal as the body.[3]

For Descartes, the brain – other than the pineal gland where he holds the thinking soul to reside – acts in an autonomous unthinking fashion. The cord for him serves only as a path from the body to and from the brain where the reflex is controlled. The problem is how reflex actions that are apparently purposeful can occur in animals without a brain. *Robert Whytt* (1714–1766) invoked a special soul-like power, a *sentient principle*, present in the brain and spinal cord that can account for reflex acts.[4] Reflex actions were not considered to be the province only of the spinal cord; reflex control is found in various brain sites.[5] Whytt gave as an example of reflexes, the responses of the iris to constrict with light and the accommodation of the pupil when shifting one's view from a nearby object to a distant one, which he said was from "the desire to see the distant object." The reflex dilation of the pupil in darkness and its constriction by light was said to be from the increased light intensity in the retina bringing about an unpleasant sensation, which – communicated to the *sentient principle* – acts to narrow the pupil. For Whytt, the eye reflexes and other examples he gave showed that reflexes are not the result of a conscious action. They are automatic, defensive, and purposeful.[6] The sentient principle residing in the spinal cord enables the spinal animal to perform protective reflexes independently of the brain.[7] When the brain of the frog is pithed, after a few minutes of flaccidity, it regains a nearly normal sitting posture. The spinal animal, although it irrevocably loses its spontaneity of action, is still able to show purpose-like reflexes. When its foot is pinched, the leg is quickly drawn up, apparently to evade the source of irritation; the response known as the "flexor withdrawal reflex." Whytt contrasted the invariant and automatic nature of reflex response to the more varied voluntary actions possessed by an animal with its brain intact. Whytt proposed

---

[3] (Lucretius, 1969), p. 105.
[4] The term "reflex" was later introduced by Hall. (Hall, 1836).
[5] An example shown in modern times is the placing and hopping reflexes found localized to a certain region in the frontal cortex of the cat (Bard, 1933).
[6] (French, 1969), pp. 84–85.   [7] Ibid., pp. 85–86.

that the apparently purposive, though automatic, movements under the control of the spinal cord showed that a sentient principle was present in the spinal cord. Only a small segment of the cord need remain to support reflex movement, all such responses ceasing when the spinal cord was completely destroyed.[8]

The long-held concept that animal spirits are elaborated in the brain and pass down into the cord and nerves to keep the body viable, would suggest that decapitation or destruction of the brain should be inimical to life. However, Mayow thought that this was the case only in the "more perfect animals," not so in insects and other "less perfect animals." In them

> the animal spirits are prepared not only in the brain but also in the protuberances of the spinal marrow, as it were so many cerebelli extended through the whole length of the spinal marrow, or rather they are stored in suitable repositories; and hence it comes to pass that in the cut-off portions of insects, the animal spirits are supplied, for keeping up to some extent life and motion, from the small piece of spinal marrow connected with each portion.[9]

REFLEXES AND INTEGRATION IN THE NERVOUS SYSTEM

In the nineteenth century, a great advance in understanding of the organization of the nervous system came about with the conception that the nervous system is composed of individual units – neurons. Even so, the interlacing of cells and fibers seen in the central nervous system appeared too forbidding to make sense of its complexity. The key to sorting out its mechanisms was found the Bell-Magendie Law (Chapter 9). Based on that insight, *Marshall Hall* (1790–1857) viewed reflexes to be the result of particular patterns of neural connectivity with no intervention of a sentient principle.[10] Conscious awareness of the reflex is subserved by a branch from the local afferent fibers leading from the cord to the brain.[11] The opposition of views, that reflexes are mechanistic processes versus the operation of a sentient soul present in the spinal cord, was brought out sharply in the *Pflüger-Lotze controversy*.[12] Pflüger, drawing on clinical experience, formulated a series of what he termed "laws of the nervous system" in a book published in 1853 in which he strongly promulgated the concept of a spinal soul. Liddell called this position "notorious," one that "hung for fifty years like a deadening blanket over physiology until the time of Sherrington."[13]

---

[8] Ibid., p. 86.   [9] Mayow quoted in (Fearing, 1930), p. 35.
[10] (Hall, 1826), (Hall, 1833), and (Jefferson, 1975).   [11] (Liddell, 1960).
[12] (Fearing, 1930), Chapter XI.
[13] (Liddell, 1960). Liddell had been associated with Sherrington in studies of reflex mechanisms, for the most part made in spinal preparations of dogs and cats.

The issues were clearly set out by Fearing:

> Does the behavior of the spinal animal (a) give evidence of the existence of consciousness, and (b) show the marks of 'purposiveness'? In connection with the first part of the question, the debate centered around (1) the metaphysical problem of whether or not consciousness or the 'soul' was divisible, i.e., was exclusively associated with the cerebral hemispheres, and (2) the physiological problem of whether or not any of the movements of the animal were spontaneous, i.e., took place in the absence of any external stimulation as a result, presumably, of an independent act of volition.[14]

Pflüger thought that the spinal cord of a cat when divided exhibits the presence of two "souls." The front part connected to the brain shows spontaneous acts under the control of will (it cries, jumps, bites, and scratches), whereas the posterior part controlled by the cord shows reflex behavior (it is sensitive, able to move in an apparently appropriate manner when stimulated, but unmoving when it is not stimulated). The apparent purposiveness of some reflexes can indeed be impressive to see. A decapitated frog suspended vertically with its legs dependent will, if a piece of paper dipped in an irritant such as acetic acid is placed on one leg, bring up its other leg to perform a series of wiping movements in an apparent attempt to brush the irritant away. Lotze, on the other hand, contended that although purposeful-like reactions appear to be under some conscious control of the spinal cord, they are only representative of the operations of an unconscious mechanism.[15] He gave examples showing that apparently protective reflexes may not always protect the animal, but that the experimental situation can be arranged so that the reflex action in response to a noxious stimulus increases damage to the animal.

Yet, the question remained as to whether, along with its mechanistic connectivity, some kind of conscious awareness is present in the spinal cord. The eminent English physiologist Michael Foster compared consciousness in the spinal cord with a flame that is lit up by a stimulus, as "a sort of momentary flash of consciousness coming out of the darkness and dying away to darkness again."[16] The prominent Italian physiologist Luciani was "able to conceive the possibility of a rudimentary spinal consciousness which may accompany the reactions of a spinal animal."[17] He pointed out that:

> Those who take the manifestations of perception and memory as the distinguishing signs of consciousness, and absolutely deny the psychical character of coordinated reflexes, do not reflect that the spinal cord is not claimed as the seat of the

---

[14] (Fearing, 1930), p. 165.   [15] Ibid., p. 87.
[16] (Luciani, 1915), p. 340. Quoted from Foster's *Text-book of Physiology*, 1897, part iii, p. 983.
[17] (Luciani, 1915), p. 340. See (Fearing, 1930), pp. 293–294.

higher intellectual functions, but only that of a simple *rudimentary* intelligence due
to the synthesis of a small group of elementary sensations... possessing transi-
torily flashes of consciousness, arising from a psychical synthesis of elementary
sensations.[18]

And further stated that:

Many of those who see in the co-ordinated spinal reflexes inherited, instinctive, but
unconscious acts, do not recognize that in admitting these they implicitly admit a
sort of fossilized intelligence for the cord, i.e., to adopt Hering's felicitous expression – unconscious memory of primitive psychical processes. The entire 'soul' of a
brainless Amphioxus is a *spinal* soul. How much of this would persist as such, and
how much (to repeat the metaphor) [remains] in a fossil state in the spinal cord of
the higher vertebrates?[19]

ENCEPHALIZATION: THE CHANGED RELATION OF THE BRAIN TO
SPINAL REFLEX BEHAVIOR IN THE COURSE OF EVOLUTION

Sherrington took an evolutionary point of view, that was given by Luciani,
in his use of the term "fossilized intelligence" when he considered that per-
formances that in earlier ages entailed mind, no longer did so after they
had become genetically fixed.[20] He analogized this to our experience in the
formation of a habit, as in learning to ride a bicycle. At first the performance
is attended with conscious effort and concern, and then later it is carried
out subconsciously. This pattern of development can be seen in the course
of phylogenetic development. The linearly organized segmented worms are
composed of a series of semiautonomous body segments, each with sensory, ganglion, and motor elements. A nerve chain controlling the body
segments extends down from the head and acts to bring the segments into
coordination in its locomotion. At the head end where sensory information
from the environment is first encountered, the ganglion undergoes an increase in the neurons and their synapses subserving these sensory inputs in
the course of evolutionary development. The lower portions undergo modification, losing much of their autonomy, becoming more machine-like and
dependent on the brain as it takes over more control. This gathering of
ever more functionality in the brain along with modifications that restrict
the range of activities of the lower part of the nervous system, is known
as *encephalization*. This process in the more recently evolved species, the
higher anthropoids, and man is shown by a greater prominence of the forward part of the brain, especially in the increased extent of the anterior
cerebrum (Chapter 15).

The behavior controlled by the nervous system has various levels of
complexity, ranging from the more stereotyped to the more complex and

---

[18] Ibid., p. 340.   [19] Ibid., p. 341.
[20] (Sherrington, 1947). Quoted in (Ochs, 1965b) on encephalization, pp. 271–275.

variable functions. *John Hughlings Jackson* (1835–1911), on the basis of his clinical experience and influenced by the philosophy of Spencer, characterized levels of behavior controlled by the human nervous system on the basis of the degree of organization and the *dissolution* of function in neural diseases:

> The doctrine of evolution implies the passage of the most organized to the least organized, or, in other terms, from the most general to the most special. Roughly, we say that there is a gradual 'adding on' of the more and more special, a continual adding on of new organizations. But this 'adding on is at the same time a 'keeping down.' The higher nervous arrangements evolved out of the lower keep down those lower, just as a government evolved out of a nation controls as well as directs that nation.[21]

The lowest level, of completely stereotyped motor actions, is exemplified by reflex action, to which on the sensory side are assigned sensations. Above this level are those combinations of reflexes that occur in such rhythmic movements as walking, where a number of reflexes are combined into a pattern of action. To this level, perceptions are assigned. At the next level are those complex motor patterns under the control of volition which, for example, include those complex muscular activities required for speaking. Corresponding to this highest level on the sensory side is mentation, ideation, judgment, and consciousness.[22]

The view that minds evolve along with increased organization of the nervous system has been supported by Sherrington.[23] He took a dualistic position, differentiating the "I," the abstract "mind," from the world of "energy," matter. "Mind refuses to be energy, just as it refuses to be matter. Energy's aspect is of motion and predicates extension which is denied to mind. But how do they interact? . . . An instance where they may approximate is at the mental process and cerebral process. There on the one side electrical potentials with thermal and chemical action, compose a physiological entity held together by energy relations; on the other a suite of mental experience; an activity no doubt but in what, if any, relation to energy[?]" But whereas the dilemma of the separate natures of mind and brain exits and how they interact, we proceed as if they are one:

> Mind has survival value. Nature in her process of Evolution proceeds as if believing in [their having] a working relationship. The finite mind, the embodied mind is colligate with the organization of the body. It shows a form of progressive organization [in] which this type of energy system and its behavior exhibit

21 (Jackson, 1931), Volume II, p. 58, et seq.
22 These views were used by James in his treatise on psychology at the turn of the century (James, 1890).
23 (Sherrington, 1941). Ekehorn has given an extensive commentary on Sherrington's thought on the subject (Ekehorn, 1949).

integration. That is, with increasing complexity, there is nevertheless, increasing unity.... In the process of evolution...further organization of the 'living' energy system develops along with the body, the integrated mind. The mind and body are two series of events concomitant in time as parts of a single series.[24]

INSTINCTIVE (INNATE) BEHAVIOR

As in the case of the spinal cord of decapitated animals, where the question was whether some degree of mindedness is present, such complex activities as the spinning of webs by spiders, nest building of birds, and so on – behaviors that have generally been viewed as due to *instinct* – raised the question of whether this showed mindedness in them. Studies of the behavior patterns of free-ranging birds and fish by ethologists, notably by Lorentz and Tinbergen, demonstrated reflex-like behavior in them, which they consider to come about by a concatenated sets of reflexes.[25] When a specific and appropriate stimulus is presented to an animal, such as food to a hungry bird, a series of stereotyped responses ensues. It opens its beak, grasps the food, swallows, with a pattern typical of the species. In response to sexual stimulation, receptive animals show a series of stereotyped mating behaviors, such as the "dance" courtship of birds peculiar to a given species. Such specific instinctive patterns of behavior were said to be *released* by the appropriate stimulus. This term implies that the behavior, although complex, is reflex – a self-contained functional unit with its neural apparatus so constituted that, on presentation of the appropriate stimulus, it is discharged into the observed motor performance. Careful observation is required to separate responses that are inborn or *innate* from those that are learned. For example, birds raised in isolation have songs typical of their species; but, in some species, birds have been shown to learn songs that can be distinguished from innate songs (Chapter 15).

To account for the complexity of innate behavior, the reflexes are considered to be chained, the release of one reflex behavior bringing the animal into position to respond to another reflex, a new situation with the possibility for a further release, and so on.[26] Tinbergen gives as an example of this chaining the hunting behavior of the peregrine falcon. When hungry, it

[24] In the forward to the 1947 reissue of his classic work, *The Integrative Action of the Nervous System*, Sherrington reveals his mature thought on the body-mind problem contraposing the views of those who believe only in a mental element, that all that exists is mind, with those who believe that all can be accounted for on physical principles, on "energy." He points to the difficulties incurred in the positions of both philosophical camps. In the end he opts for a dualistic view saying, "that our being should consist of *two* fundamental elements offers I suppose no greater inherent improbability than that it should rest on one only." p. xxiv.
[25] (Tinbergen, 1974), (Lorenz, 1982), and (Tinbergen, 1989).
[26] (Tinbergen, 1989) Chapter 5, pp 101–127, and see Chapter 16.

begins a hunting drive flying with a characteristic search path over a hunting territory. The sight of prey releases diving and harrying maneuvers. Proximity to the prey releases catching, then killing, then eating. In this example, hunger is the *drive* that leads to increased activity and the series of reflex behaviors directed to obtaining food. If sufficiently intense, that behavior may be realized without a food object actually being present. In young starlings that have been starved, the whole sequence of watching, killing, and swallowing may occur without food present, inappropriate behavior called *vacuum activities*. Even in organisms as evolutionarily low as the *Coelenterata*, vacuum behavior has been seen. The animal uses its tentacles to wave food in the water into its mouth. If sufficiently starved, tentacle responses can occur spontaneously without food present. Such inappropriate feeding behavior can be stopped by feeding the animals or by injecting mussel juice into them.[27]

Complex behavior has been explained by the *reflexologic school* to come about as the result of a set of learning stages during embryonic development. It is known, for example, that the salamander, *Ambystoma*, goes through a series of characteristic body reactions in its early embryonic development. They first show simple bending of one side of their bodies to form a 'C.' Later on, a reverse bending occurs, with their bodies assuming the form of an 'S.' Still, later, this S shape is repeated in rapid sequence as part of the animal's swimming movement. The reflexologic school holds that each of the simpler reflex stages must be performed before the next stage can be learned. In opposition to this concept, Coghill considered that the "total pattern" responsible for swimming is laid down in their nervous system, the C or S behaviors representing "partial patterns."[28] The sensory and motor systems may have different rates of growth, but when all systems are present, the motor responses are complete and orderly from the onset. The experimental studies of Weiss gave support to this "whole-pattern" view.[29] He kept *Ambystoma* embryos continuously anesthetized for days during the time when they are presumed to be forming their swimming behavior by a learned sequence of reflexes. When the anesthetic wore off, the animals at once performed the complex motor behavior shown by normal animals at that stage of development without having passed through a sequence of learned primitive reflexes. Weiss pointed out that salamanders are so tightly enclosed in their eggs that primitive behavior is not possible before their free-living life commences.[30] Another example he gave was that of the butterfly enclosed in its pupal case with no possibility of any movement of its wings. On emergence from the cocoon, it unfolds its wings and flies off with well-coordinated movements. Similar studies of reflexes of the human

[27] (Parker, 1915).   [28] (Coghill, 1929).   [29] (Weiss, 1939), pp. 566–572.
[30] (Coghill, 1929) and (Weiss, 1939).

fetus and newborn also support the concept of behavioral mechanisms laid down as a whole to give rise to an integrated performance from the onset.[31]

The complex behavior shown is an expression of inborn patterns of neuronal connectivity that are genetically determined for the species. It depends on certain specific stimulus patterns that are to be met with in the environment. This is shown by the *imprinting* observed in ducklings soon after hatching. On perceiving the mother duck, it normally walks after her. In the absence of the mother, the duckling will follow a model resembling the mother duck; its readiness to do so increases the closer the shape, size, and color of the model approach that of the mother duck. This propensity to follow a model of a mother duck is very strong for only a few days after hatching, during the *critical period*, after which it falls off and is then gone.[32] A similar innate readiness to react to particular sensory inputs without previous experience is the fear reaction shown by newly hatched birds when a model of a falcon having the general shape of their natural predator is passed over them.[33] Those studies indicate that biological value has been placed on specific sensory stimulation patterns in the environment that releases the behavior appropriate for the survival of that species. This inborn propensity to respond to specific sensory input differs from the concept of a *tabula rasa*, which Locke considered to represent the nervous system at birth, the "blank sheet" on which experiences will be written.[34] The ethologists have shown that an inborn neural organization is present in the animal that anticipates specific objects it is designed to meet with in its environment. The inborn propensity to respond to certain experiences has some relation to the philosophical position of Kant, who considered that a priori conditions of time and space are required before objects can be experienced as meaningful.[35]

EVOLUTION OF THE NERVOUS SYSTEM: A PROCESS OF DIFFERENTIATION AND DEVELOPMENT

In the previous sections dealing with encephalization, the view of Hughlings Jackson that "there is a gradual 'adding on' of the more special, a continued adding on of new organization," indicates that newly developed functions are not superimposed, *de novo* but develop from structures and functions already present in their ancestors. If this line of thought is followed along the chain of evolution step by step, it implies that even the most primitive organisms could have an anlage of the functions of higher organisms. The progression of neural development in the course of evolution is shown in the most elementary metazoans (Figure 14.1). In the

---

[31] (Hooker, 1952).   [32] (Lorenz, 1982).   [33] (Locke, 1959), and see Chapter 16.
[34] (Tinbergen, 1989).
[35] See discussion of Kant's philosophical view at the end of Chapter 16.

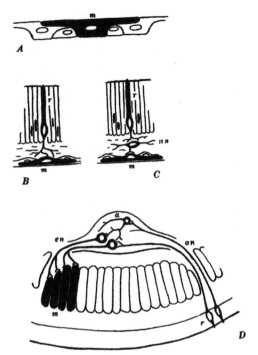

14.1. Evolutionary development in lower organisms. **(A)** **(m)** represents a muscle cell of the sponge. **(B)** **(r)** represents a receptor cell in the sea anemone with its fibers synapsing directly on muscle cells. **(C)** The interneuron cell (**lt n**) can form a nerve net. **(D)** A reflex arc (**a**), seen in a cross-section of an earthworm with receptors at the body surface, **(r)**, the afferent neuron **(an)**, and the efferent neuron **(en)**, leading to the muscle effector **(m)**. In the ganglion, two types of connectivity are shown: one directly from the afferent neuron **(an)** to the efferent neuron **(en)** synapsing on the muscle cells, and the other with a ganglion cell interposed through which connectivity to other neurons in and between body segments carries out coordinated body activities. Modified from (Bard, 1956), Figure 316.

single-celled amoeba local alterations of its membrane initiate extensions and retractions of its pseudopodium to alter the direction of its movement. In the paramecium and other protozoa, cilia and flagella bend in a coordinated sequence, allowing for their swifter movements.[36] These primitive functions in the single-celled organisms are handed over to selected cells in the metazoa to carry out those specialized functions. In the sponges, cells found around their pores contract in response to mechanical stimulation and to substances in their environment that is part of their feeding behavior. In the Coelenterata, the sea anemone, hydra, and jellyfish receptor cells have evolved, which form connections to specialized contractile muscle-like cells. An intermediary cell has also arisen to connect the sensory and

[36] (Jennings, 1962).

motor cells and serve to organize the overall movements of the animal. These intermediary cells evolve into neural networks that integrate sensory information and motor output in the very complex central nervous system of higher forms.

In that evolution, a most significant development was the elongation of the nerve fibers that permitted neural functions to be distributed over widespread regions of the nervous system giving room for the development of specialized functions. The elongation could not have occurred without the attending mechanism of axoplasmic transport in them; the participation of this process related to the higher functions of the brain is described in the next chapter.

15

NEURAL EVENTS RELATED TO
LEARNING AND MEMORY

The machine-like reflex responses of spinal animals and the instinctive behavior seen in lower species contrast with the adaptability of the behavior controlled by the brain in the higher species, especially in man where higher cognitive functions and willed behavior predominate (Chapter 14). In some of the earliest speculations, the brain was held to be the site where a higher spiritual entity, the soul, was responsible for cognition and willed activities. As more was known of the complexities of the brain, the Alexandrians Herophilus and Erasistratus, and above all Galen, assigned the higher functions of imagination, reasoning, and memory to the passage of animal spirits in the ventricles. This localization of functions was enshrined in the "cell theory" which held sway throughout the Middle Ages (Chapter 2). As the anatomy of the brain became better known, higher functions were assigned to various brain structures. The cerebrum, with its complex gyrations and the greater expanse of the cortex over the surface of the cerebrum in man, became identified with the higher functions of reasoning, learning, and memory. When the neuron and its interactions were recognized as the basis of nervous integration, changes in neuronal structure, particularly of the dendrites where synaptic interactions on them were seen to occur, was held to account for learning and memory.

LOCALIZATION OF FUNCTION IN THE CORTEX
In the early decades of the nineteenth century, Gall advanced the view that the cerebral cortex contains some twenty-seven (later more) separate "organs" or faculties to which he assigned faculties such as benevolence, veneration, hope, love, number, time, self-esteem, and so on. These, he held, could be determined by the shape and prominence of the skull over the areas responsible for those faculties.[1] The theoretical basis given for

---

[1] (Gall and Spurzheim, 1967), (Ackerknecht and Vallois, 1956), and (Zola-Morgan, 1995).

such regional variations in the brain and skull was that a greater growth of fibers into a given cortical region causes a greater growth of the cortex in that site with, in turn, an increased protuberance of the overlying skull to accommodate the enlarged cortex.[2] By palpating the scalp, the prominence of the protrusions showing the greater development of one region or other and the strength of that faculty in individuals were popularized by Gall and his associate Spurzheim under the name "phrenology." It had considerable vogue throughout much of the nineteenth century. Although phrenology was eventually rejected by the scientific community as spurious, it had given an impetus to the search for the localization of function in the brain. Bouillaud reported in 1825 that lesions in the brain could interfere with speech without impairing higher intellectual functions,[3] and Broca localized speech in an area in the left frontal region.[4] A map of a motor region controlling the musculature of different parts of the body was later found by electrical stimulation of the cortex[5]; and sensory modalities, defined by their loss following lesions in specified territories of the cerebral cortex, furthered the acceptance of the localization of functions in the cortex. The numerous cases of bullet and shrapnel wounds of the brain incurred in war were used to relate a functional loss or derangement to a region of the cortex.[6]

Opposition to the concept of a localization of functions came from the antilocalizationists, most prominently from Flourens.[7] At first, he had sided with Gall, but then vigorously opposed phrenology and came to view intelligent behavior as an invisible function of the activity of the entire cerebrum – a "summation of the energy of the whole organ . . . a dynamic action." As Lashley, an antilocalizationist, later stated:

> The cerebral hemispheres are the sole organs for the perceptions and volitions. . . . All perceptions and volitions have the same distribution in the hemispheres; the faculties of perceiving, understanding, and willing constitute a single function which is essentially unitary. . . . Excitation of one point in the nervous system involves all others; there is community of reaction, of changes, of energy. Unity is the great principle which rules, is universal, dominates all. The nervous system forms a single unified system.[8]

Goltz and Loeb, among other prominent investigators, also took the view that the brain functions as a whole. Goltz decorticated his pet dog whose

---

[2] The differential growth of brain regions must be considered to be genetic, occurring in embryonic or in early maturation, because the bony skull could hardly be resculpted in later life.

[3] (Bouillaud, 1825).   [4] (Broca, 1861). Now known as Broca's speech area.

[5] (Fritsch and Hitzig, 1963).   [6] (Markowitsch, 2000).

[7] (Flourens, 1846).   [8] (Lashley, 1963), p. 4.

behavior he knew intimately and found that the dog might show displeasure by growling or biting at painful provocation, but was otherwise passive, experienced no joy, and was without a trace of perception – "all trace of memory seemed lacking to the creature."[9] Loeb proposed a "resonance" to account for the unity of brain function.[10] Other theories involved electrical fields that produce integration by the changes of excitability of individual neurons under its influence. The appeal of such a viewpoint was that it could account for the unity of consciousness. However, there was no gainsaying the existence of sensory and motor regions. The difficult question was whether intelligence could be localized to some special region in the cortex.

Rather than trying to assess intelligence by inference of an animal's overall behavior, a more objective assessment was afforded by use of experimental studies designed to determine specific behaviors. Lashley developed the use of mazes to determine where learning and memory retention could be assigned in the rat cerebrum after making lesions in different regions.[11] He came to the conclusion that the capacity to learn a path through a maze or its memory was reduced in proportion to the amount of cortical destruction rather than to the loss of any particular field of the cortex. An early indication of the importance of the amount of cortex in behavior was suggested by Darwin in the nineteenth century who reported that the brains of domesticated animals were smaller in comparison with those of animals in the wild. This could perhaps be the result of the greater sensory stimulation and motor interactions met with of necessity in a wild environment.[12] This inference later received support when studies of the cerebral cortex showed that it could increase by as much as 15 to 30 percent in thickness in rats raised in an enriched environment, compared to those in an impoverished environment.[13]

Such cortical increases were not ascribed to increases in the number of neurons. It was the received doctrine that, once developed, the neurons remain for the life of the organism with no new ones arising. This was the view of Ramón y Cajal who, despite careful searching for them, saw no cells undergoing mitosis in the mature brain. A succession of similar such studies over the years showed the same result until, with the introduction in the 1960s of a new technique to label mitotic cells, evidence for neurogenesis in the adult brain was advanced. For the latter part of the nineteenth century and throughout most of the twentieth century, theories of neuronal changes related to learning and memory were based on changes in the dendritic extensions of the neurons and the fibers synapsing on them.

[9] Quoted from (Sherrington, 1947), p. 267.
[10] Quoted from (Lashley, 1963), p. 5.   [11] Ibid., p. 175.
[12] (Jacobson, 1978), p. 203.   [13] Ibid., pp. 203–204.

CHANGES IN THE CONNECTIVITY (SYNAPSIS) OF NEURONS IN THE CEREBRAL CORTEX RELATED TO HIGHER FUNCTION

The multiplicity of neurons packed into the substance of the cortex, and the exuberance and complexity of their fiber interconnections forming the *neuropil* posed a forbidding obstacle to an analysis of neuronal function until the introduction by Golgi of his method of silver staining.[14] With it, individual neurons with their ramifications of dendrites and axons could be pictured in apparent isolation, allowing inferences to be made as to the role of neurons and their processes in relation to cerebral function (Figure 15.1). Although Golgi theorized that the dendrites served for the nutrition of its cell body,[15] Ramón y Cajal, in opposition, viewed the large tree-like extension of the apical dendrites as sites on which numerous axons from other cells terminate.[16] The passage of excitation from the axon terminals of neurons to the dendrites and cell bodies of other neurons with, in turn, their axons ending on other neurons, and so on, were pictured as the basis for the processing of sensory information. Theories were advanced early on at the turn of the nineteenth century to account for sleep and wakefulness; and the effects of narcosis, neurotoxins, disease, learning, and so forth, on the basis of an alteration of the closeness of the contact of fibers with cells and their dendrites, with the close contact between them later referred to by Sherrington as the *synapse*.[17]

Changes in connectivity causing changes in function were proposed early on by Rabl-Rückhard. He proposed that changes were brought about by the ameboid movement of axon terminals approaching or withdrawing from cells and dendrites. He held this to be the basis of the formation of memories, forgetting, dreaming, hypnotic trances and alertness, and so on – the concept termed *amoeboidism*.[18] Lépine also considered that differences in the degree of intimacy with which fibers make contact with the dendrites or cells affect the state of consciousness.[19] He envisioned the terminals as

---

[14] (Pannese, 1996).   [15] (Golgi, 1967).

[16] (Ramón y Cajal, 1967), and (DeFelipe and Jones, 1988).

[17] The term was suggested to Sherrington by a Greek linguist scholar. The root of the term given in Liddell and Scott's Greek-English lexicon is "to knit or join together, to be closely engaged or entangled, to be attached to or combined with." In studies made before Sherrington's, the term was not used but it is given here to express the implied concept.

[18] (Rabl-Rückard, 1890). The concept of Rabl-Ruckardt and similar ones was discussed at length by Soury in a review (Soury, 1898) and in his history of the nervous system (Soury, 1899), pp. 1754–1757. He notes that the theory of amoeboidism implies that the protoplasmic expansion of the nerve cell is constantly "animated," i.e., is not a fixed structure. A summarization of amoeboidism and allied theories was also given by (Van Gehuchten, 1900), pp. 267–279, and Ramón y Cajal (Ramón y Cajal, 1995), pp. 721–726.

[19] (Lépine, 1895).

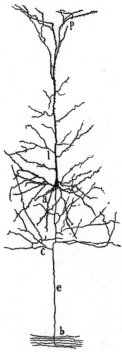

15.1. A pyramidal cell of the cortex of a mouse drawn by Ramón y Cajal from a Golgi-stained preparation. Along the shaft of the dendritic extension and branches (**p,l,c**) of a cell (**a**) extending through most of the cortical thickness, small projections are seen: the *dendritic spines* (cf. Figure 15.3). Its axon (**e**) is also greatly branched to synapse locally, with a longer length entering the white matter (**b**) under the cortex to pass to other farther cortical regions and to subcortical regions. From (Ramón y Cajal, 1995), Volume 2, Figure 344.

having amoeba-like properties much as Rabl-Rückhard did. Other mechanisms proposed to account for the approach or withdrawal of terminals at their contact with cells invoked a swelling and shrinking of the terminals, or alteration due to changes in the metabolism of the terminals resulting from their activity.[20] An alternate hypothesis advanced by Ariëns Kappers was that an electrical potential difference existing between the terminating fibers and cell bodies and their dendrites acts to bring them together – a process he called *neurobiotaxis*.[21]

Changes in the form of the dendrites were also invoked to account for a modulation of function. Whereas apical dendrites are generally seen to be

---

[20] (Lugaro, 1898). See (Van Gehuchten, 1900) pp. 278–279.
[21] (Ariëns-Kappers, 1929), and (Ariëns-Kappers, Huber, and Crosby, 1936).

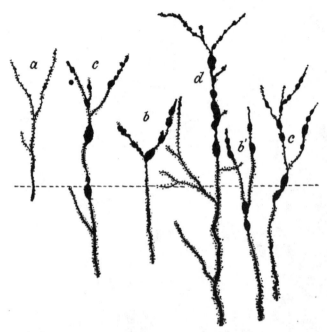

15.2. Terminal tufts of the apical dendrites of pyramidal cells (a,c,b,d,b′,c) above the dashed line, drawn between the first and second layers of the cortex, show beading in a case of "alcoholic insanity." From (Andriezen, 1884), Figure 34.

smoothly tapering from their origin in the cell body, varicosities (beading)[22] were reported in the dendrites after the administration of a wide variety of chemical agents – alcohol, ricin, morphine, chloral hydrate, chloroform, and so on – or by electrical stimulation (Figure 15.2). Demoor hypothesized that, as a result of this form change, the closeness of contact was altered to produce functional changes.[23] Such beading cannot be ascribed to an artifact brought about by differences in the early use of Golgi staining; the beaded form change of apical dendrites has been reported in recent years by accomplished neuroanatomists using the Golgi technique.[24]

Another feature of apical dendrites is the numerous small projections termed "spines" or "spinous processes" present along their lengths (Figure 15.3). Since their first description by Ramón y Cajal, variations in their shape

[22] The form change is similar to that of the beading of peripheral nerve fibers described in Chapter 8.

[23] (Demoor, 1898), p. 4. Lugaro reported that varicosities could be reduced by the manner in which animals are prepared. injecting them intracarotidly with Cox solution or morphine reduced them considerably (Van Gehuchten, 1900) pp. 278–279. But those varicosities that remained in spite of this posed a problem for Van Gehuchten. See p. 280.

[24] (Purpura, 1974), (Scheibel et al., 1976), and (Gross, 2001).

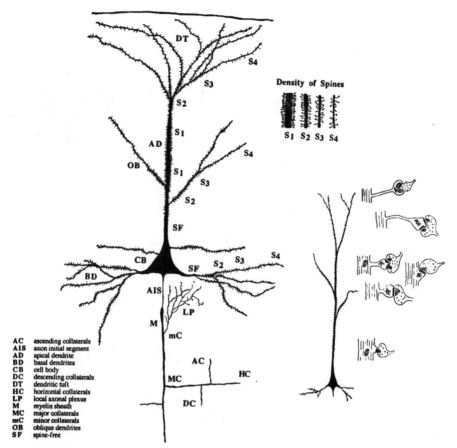

15.3. Detailed representation of the spines of a pyramidal cell. The dendrite shaft (**AD**) has the greatest density of spines that varies with its position along the dendrite and its thickness, as shown on the right for spines **S1–S4**. Spine-free (**SF**) regions are seen near the cell body. Different forms of spines with bouton terminals synapsing on them are drawn on the lower right. From (DeFelipe and Fariñas, 1992), Figures 3 and 18.

and number have been reported in pathological conditions. Their number was seen to be strikingly reduced in cases of schizophrenia[25] and in aged brains (Figure 15.4).[26] A thinning of the cortex was also found associated with diminished mental function. A decreased thickness of the cortex of neonatal rats placed in an impoverished environment was also seen and was

---

[25] (Glantz and Lewis, 2000).

[26] In this figure, cortical pyramidal cells and limbic cells in the aged brain are shown. Learning is also associated with changes in the cells of limbic structures, particularly the hippocampus as will be discussed in a later section.

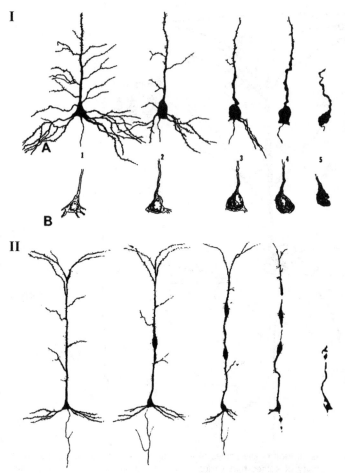

15.4. (I) The upper cells show that the dendritic processes of cortical pyramidal cells (labeled 1–5) stained with Golgi impregnation **(A)** undergoes progressive changes with age. The comparable Bielschowsky-stained cell bodies **(B)** also show degenerative changes (Scheibel et al., 1976), Figure 5. **(II)** Progressive changes in pyramidal cells of a subcortical region, the limbic system, with senescence (Scheibel et al., 1976), Figure 6.

related to a decreased synaptic connectivity.[27] The decrease in thickness of the visual cortex found in kittens raised in darkness, was due to a diminution of the neuropil.[28]

The relation of neural interconnectivity to behavior was summed up by Ramón y Cajal. He theorized that it is by the rapid growth of interneuron connections in which the "ability of neurons to grow and create new

[27] (Diamond et al., 1972).   [28] (Takacs et al., 1992).

associations in the adult that explains human adaptability and the facility to change ideational system ... [and where] the decline of such [formations in the aged and in those with] brains rigid from lack of education explains [their having] unshakable convictions ... and even misoneism [lack or hatred of what is new]."[29] Ramón y Cajal noted that such growth required the need for nutrients to be supplied to the processes, and he pointed out the involvement of neurofibrils and a possible role for them in that process.[30] These observations of Ramón y Cajal are prescient with respect to our present understanding of axoplasmic transport, and how it participates in the changes of form and function of neurons related to higher functions, as discussed in the following sections of this chapter.

EXPERIMENTAL PROCEDURES THAT RELATE LEARNING TO SYNAPTIC CHANGES

To further objectively relate behavioral changes to the properties of neurons and their connections, Konorski experimented with the *orientation reaction* – the pricking up of the ears of an animal, such as a dog, to an unexpected sound.[31] The reaction abates with repetition of the sound indicating the primitive form of learning referred to as *habituation*. He proposed this to be a "plastic" change with the formation of new relationships between neuronal processes that become coupled on the basis of the neurobiotaxis principle hypothesized by Ariëns Kappers.[32]

Ivan Pavlov (1849–1936) introduced the technique of *conditioning* to examine the more complex changes he believed took place in the cerebral cortex during learning. Hungry dogs were given meat or meat juice via a cannula in the mouth, just preceded by an auditory stimulus. The food, which elicits salivation, was paired with a sound until, after a sufficient number of such pairings, the sound itself gave rise to salivation. Pavlov termed the food an *unconditioned stimulus* (US) because it will ordinarily elicit the salivary response, whereas the sound is a *conditioned stimulus* (CS), one that ordinarily does not elicit salivation. The elicitation of the conditioned response was termed a *conditional* or *classical reflex*. Conditioning is not limited to responses to food as an unconditioned stimulus. Various unconditioned and conditioning stimuli can be used. In one such experimental design, a puff of air to a rabbit's eye is the unconditioned stimulus eliciting a reflex contraction of the nictitating membrane. When preceded or given along with a sound as the conditioned stimulus, after a sufficient number of such pairing trials the nictitating membrane responds to the sound alone. This is to be contrasted to another type of learning, *single trial learning*, e.g. where a rat is placed on a small platform facing the dark hole of

[29] (Ramón y Cajal, 1995), Volume 2, pp. 724–726.   [30] Ibid., pp. 725–726.
[31] (Konorski, 1948).   [32] (Ariëns-Kappers, 1929), and (Ariëns-Kappers et al., 1936).

a box into which it instinctively enters. When it does so, it is given a strong foot-shock. When placed on the platform the next day, the shocked animal will not enter the box again. Its remembrance of the single experience lasts many days, weeks, or months. This is an example of *one-trial passive avoidance conditioning*.[33] If, however, soon after being shocked in the box, the animal is deeply anesthetized,[34] or its brain electroshocked,[35] it will show no memory of having been foot-shocked and when placed on the platform the next day it will enter the box as readily as a naive animal would. The time between foot-shock in the box and the effect of an electroshock to the brain or the administration of a deep ether narcosis to disrupt memory formation is the *consolidation* period. It lasts approximately one hour and, after this period of consolidation is over, administration of electroshock to the brain or deep narcosis is ineffective in disrupting the memory engram. The animal will not enter the box on the next day and on following days, showing its retention of the memory of the foot-shock.

The effect of deep narcosis to completely silence all ongoing electrical activity has a bearing on an alternate hypothesis that holds memories not to involve a lasting material change in neurons. Instead, the theory proposed was that learned patterns of behavior are maintained by activity continually recirculating in networks of chained loops of neurons.[36] Such a circulation of activity in neural networks would be completely interrupted by deep narcosis, nevertheless, experience shows that, on recovery from narcosis, memories are still retained.

THE CELL-ASSEMBLY THEORY OF HEBB
A theoretical model of learning based on changes in synaptic properties advanced by Donald Hebb in the mid-twentieth century gained considerable support by psychologists.[37] More recently, computer modelers of artificial neural networks that show "learning" have described a class of such networks as *Hebbian*. In opposition to *connectionist* theories of learning in which the nervous system is pictured as a complex switchboard with incoming sensory impulses entering the cortex along paths rigidly connected from cell to cell until the outcome is directed to a motor output, Hebb espoused a *field theory*. He viewed sensory input to the cortex becoming integrated in some statistical interactive fashion in a group of cells in which

---

[33] The different procedures used for behavioral studies, classical (Pavlovian), operant (instrumental), one-trial, have been characterized by (Keller and Schoenfeldt, 1950).
[34] (Abt, Essman, and Jarvik, 1961).   [35] (McGaugh and Herz, 1972).
[36] In theories presented by Rashevsky, Householder and Landahl in the Mathematical Biophysics Program at the University of Chicago. The general approach of the group is given by (Rashevsky, 1938), Chapter XXII.
[37] (Hebb, 1949).

memory traces develop, in a *cell assembly*. Activity in any particular neuron in that group is not specifically determined; The distribution of excitation over the field of neurons determines the actions of its individual neurons. These cell-assembly groups interact with other such assemblies in the cortex and elsewhere in the brain to constitute a *phase sequence*. Each assembly's action may be aroused by a preceding assembly action, by a sensory event, or by a thought process – the sequence ultimately determining a specific motor outcome. The cell assembly is the basis of psychological properties, such as "attention," which gives a direction to perception and serves as the basis for learning. The interaction of the thought process with the neural machinery remains a problem for Hebb when he asks:

> What is the nature of such relatively autonomous activities in the cerebrum? We know a great deal about the afferent pathways to the cortex, about the efferent paths from it, and about many structures linking the two. But the links are complex and we know nothing about what goes on between the arrival of excitation at a sensory projection area and its later departure from the motor area of the cortex.[38]

Hebb considered that the way to bridge the gap was to further understand the processes going on in the brain during learning. He postulated that reverberatory activity takes place in the cell assembly to form a *trace* before the cell changes are consolidated and made permanent. Formally stated, the premise made was that: "When an axon of cell A is near enough to excite a cell B and repeatedly or persistently takes part in firing it, some growth process or metabolic change takes place in one or both cells such that A's efficiency, as one of the cells firing B, is increased." Hebb considered that the most probable way in which a cell's increased tendency to fire was by an increase in the *synaptic knobs*, the *boutons* at the ends of the fibers, synapsing on neuron cell bodies and dendrites. By their expansion, the boutons increase the area of contact between them and the cells to increase their efficacy. Additionally, new boutons and synaptic contacts could develop. These were proposed to be induced in axons passing by an active cell even at a distance "in the range of a centimeter or so is still possible."[39] An action over such distances was suggested by the picture of synaptic terminations on cell bodies given by Lorente de Nó, where "irregular thickenings in the axon" appeared as it threads its way through a thicket of dendrites and cell bodies.

The point in the axon where the thickening occurs is not determined by the structure of the cell of which it is a part, but external to it.[40] The means by which activity in cells affects nearby cells and axons was by an "electrotonic" effect. He referred to the findings of Arvanitaki, who showed excitability changes in giant nerve fibers resulting from the electrical activity

[38] Ibid., pp. xvi ff.   [39] Ibid., p. 63.   [40] Ibid., pp. 62–66.

in nearby fibers.[41] At the time that Hebb wrote, synaptic activation was generally considered to be accomplished by electrical currents passing from the presynaptic bouton to the postsynaptic cell. This theory was proposed by Eccles for excitation at the neuromuscular junction and in synapses in the spinal cord was widely accepted, only to be later overturned by Eccles himself. Using microelectrodes to record from motoneuron cell bodies in the spinal cord (Chapter 7), he found a synaptic delay due to a neurotransmitter action, evidence against an activation by electrical currents.[42] A wide range of various neurotransmitters subsequently found in the brain has opened up new understanding of how learning and memory comes about, as will be seen in the following sections.

LONG-TERM POTENTIATION IN HIPPOCAMPAL NEURONS
Although it is generally believed that memory engrams are retained in the cortex, the hippocampus plays an important role in the early phase of memory, in the consolidation of memory traces. This was indicated by Scoville and Milner in 1954 in a patient known as H.M., who had been subjected to bilateral temporal lobectomy to alleviate persisting convulsions. He showed a lack of retention of recent experiences, even though old memories were retained.[43] Examining the phenomenon further, they found this memory defect to be produced also in patients in whom surgical ablation had gone far enough to remove or damage portions of both the anterior hippocampi. Damage of the hippocampus on one side only did not produce this defect of memory. The investigators concluded that the hippocampus and hippocampal cortex are critically concerned with the retention of recent experience, though it was not known if the amygdala, which was removed together with the hippocampus, plays a part in the loss of recent memory.[44]

The hippocampus and amygdala are part of a *limbic system* of subcortical regions connected with pathways ringing the medial aspect of the two large brain hemispheres.[45] These structures further connect to a number of other brain structures, to the cingulate cortex above and the hypothamus below which controls a number of vital autonomic functions that Papez proposed as constituting a circuit responsible for emotions. This was contrasted with the lateral neocortex held responsible for cognitive functions. This division of function is not clear-cut. The limbic system has important connections to the neocortex in which long-lasting memories are stored and utilized in cognition.

A cellular change related to consolidation of memory was seen in the hippocampus by Bliss and Lømo by a long-lasting increase in the amplitude of

[41] (Arvanitaki, 1938).   [42] (Eccles, 1953), and (Eccles, 1964).
[43] (Scoville and Milner, 2000).   [44] (Milner, 1972).   [45] (Swanson, 1999).

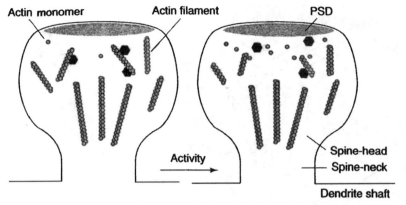

15.5. Internal components of spines that are hypothesized to alter the shape of the spines after synaptic activity. In this model, actin filaments become disassembled by the action of the neurotransmitter glutamate. Ca^{2+} entering the spine activates a calmodulin-dependent protein kinase to depolymerize the polymeric F form of actin to its monomeric G form, thereby changing the shape of the dendritic spine. The synaptic site is at the region showing a post synaptic density (PSD). From (Halpain, 2000), Figure 2.

evoked responses after a short burst of tetanic stimulation.[46] Such increased responses lasting for hours, days, or weeks were termed *long-term potentiation* (LTP).[47] Van Harreveld and Fifková found a dramatic change in the form of the spines on the dendrites of the dentate granule cells in the hippocampus to accompany the LTP.[48] Though variations in spine shapes are often found (cf. those shown for cortical pyramidal cells in Figure 15.3), they generally look somewhat like toadstools, bulbous heads connected via a narrow neck to the dendrite. Stimulation inducing LTP caused the length of the spines to shorten and their necks widen. The shape change was thought to result from the action of actin in the dendritic spines on analogy to skeletal muscle contraction effected by actomyosin.[49] Fifková demonstrated the presence of actin within the spine[50] and pointed out that its disposition in the head of the spines extending into the neck was favorably positioned to cause the spine to shorten.[51] Upon synaptic excitation of the spine, calcium enters it changing actin from its inactive monomeric globular G form to the active polymeric F form, which is able to contract and shorten the spine (Figure 15.5).[52] This form change was considered to increase the effectiveness of synaptic transmission by decreasing the electrical resistivity in the narrow

---

[46] (Bliss and Lømo, 1973).
[47] (Bliss and Gardner-Medwin, 1973), and (Douglas and Goddard, 1975).
[48] (Van Harreveld and Fifková, 1975).   [49] (Crick, 1982).   [50] (Fifková, 1987).
[51] (Fifková and Morales, 1992).
[52] (Piccolino, 2000). How much actin is in the F form and in the G form, a question relevant to the lability of the form of the spines, is unknown.

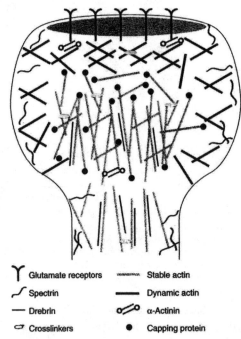

Υ Glutamate receptors   ⚊⚊⚊ Stable actin

$\mathcal{S}$ Spectrin   ⚊⚊⚊ Dynamic actin

⚊ Drebrin   o⚊⚊o α-Actinin

⚬ Crosslinkers   ● Capping protein

15.6. Model of dendritic spine containing actin and spectrin as possible participants in shape changes resulting from synaptic activation. The stable actin is capped with a protein. On synaptic activation, it is uncapped and dynamic changes of actin occur, mainly at the periphery of the spine. The spectrin associated with the membrane, as it is in neurons elsewhere, is part of the membrane skeleton mechanism that has control over the shape of the cell and its processes. From (Halpain, 2000), Figure 3.

neck of the dendritic spine when it was widened, allowing a greater flow of synaptic current to spread further out into the dendrite.[53]

However, the change in resistivity due to the spine change was calculated by Lynch to be too small, and he proposed another mechanism.[54] He theorized that, on synaptic excitation when the neurotransmitter *glutamate* acts on its receptor *N-methyl-D-aspartate* (NMDA) in the spine head, calcium entering activates the enzyme *calpain* that degrades spectrin in the membrane (Figure 15.6). This destabilizes the membrane, allowing it to expand and expose hidden NMDA receptors to thereby increase its responsiveness to the neurotransmitter.[55] In addition, a growth of new spines is promoted adding to the potentiation. Polyribosomes present at the base of the dendritic spines[56] could synthesize polypeptides that participate in the

[53] (Rall and Rinzel, 1971), and (Rall, 1995).   [54] (Lynch, 1986).   [55] (Schliwa, 1986).
[56] (Steward and Levy, 1982).

increased outgrowth of spines and contribute to the consolidation of memory traces associated with the spines.

The form change proposed to be due to spectrin[57] was based on evidence that this large molecule is integrally associated with the membrane, as part of the *membrane skeleton* that, with other molecular components, is responsible for the shape changes of cells.[58] Spectrin, along with actin in the membrane skeleton, was held responsible for the beading of fibers (Chapter 8). Using high-resolution confocal imaging of dendritic spines in the cortex in vivo, the spines were indeed seen to change form on stimulation, along with changes in their intracellular concentration of calcium.[59] A moderate rise of calcium gave rise to an elongation of the spines, whereas a larger concentration of calcium caused shrinkage and eventual collapse. An analogous process was proposed for neurons in the hippocampus, where an activation of NMDA receptors was associated with a formation of new spines and the pruning of others.[60] That the new spines were likely to be functional was indicated by the presence of presynaptic terminals on them.

INTEGRATION IN DENDRITES AND THEIR SPINES

Synaptic activation of dendrites does not always initiate propagated impulses in pyramidal neurons due to the relatively long distance between synaptic sites on their far reaches of the dendrites from the cell body and initial segment of the axon just below it where excitation of propagated responses occurs. The passive electrotonic spread of synaptic potentials down to the initial segment would be much too attenuated to be effective unless there was a summation of electrotonic currents by synchronous synaptic activity over a large enough portion of the dendritic tree. Such a global summation was indicated by the theoretical analyses of Willifred Rall and his collaborators who modeled the complex branching of the dendrites by a single cable, taking into account the branching of the dendrites, their sizes and numbers.[61] In their analyses the presence of nonlinear behavior indicated the possibility that regenerative activity could act to augment that spread.[62] This was investigated by electrically stimulating the cortical surface with brief pulses and recording smooth 10–20 msec depolarizations that were termed the *dendritic potentials* at distances up to 10 mm.[63] These were generally taken to be synaptic potentials. These negative (N) wave *direct cortical responses* (DCRs) did not, however, show the summation expected

---

[57] (Halpain, 2000).
[58] (Marchesi, 1985), (Stokke, Mikkelsen, and Elgsaeter, 1985), and (Devarajan and Morrow, 1996).
[59] (Spillane, 1981).   [60] (Golding and Tonge, 1993).   [61] (Rall, 1995).
[62] (Segev and Rall, 1998).   [63] (Chang, 1951).

of synaptic potentials. Instead, spatio-temporal interaction studies showed refractoriness, the occlusive behavior indicative of regenerative responses.[64] Evidence for regenerative active responses in dendrites involving calcium ion channels was given by Rodolfo Llinás.[65] Such regenerative action can augment the spread of depolarization in distant dendritic regions.[66] More recently the action was traced to the dendritic spines.[67] Two-photon excitation imaging of active individual spines showed the small spines (filopodial and thin) and large spines (stubby, fenestrated and mushroom-shaped) to have different responses to the transmitter glutamate.[68] The large ones are more stable, the small ones more labile suggesting the hypothesis that the larger ones are related to "memory," the smaller ones to "learning." The picture that emerges is that the mixture of electotonic spread and regenerative processes with various patterns of excitatory and inhibitory inputs strategically placed on the spines at various regions of the dendritic tree can confer logical AND-NOT and coincidence detection properties on the neuron in the course of the reading in and reading out of information through it. Such properties could very well serve in the process of learning and memory.

THE MOLECULAR BASIS OF MEMORY CONSOLIDATION
Considering the large number of memories that are retained by lower animals, and the much greater number present in humans that accumulates over the course of many years, theoretical reasoning suggested that the formation of so many memories could be accounted for by individual proteins, on consideration of the astronomical number of different proteins that could be formed from the twenty amino acids composing them. The hypothesis that memory formation involves the synthesis of proteins was given support by the failure of consolidation seen when a protein synthesis blocking agent, puromycin or cycloheximide, was given during or shortly after training sessions. When administered after consolidation had taken place, they were ineffective.[69]

Because proteins have a turnover with short half-lives of weeks or even days,[70] the newly formed memory proteins would require a change in nuclear DNA in order to perpetuate the memory proteins for months or years. A DNA change was seen in neurons early in learning with the appearance

---

[64] (Ochs and Booker, 1961), (Suzuki and Ochs, 1964), (Phillis and Ochs, 1971), and (Arikuni and Ochs, 1973). By cutting through the cortical layers between stimulated and recording sites except for the uppermost molecular layer, these were shown to carry fibers passing laterally to synapse on the upper reaches of the vertically oriented pyramidal cells to give rise in them to the N wave DCR (Ochs and Suzuki, 1965).
[65] (Llinás et al., 1968), and (Llinás and Sugimori, 1980).
[66] (Shepherd et al., 1985).   [67] (Segev et al., 1998).   [68] (Kasai et al., 2003).
[69] (Agranoff, Davis, and Brink, 1966) and (Davis and Squire, 1984).
[70] (Lajtha, 1964).

of the proto-oncogene c-fos, one of a group of similarly acting DNA genetic domains that have been related to learning.[71] c-fos triggers a translation of the protein *fos* that binds to DNA sites regulating transcription. The *c*-fos changes, however, are not permanent, and the link from *c*-fos to enduring DNA changes responsible for the long-lasting proteins of the memories has yet to be determined.[72]

Insight into the molecular nature of the learning process was gained by Kandel and colleagues using the simpler neuronal system in the invertebrate marine snail *Aplysia californica*.[73] In this animal, memory mechanisms were determined by electrical responses of neurons and molecular changes in them in relation to behavioral changes. The animal is opened to display its siphon by which fluid is ejected from it (Figure 15.7A). Stimulation of the siphon with a brush stroke stimulates the expression of fluid. Microelectrodes are inserted into the cell bodies of the sensory neuron of the siphon and that of the motoneuron to the gill on which it synapses to record their synaptic potentials. Stimulation of the siphon, if given repeatedly, shows *adaptation*, the synaptic response in the motoneuron then seen to be much diminished. With repeated sessions of stimulation, the habituation could last for weeks. This primitive learning is due to a decrease of the synaptic discharge from the sensory neuron synapsing on it, as well as other interneurons synapsing on the motoneuron. Neurons from other body regions that synapse directly and via interneurons on the gill motoneuron act to modify its discharge. That from the tail when strongly innervated by an electrical stimulus can *sensitize* the motoneuron, producing an increased synaptic response by facilitating interneurons that terminate directly on the motoneuron and on the presynaptic terminal of the sensory neuron ending on the motoneuron, at an axo-axonal junction. Classical Pavlovian conditioning showed such sensitization by pairing stimuli to the siphon as the CS with US given to the tail.[74] After a number of such pairings, conditioning is shown by the increase of the synaptic response of the motoneuron. This kind of learning may be short-lived or with strong conditioning may last days or weeks. The long-lasting changes were shown to be due to a series of molecular changes in the presynaptic terminals acted on by facilitating neurons and changes in the DNA of the nucleus in the cell body of the sensory neuron (Figure 15.7B).

On stimulation of the tail, a facilitating interneuron ending as an axo-axonal junction on the sensory nerve terminal causes a release in it of the transmitter serotonin (5HT). On binding to its receptor on the terminal, it results in generation of cyclic AMP (cAMP) and in turn phosphokinase A (PKA), which produces an enhanced outflow of transmitter released onto the motoneuron. In addition, long-term memory is brought about by a

---

[71] (Morgan and Curran, 1989).   [72] (Goelet et al., 1986).
[73] (Kandel, Schwartz, and Jessell, 2000), Chapter 63.   [74] (Kandel et al., 2000).

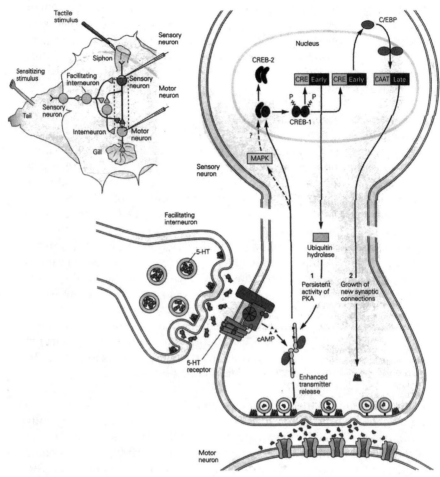

15.7. Classical conditioning of neurons of the snail *Aplysia californica*. The animal is pictured in the upper left (**A**), with its neurons exposed for experimental study of the gill withdrawal reflex. An unconditioned stimulus given to the tail area activates a sensory neuron that synapses on a facilitatory neuron that in turn synapses onto the motor neuron controlling the gill area. A sensory neuron from the mantle to which a conditioned stimulus is given also synapses on the motor neuron. On the right, details of the molecular mechanisms responsible for conditioning are given in (**B**). The neurotransmitter serotonin (5-HT) released from the facilitatory inter neuron activates adenylyl cyclase, giving rise to cAMP. With the preceding activity by the conditioning stimulus, the calcium channel is activated, thus allowing calcium to enter and bind to calmodulin. It acts on adenylyl cyclase to enhance the synthesis of cAMP leading to the release of a greater amount of neurotransmitter than ordinarily would occur in the unconditioned state. Its upper passage to the nucleus (by diffusion or axoplasmic transport?) along with mutogenic activated kinase (MAPK) leads to genetic changes via a cAMP-response element (CREB) protein maintaining the changes induced by conditioning, in which CRE, transcription factor c/EBP, DNA response element CAAT and other factors participate. The downflow from the nucleus gives (1) persistent activity of PKA with enhanced transmitter release, and (2) growth of new synaptic connections. Modified from (Kandel, Schwartz, and Jessell, 2000), Figure 63.5.

genetic change in the nucleus of the sensory neuron. This occurs by a retrograde ascent of PKA, presumably by a transport system like that of axonal transport, to the nucleus of the sensory neuron. There, it phosphorylates one or more transcriptional activators binding to cAMP-response elements (CREs) located on a region of cAMP-inducible genes. The transcriptional activators belong to the protein family of CRE binding proteins (CREBs) activating the genes encoding two classes of protein – one a persistent activator of PKA that enhances transmitter release and a second one responsible for the growth of new synaptic connections. These are carried down into the terminal of the sensory neuron to have their action there.

In their studies carried out in the neurons of hippocampal slices, Tsien and colleagues found that the calcium, which enters the neurons after synaptic activation, leads to a buildup of calcium/calmodulin close to the calcium channel. The calmodulin is then translocated to the nucleus as a signal regulating gene expression via CREB phosphorylation.[75] They postulated that "tags" were present as markers for guidance of calmodulin to the requisite site in the nucleus. Whether such translocation occurs via diffusion or by a transport mechanism, such as one based on microtubules or actin, has not been resolved. In the dendritic spines where changes in them are considered as the basis of learning and memory, specific proteins to maintain their altered state would entail a lasting change in the DNA of the nucleus. In the neurons of the hippocampus, the spines are a far distance from the nucleus, and a transport mechanism would be required to transport the proteins to the specific spines that have participated in learning. For this, routing could be involved (Chapter 11). A subset of microtubules could carry signals from the activated synaptic sites to the cell bodies to promote the nucleic acid transformation, and the new proteins they produce routed back to those spines. Microtubules are seen in electron micrographs to turn in at the base of the dendritic spines.[76] However, given the very large number of spines on apical dendrites, compared with the much smaller number of microtubules present in the dendrites, each spine could hardly be supplied by its own subset of microtubules. A selective supply to the spines could be accomplished by tags, signals specifying where the off-loading of specific proteins to particular spines is to occur. The tags may be polypeptides produced by the polyribosomes present at the base of the dendritic spines.[77]

mRNA PROVIDING SELECTIVE CHANGES IN SPINES AT SYNAPTIC SITES RELATED TO LEARNING AND MEMORY

The polyribosomes and associated membranous elements in the dendrites, located particularly at the base of the dendritic spines are where mRNAs and

---

[75] (Deisseroth, Bito, and Tsien, 1996).   [76] (Van Harreveld and Fifková, 1975).
[77] (Steward and Levy, 1982).

the machinery for translation of specific proteins are also found.[78] Such localization of the machinery for translation of specific proteins is seen in a number of different neurons, in dentate granule cells, hippocampal pyramidal cells, cortical neuron, and cerebellar Purkinje cells. The synthesis of proteins in response to synaptic activity in the spines could act to modify the long term action of the synapse, and/or could provide tags or signals transported to the neuron cell body to provide a continued supply of specific mRNAs to sustained long-term memories. Some transcription components are also found associated with the spines. Elements involved in synthesis were shown to be transported in the dendrites, an A2 response element and heterogeneous nuclear ribonucleoprotein (hnRNP) A2, their transport by microtubules indicated by the block seen with colchicine.[79] Translocation of granules colocalized with mRNA seen in combined light and electron microscopic studies was shown to move within living neurons, most likely along microtubules, the movement interrupted with colchicine.[80] An RNA binding protein, zipcode binding protein 1 (ZBP1) in the form of granules was seen using digital imaging of hippocampal neurons to move from the cell body into dendrites in a proximo-distal direction following KCl depolarization.[81] High speed imaging showed both anterograde and retrograde movements in dendrites and as well in the spines.[82] The RNA binding proteins could serve as regulators of mRNA transport to synapses that have become activated.

Such machinery for local production of proteins at the dendritic spines and other active sites on the dendrites could serve to selectively strengthen synapses on given groups of synapses on the dendritic tree strategically placed so as to enhance some synaptic inputs over others. This work is in a relatively early stage and the presence of such a diverse population of mRNA and associated components for the synthesis of numbers of various proteins related to the pattern of spine activation in the dendrites will no doubt throw more light on the molecular processes underlying learning and memory processes.

CELL CHANGES IN SPECIFIC REGIONS OF BRAIN AFTER SINGLE TRIAL LEARNING

The greater complexity of the brain in the higher forms brings about questions as to how components involved in memory formation in one brain site are carried to another site where they are maintained. This was examined in the brains of newly hatched chicks using *one-trial passive avoidance conditioning*.[83] The chicks peck spontaneously at small, brightly colored

---

[78] (Steward and Schuman, 2001).   [79] (Shan et al., 2003).
[80] (Malinow and Malenka, 2002).   [81] (Tiruchinapalli et al., 2003).
[82] (Knowles et al., 1996).
[83] (Morgan and Curran, 1991), (Rose, 1991), and (Morgan and Curran, 1989).

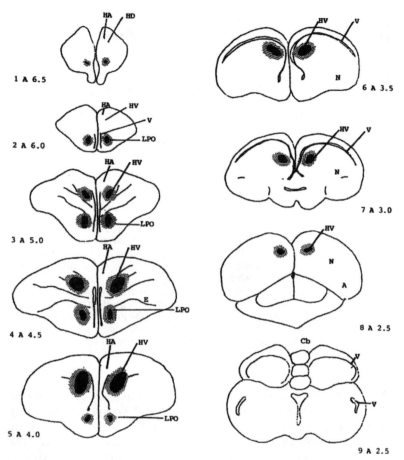

15.8. Specific areas in the chick brain involved in passive avoidance memory. Beads coated with a bitter substance and distinctively colored are rejected by the chick after a single trial. Lesions made in the regions shown in a series of coronal sections taken through the chick brain indicated by the black and shaded regions interfere with avoidance conditioning. The regions involved are the IMHV and below it the **LPO** (lobus parolfactorius) nuclei. Although the IMHV on the left side is critically involved, that on the right is not (cf. Figure 15.9). From (Rose, 1991), Figure 3.

beads. If beads with a given color had been dipped in a bitter-tasting liquid, the chick will peck them once and when tested the next day will avoid beads of that color. When chick brains are taken for analysis within 30 minutes of such a trial, two brain regions were found to have an increased accumulation of labeled deoxyglucose in them, indicating an increased neural activity.[84] The sites were the *intermediate medial hyperstriatum ventral* (IMHV) and the *lobus paraolfactorius* on the left side of the brain (Figure 15.8). A

[84] (Sokoloff, et al., 1977).

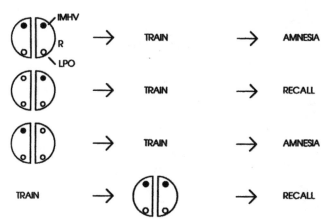

15.9. Cartoon showing the effect of IMHV lesions on recall of passive avoidance conditioning to a bitter bead. Chicks trained to peck at a bitter-colored bead will not retain the memory of it if lesioned within less than one hour after the experience and will peck at it again the next day. When lesioned after one hour, consolidation had taken place and when tested the next day will avoid that colored bead. Black circles (●) indicate lesioned areas; open circles (O) indicate unlesioned sites. With both right and left IMHV regions lesioned, the chicks were amnesic, no longer avoiding the colored bitter bead. With the right IMHV area lesioned, the chick could still recall the training. But, with the left IMHV lesioned, the animal was amnesic, showing that it is only the left IMHV region that is related to memory formation. LPO, lobus parolfactorius was not involved. See Figure 15.8 for actual loci of the lesions. From (Rose, 1991), Figure 4.

great increase in the density of the dendritic spines in large multipolar neurons was subsequently found in those regions. Isolation of synaptic membranes from those sites showed an increase of the presynaptic 43 kD protein, the *growth-associated protein* (GAP-43) and other GAP proteins activated by phosphokinase C (PKC). Injection of PKC inhibitors into the left IMHV region just before or just after avoidance training blocked the retention of the conditioned behavior: The chicks did not avoid pecking at beads of the color they had earlier found bitter. By making lesions in one or another of the IMHV regions, it was seen that the left IMHV region was primarily involved in forming the memory with a transfer of learning from it to the IMHV on the right side of the brain (Figure 15.9).

In the mammalian brain, Rolls described the transfer of a learned engram from the hippocampus to the cerebral cortex and other subcortical regions, also to the cerebellum that had been implicated in learning and/or memory storage.[85] He pictured the CA3 region of the hippocampus as an auto-association system allocating coding for complex inputs stored in neurons of the neocortex and elsewhere.[86] The CA3 region of the hippocampus

[85] (Thompson and Yu, 1987).   [86] (Rolls, 1990).

acts to guide and supervise cortical learning by back-projection from it to the cortex to modify sensory input to it. The CA3 region, by detecting useful conjunctions globally, directs storage from the earliest stages – dynamically adjusting processing to facilitate optimal satisfaction of the multiple constraints working to direct the attention of the animal. Furthermore, it provides filters representative of input for later processing, with back-projections serving for recall from stable memories deposited in the cerebral cortex.

HEMISPHERIC LEARNING IN THE SPLIT-BRAIN PREPARATION

The optic nerves in the brain of higher mammals are partially crossed with half their fibers passing to the visual cortex on the opposite side and half to the hemisphere on the same side (Figure 15.10).[87] By making an antero-posterior cut down through the midline between the hemispheres, through the corpus callosum plus other crossing fiber paths, the anterior and posterior commissures and the massa intermedia, the *split-brain* preparation so produced was used by Sperry and colleagues to study visual functions separated in the two hemispheres.[88] With one eye covered, the split-brain animal was trained to a visual image presented to the open eye that was retained in the visual region of the hemisphere on the same side. When that eye was covered and the other eye open, the animal acted in a naive fashion. Normal animals, trained with one eye covered and then after training tested with the previously covered eye open, were able to recognize the image – indicating that transfer of learning occurs via the interconnections between the hemispheres. Following experiments made by cutting the various interconnecting tracts separately, transfer of the learned behavior in the cat from one hemisphere to the other was shown to occur via the corpus callosum; in the monkey, the anterior commissure also transfers visual information.

RETENTION OF LEARNING AND ITS TRANSFER IN MAMMALIAN BRAIN USING SPREADING DEPRESSION

The localization of function in the cortex has classically been studied by the loss seen after making a lesion. The effects of a lesion are, however, complicated by the effects it produces in other regions and the need for recovery from the surgery during which adaptative changes can take place in other regions. And, in the split brain experiments, transfer of the learned engram

---

[87] The figure of the visual connections given are those of the monkey. Compare this figure that shows part of the optic nerve terminating on the same side and part crossing to the other side with the completely crossed optic nerves present in the rat and bird shown in Figure 11.17.

[88] (Sperry, 1962), and (Sperry, 1961).

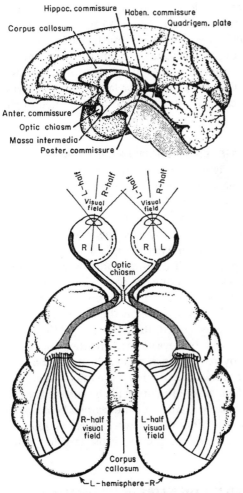

15.10. Spit-brain preparation of the monkey brain. The upper figure shows the medial surface of the hemisphere of a brain cut sagittally in the antero-posterior midline. Some of the main interhemispheric tracts connecting the hemispheres are labeled. The bottom figure shows the brain from the dorsal view. The main tract connecting the hemispheres, the corpus callosum, is either cut or left intact to show its effect on transfer. The optic chiasm is cut in the antero-posterior direction so that visual field training is carried to only the cortex of the hemisphere on that side. **L**, left; **R**, right. From (Sperry, 1961).

had to be inferred by its loss following section of the interconnecting path. It became possible to disable one cortical hemisphere temporarily to investigate memory retention in one cortex and then show its transfer to the naive cortex using *spreading depression* (SD), the phenomenon discovered

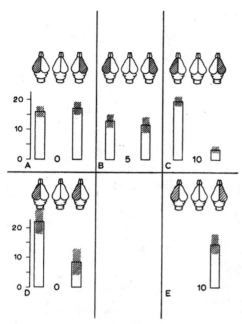

15.11. Unilateral conditioning and interhemispheric transfer. One of the rat's hemisphere was made temporarily inoperative with spreading depression produced by potassium chloride applied to the left hemisphere (**A**), this indicated by hatching. The number of conditioning trials to criterion is indicated by the bar graph. Fifteen trials were required for the rat to reach criterion. The next day, the same number of trials was required when spreading depression was initiated in the right cortex, showing that it was naive. When five trials were permitted without spreading depression present (**B**), this had no effect on the next day's test. However, when 10 trials were given with both sides open (**C**), a transfer of the training to the naive side was achieved. This was shown when the trained side was depressed, and fewer trials were required by the previously naive cortex (**D**). Spreading depression without intervening training has little effect (**E**). From (Bureš et al., 1974). Redrawn in (Ochs, 1965), Figure 25.14.

by Leão.[89] SD is elicited by a brief, electrical depolarizing current or application of potassium chloride to a point on the cortex. From that site, a wave moves through the cortex at a slow rate of 3–5 mm/min extinguishing the EEG, evoked responses, and in general all neural activity as it occupies a region of cortex. The spread occurs much as waves expanding out in a circle from a pebble dropped into a pool. Recovery of function occurring within 10 to 20 minutes after the wave of SD has passed by; but, with a higher concentration of potassium chloride, a series of SD waves is evoked that can keep the cortex depressed for longer times. Bureš and colleagues made

[89] (Leão, 1944), (Ochs, 1962), (Van Harreveld, 1966), and (Bureš, Burešova, and Křivánek, 1974).

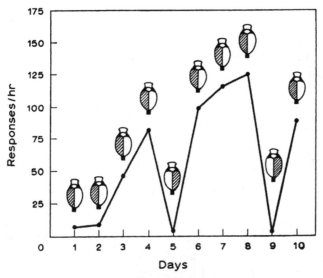

15.12. Single-trial learning engram in one hemisphere obtained using spreading depression. The rat was prepared with cups inserted in its skull so that spreading depression could be induced with potassium chloride on one side or the other of the brain. The animal was put in a Skinner box for one-hour sessions with one side depressed (hatched). If the animal presses the bar inside the box, a food pellet is delivered. On days 1 and 2, the animal makes only a few accidental bar-presses; but, on days 3 and 4, learning had occurred as indicated by high bar-press rates for food reinforcement. On day 5, the trained side (that which had not been depressed) was depressed, and the rate fell to that of a naive animal. On days 6, 7, and 8, the original trained side was again made functional and a return to high rates of bar-pressing was seen. On day 9, the untrained side was again tested and few responses were seen. On day 10, after return to the original side, the rate was high. From (Steele-Russell and Ochs, 1961). Redrawn in (Ochs, 1965), Figure 25.15.

use of the method to examine conditioned learning in the cortex.[90] Rats were classically conditioned with one cortex depressed to prevent learning in it, whereas the other cortex kept "open" could be conditioned (Figure 15.11). That learning had taken place in it was shown by depressing the trained cortex and finding that the animal no longer showed conditioned behavior. Then, without SD in either cortex, the trained cortex was able to "teach" the naive cortex, transfering of the learned engram to it. That transfer had occurred was shown when the cortex originally trained was depressed, and the animal then showed the learned behavior.

Further characterization of learned behavior was revealed using *operant conditioning* and SD.[91] Rats were put into a Skinner box with a bar projecting

[90] (Bureš et al., 1974).
[91] (Steele-Russell and Ochs, 1961), and (Steele-Russell and Ochs, 1963).

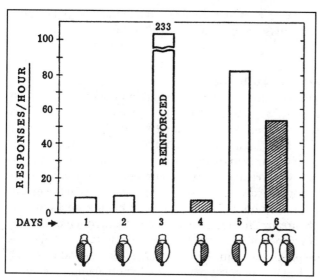

15.13. One-trial interhemispheric transfer of an engram. An engram is learned on one side after days 1 and 2 as shown by the high rates of bar-presses on day 3. With spreading depression on the learned side a failure to bar-press was shown on day 4. On day 5, high rates are found on the learned side even without food reinforcement. On day 6, transfer was achieved by letting the animal make a bar-press and receiving a food pellet with both cortices not depressed. Then, one hour later, the trained side was depressed, and the previously untrained side showed a high rate of unreinforced responses showing that transfer had occurred. From (Steele-Russell and Ochs, 1961).

into it which, if struck by the rat, causes a pellet of food to drop into a hopper. A hungry rat placed in the box will, when by chance it strikes the bar and a pellet of food appears, make the connection and thereafter continue to bar-press for pellets at rates that can be quite high (Figure 15.12). With one hemisphere inactivated by SD, rats learned to bar-press for food with the "open" cortex. When the trained cortex was inactivated by SD, the animal acted as a naive one would. Without SD in either hemisphere, the animal still behaved as a naive animal until it pressed the bar and received a food pellet reward. The learned behavior then spread from the trained to the untrained hemisphere (Figure 15.13). This was shown by depressing the trained cortex with SD and finding that the animal bar-pressed for food as it did with the previously trained cortex. Transfer required the process of reinforcement with the food reward. Whether transfer takes place from the trained cortex to the naive cortex via the callosal fibers interconnecting the two hemispheres, or through some other path – perhaps by an interaction with the hippocampus – is unknown. That axonal transport is likely to be involved in conditioned learning was indicated in other experiments by the block of consolidation seen when colchicine was administered immediately

A HISTORY OF NERVE FUNCTIONS

after training sessions with a resultant failure to retain the learned behavior. Colchicine was ineffective when administered after the phase of consolidation.[92]

NEUROGENESIS IN THE ADULT BRAIN

In the postnatal period of the rabbit, a proliferation of neurons, their dendrites and axonal processes, takes place over a period of a month or so postpartum as its cortex increases in thickness toward its mature size.[93] A similar early postnatal development is seen in the rat. In the cat, the proliferation takes longer; and in man, many months. Once the developmental period had ended, it was accepted as a rule that no more neurons arise, this based on the lack of evidence for the cells undergoing mitosis in the mature animal. Learning and memory storage in the adult brain were ascribed to changes in the cells already present, in their synaptic components. The doctrine that neogenesis does not occur in the mature animal was challenged in the 1960s by Joseph Altman using [^{3}H]thymidine to label dividing cells.[94] He saw thymidine labeling of small cells in the hypothalamus and subsequently elsewhere in the brain, including the cortex. Opposition raised to those findings was strong: It was held that the dividing cells seen were glia and not neurons. Support for neurogenesis in the adult brain came from Kaplan, who showed in electron micrographs that the cells in question were indeed neurons.[95] The labeled cells in the neocortex, olfactory bulb, and hippocampus of the adult rodent were seen to have axonal processes and dendritic extensions typical of neurons, and he even saw synaptic sites present on them. Nevertheless, this evidence was not enough to reverse the judgment against a neurogenesis occurring in the adult brain.

A clear example of neurogenesis was subsequently discovered by Nottebohm in the adult male canary in a region of their brain that increased in size when the birds learned new songs.[96] Rather than learning their songs once and for all (e.g., during a "critical period" before reaching sexual maturity), male canaries can learn songs in maturity. The birds might live for eight years or so, and each spring they learn songs anew for use in the breeding season, losing them later in the year. A young canary kept in a cage with its father will develop a song similar to his, or learn recorded songs, whereas a deafened bird will only give a simple repertoire of sounds. The region in the brain that increases in size on acquisition of learned songs, the HVc, can increase by as much as four-fold. The cells in this region receive auditory signals, and its fibers project to the syrinx in the trachea, the bird's vocal apparatus. The HVc continues to grow while the bird is in

---

[92] (Cronly-Dillon, Carden, and Birks, 1974).   [93] (Schadé and Baxter, 1960).
[94] (Altman, 1962), (Altman, 1963), and (Altman, 1967).   [95] (Kaplan, 1985).
[96] (Nottebohm, 1985).

the phase of song learning and stops growing after 7 to 8 months when the bird achieves the stable adult song. When learning the song, the cells of the HVc nucleus were found to be labeled with [³H]thymidine, indicating that it contained dividing neurons. Microelectrode records taken from the cells showed them to indeed be neurons by their electrical responses to auditory input. Furthermore, horseradish peroxidase (HRP) – a marker of retrograde transport (Chapter 11) injected into parts of the brain known to receive terminal fiber projections from the HVc – was taken up by the terminals and transported to their cell bodies. To add to the evidence for neurogenesis in the adult mammal, McEwen found that stress, or the administration of hormones producing a display of sexual characteristics, also caused neurogenesis in the hippocampus of the rat.[97]

Although the evidence so far given indicated the occurrence of neurogenesis in the canary and in regions of the brain of the rat, cat, and other lower mammals, the question remained as to whether neurogenesis takes place in higher mammals, especially in primates evolutionarily close to man. Despite an extensive search, Rakic could not find evidence for mitotic cells in the adult primate brain.[98] The few cells he did find, he considered to be glial and not neuronal. The situation changed with the introduction of a new method to show neurogenesis, the use of 5-bromo-3'-deoxyuridine (BrdU). As does [³H]thymidine, it becomes inserted into the DNA of cells undergoing mitosis. Its great advantage over [³H]thymidine labeling is that the BrdU-labeled cells can be examined with neuron-specific and glial-specific antibodies, allowing neurons to be differentiated from glial cells. By its use, the hypothalamus – especially the dentate region that is involved in early learning – was shown to contain newly developed neurons.[99] Gould found neurogenesis to occur not only in the hypothalmus of the rat, but also elsewhere in the brain and in other species, most importantly in the prefrontal cortex of the macaque, the old world monkey evolutionarily close to man. The new neurons were seen early on in the subventricular zone lining the lateral ventricles and later, after several weeks, to have migrated into the white matter of the frontal and temporal cortex, and then enter the cortex. More recently Gage and colleagues used this technique to show that neurogenesis occurs in the human brain.[100] They found on autopsy of the brains of terminally ill cancer patients given BrdU to label cancer cells in their body organs that it had also been taken up by brain cells.

The finding that neurogenesis can occur in the adult human brain is of intense interest with respect to the mechanisms of learning and memory

[97] (McEwen, 1992).   [98] (Rakic, 1985).
[99] (Rousseau, 1989), and (Gould et al., 2001).
[100] (Rousseau, 1989), (Eriksson et al., 1998), (Gould et al., 2001), and (Palmer et al., 2001.)

storage. How the new neurons interact with those neurons already present, and whether and how memories in them can relate to those previously stored in other cells are questions that come to mind. On a clinical level, investigation of neurogenesis could lead to possible treatment of such devastating brain disorders as Alzheimer's disease, Parkinsonism, and the loss of function after a stroke.

PHANTOM LIMB: THE PERCEPTION OF BODY REGIONS IN RELATIONSHIP TO THE BRAIN

Insight into the relationship of mind to the nervous system is afforded by the phenomenon of *phantom limb*. In this condition, first described in the medical literature in 1551 by Paré, the great French military surgeon, an amputated limb was felt by the patient as still being present.[101] Hartley wrote in 1749 that "a person who has lost a limb [may] often feel a pain, which seems to proceed from the amputated limb; probably because the region of the brain corresponding to that limb, is still affected."[102] A central site for the phantom limb was also proposed in 1798 by Aaron Lemos.[103] He theorized that the movement of one leg is so strongly associated with the other leg that, after amputation of one leg, the mind attempts to supply what is missing to restore the original association. However, there are reports in the literature of patients born without limbs who have the experience of a phantom limb. This indicates that the phenomenon is not the loss of a learned association, but rather that the sense of the limb is innate.

The phantom limb phenomenon suffered relative neglect in the medical literature, for the most part discounted as hallucinatory or a mental aberration until, in the latter part of the nineteenth century, it was brought forcibly to the attention of physicians and the reading public by S. Weir Mitchell (1829–1914).[104] He and his colleagues, George Morehouse and William W. Keen, cared for Civil War soldiers in a special hospital set up to tend to the large number of cases of nerve injury.[105] They found that a majority of amputee patients experienced a phantom limb and in a number of them severe persisting pain. The phantom limbs in most cases could appear to be moved at will – the arm flexed or a hand opened and closed. In some cases,

---

[101] (Price and Twombly, 1978). Intimations of the phenomenon have been found couched in folkloric and miraculous terms in medieval and early renaissance literature. A still earlier awareness of the phenomenon is indicated by Lucretius in his poem in which similar observations of severed limbs were given in part in Chapter 14.

[102] (Hartley, 1967), p. 32.   [103] (Price and Twombly, 1972).

[104] Mitchell was not only a leading neurologist of the time, but also enjoyed a reputation as a widely read author of popular novels. His story of Dedlow, who suffered loss of all four limbs and experienced them as phantom limbs, based on his clinical experiences (Ochs, 1997), was a great popular success and still worth reading.

[105] (Mitchell, 1965).

the position of the phantom was immovable. Livingston gives a graphic description of such a case in which the hand was felt to be clenched, unable to be opened and the nails of the fingers digging into the palm giving rise to an unremitting, excruciating pain.[106]

Mitchell believed that irritation of the neuroma at the cut end of the nerve in the stump was the cause of the painful phantom limb. However, injection of narcotic into the stump, or excision of the neuroma, gave only temporary relief. In more recent times, dorsal roots and fiber tracts in the cord carrying pain sensations were sectioned to alleviate pain. This helped for a time, but again the pain eventually returned. Even surgical removal of the sensorimotor cortex did not give lasting relief – the pain returned. Such results pointed to some other locus in the brain for the phantom limb. Melzack proposed that a network of neurons extending widely throughout the brain, a "neuromatrix" that had been formed, as by a Hebbian cell assembly, was responsible for phantom limb pain.[107]

Nevertheless, participation of the specific sensory receptor areas in the cerebral cortex does play a role in the changed perception of the phantom limb. The delimitation of sensory regions in the cortex had classically been determined by the sensory loss observed after making lesions in those regions of the cortex.[108] Additionally, those regions of primary sense reception were shown by the evoked electrical responses recorded in them in response to optic, auditory, or somesthetic stimulation.[109] On stimulation of the skin at points over the body surface, cortical responses were mapped in animals ranging from the rat to man. The primary sensory receptor area so determined, the somesthetic region, represents the body surface by a figurine or "homunculus." In the monkey and man, it was found to run in a band from the top to bottom of each of the hemispheres located just posterior to the central sulcus.[110] The size of the cortical representation for a given part of the body surface, depicted on the cortical surface, depends on the density

---

[106] (Livingston, 1943).   [107] (Melzack, 1990).

[108] Minkowski, at the turn of the twentieth century, showed the presence of the visual region in the cortex by the use of localized lesions.

[109] The pioneer use of the evoked response technique by Marshall laid the foundation for the topological representation of sensory modalities in the cortex in early such studies extended by Bard, Woolsey, Mountcastle, Hubel and Wiesel, and others (Bard, 1956).

[110] The primary somesthetic area SI that was seen as uniform area in earlier studies was shown later in the macaque cortex to be composed of five parallel strips alongside one another, each strip generally representing a similar topographical representation of the body. Neuronal interconnectivity between strips and other sensory regions outside SI (e.g., of the second somatosensory region SII) were found. Their connectivity was shown by injection of the hand, foot, and face regions with horseradish peroxidase. Taken up by axon terminals in those regions, they were retrogradely transported to other sensory regions, such as SII, with which they interconnect.

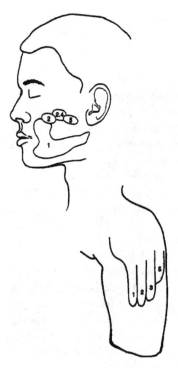

15.14. Place on the upper arm where stimulation of the face evokes a phantom hand in the upper arm of a patient whose lower arm had been amputated 10 years earlier. Fingers labeled 1 to 5 indicate those of the phantom hand on the upper arm corresponding to sites on the face from which they are perceived. From (Ramachandran and Blakeslee, 1998), Figure 2.

of innervation and the use made of that part of the body by a given species. In man and monkey, the individual fingers take up a very large proportion of the sensory area with respect to the rest of the body. In animal species where the use of the hand is much more limited, or in four-footed animals in which a foot is present instead of a hand, its area of representation in the cortex is smaller or missing altogether.

These studies were thought to indicate that the primary sensory receptor areas are fixed. It was therefore surprising to find in the cortex of adult macaque monkeys, in which their dorsal roots conveying input from the arm had been severed twelve years beforehand, that massive changes had occurred in the somesthetic cortex.[111] The neurons in that region to which sensory input from the hand normally projects were electrically silent when stimulating the skin of the paralyzed hand. That was expected because the afferent fibers from the arm to the central nervous system had been cut. However, on stimulating the adjoining face region, a marked increase of neuronal activity within the previously silent somesthetic receptor area for the hand was recorded. This indicated that a neuronal spread from the face area had entered the denervated somesthetic hand area. Inspired by

---

[111] (Pons et al., 1991).

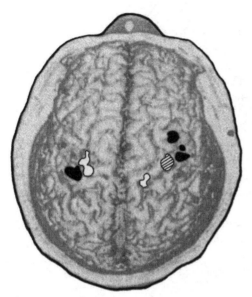

15.15. Magnetoencephalography images taken from the brain of a patient whose right arm had been amputated below the elbow. The cortex of the right hemisphere to which the left arm projects shows the usual location of the hand region (hatched), the face (black), and upper arm (open) are present. On the left hemisphere, where sensory input from the right arm is missing, activity from the face and upper arm regions has shifted to overlap the missing hand region. Modified from (Ramachandran et al., 1998), Figure 3.

these studies, Ramachandran examined humans who had suffered the loss of a limb to look for such a spread. He found in patients whose lower arm had been amputated that stimulation of their face by stroking it with a cotton swab could evoke a phantom of the missing hand (Figure 15.14).[112] In the homunculus representing the body surface of the human, the face and hand regions are seen to be closely associated, suggesting that neurons from the face area had grown into the quiescent hand region to give rise to the phantom limb. Cortical shifts in these sensory regions were looked for using the technique of magnetoencephalography, which shows the activity of cortical regions directly through the intact skull.[113] In those patients with a phantom limb, the sensory region for the face had shifted its position to overlap the hand region. So also did a shift occur in the cortical region of the upper part of the arm from which a phantom of the lost lower part of the limb could be evoked (Figure 15.15).[114]

[112] (Ramachandran and Blakeslee, 1998), Figures 2.2 and 2.3.
[113] (Rezai and Mogilner, 2002), and (Singh, 2002). Magnetic resonance imagining does so indirectly, showing an active region of the brain by the increased blood flow associated with it.
[114] Modified from (Ramachandran and Blakeslee, 1998), Figure 2.3.

The rapidity with which a phantom limb appears after amputation, in a matter of a day or less, makes it difficult to account for a growth of fibers from the face area into the hand region. The distance seems to be too great. An alternate possibility is that interconnections between these regions preexist, but are normally inhibited by the sensory information normally coming into the sensory receptor area. When sensory input to it is missing, as in the example given in which the hand input was cut off, the preexisting connections from the face area are no longer inhibited so that, as in the example given, stimulation of the face evokes a phantom hand.

Ramachandran gives other examples of patients with a phantom that can be evoked from another sensory region close to it in the cortex. After the loss of a foot, which on the homuncular map is close to the genital region, the phantom of the foot could be evoked when the genitalia are activated in sexual union.[115] Other projecting body parts, the nose and teeth, may also show phantoms after their loss. The sensory modalities are preserved in the phantom. Warm water on the face is felt as warm in the phantom limb, and spilled on the face, it was sensed as trickling down the phantom arm.

In addition to the presence of a phantom experienced after the loss of a limb or body part, a supernumerary phantom can be experienced after a cortical lesion.[116] This was reported in a right-handed woman who had a stroke in her right frontomesial cortex. She then sporadically experienced a supernumerary "ghost" left arm – one that was present along with the perception of her normal left arm. The patient was aware that this was an illusion, but at times there was confusion of the supernumerary arm with the normal arm.

A change in the function of a primary sensory region by another sense was shown experimentally by routing optic nerve input into the auditory region.[117] Something like this may occur in the profound disarray of sensory connectivity present in the autistic brain. Examination of a unique autistic patient by Merzenich and colleagues has given unusual insight into the autistic mind. This youth was able to communicate his sensory experiences.[118] He was unable to appreciate several senses simultaneously and felt forced to concentrate on one sensory modality at a time. He also showed a defect of sensation, with seconds needed to experience a sound or a light flash rather than in a normal fraction of a second. Microscopic examination of autopsies made on the brains of autistic children has shown the presence of many more neurons in the cortical columns of their sensory regions, along with the many more connections they make than is

---

[115] Ibid., pp. 33–38.   [116] (McGonigle et al., 2002).
[117] (Merzenich, 2000).   [118] (Blakeslee, 2002).

normal. Such hyperconnectivity may be the basis for the sensory defects experienced by the autistic child.

The shifts in the area representing sensory areas, with an apparent replacement of a sensory area in the cortex, may not actually originate in the cortex but in the thalamus. The thalamus is the relay station for upward projections of sensory inputs to the cortex and also a projection of fibers from the sensory regions of the cortex down into those same regions of the thalamus. When sensory input to the thalamus of one modality is lost after denervation, its thalamic neurons become sensitized, allowing neurons of another sense modality to excite them. This reorganization of the thalamus was found in macaque monkeys with a limb that had long been amputated.[119] Mapping neurons in the thalamus relaying face sensory representation showed that it had expanded into the area normally occupied by the hand region.

Revelations of a genetic basis for disruption of higher functions based on our recent acquisition of the genetic code are just at their beginning. Schizophrenia – characterized as an impairment of planning, abstract reasoning, and judgment – has recently been associated with changes in chromosome 8 in a group of Icelandic schizophrenics. The defect causes a change in the signaling protein, *neuregulin1*, leading to the accumulation of wrongly formed synapses.[120] Another study has associated chromosome 6 with schizophrenic in an Irish population.[121] The gene produces a protein, *dysbindin*, which may cause schizophrenia by another process. Both genes acting together may be required to produce the full-blown disease.

Studies of the frontal lobes in monkeys have indicated they have to do with the control of a higher, complex form of learning – that of a delayed response.[122] Frontal lobotomy, performed as a supposed therapy for various psychiatric conditions,[123] a practise fortunately now abolished, gave rise to schizoid symptoms such as difficulty in planning, performing appropriate social behavior despite knowing what those behaviors are, and lack of affect. A recent, thorough histological study of the prefrontal lobes of schizophrenics has shown a decreased thickness of their prefrontal cortex.[124] This would fit with a decreased synaptic connectivity and a diminished neuropil (Chapter 15). Although magnetic resonance imaging (MRI) has not consistently shown the decrease in the prefrontal cortex, a recent MRI study has indicated a loss of gray matter in the bilateral superior temporal gyri and anterior/hippocampal gyri.[125] There may be different loci giving somewhat similar syndromes of schizophrenic-like behavior. With the improved tools

---

[119] (Pons et al., 1991), and (Jones and Pons, 1998).   [120] (Stefansson et al., 2002).
[121] (Lynch and Staubli, 1993), and (Straub et al., 2002).
[122] (Stamm and Pribram, 1960), and (Stamm, 1961).
[123] (Fulton, 1951).   [124] (Selemon et al., 2002).   [125] (Anderson et al., 2002).

in hand and those in the offing, it is likely that the matter will be resolved before long.

Study of the aberration of perception present as a phantom limb in patients related to neural changes found in their brain and the genetic relation to behavior have philosophical implications bearing on the enigma of how the mind can be accounted for by the actions of neural networks that are discussed in the next chapter.

16

EPILOGUE: WITH OBSERVATIONS ON THE
RELATION OF THE NERVOUS SYSTEM TO MIND

Our aim in this chapter is to consider how what we now know of the nervous system may account for understanding human behavior. Rather than the early view of animal spirits and surrogates for them, our view of the nervous system is that it is composed of neurons, with a mechanism of axoplasmic transport in them as schematized for a peripheral neuron in Figure 16.1. The same axonal mechanism is present as well in all neurons in the central nervous system. An example is that of cortical neurons crossing from one hemisphere to the other via the callosal tract.[1] In addition to components required to maintain the viability of fibers and to provide the neurotransmitters acting at synaptic junctions. Other molecular signals are transported between neurons to establish and maintain the networks responsible for the integrated behavior of the organism. Networks formed in the brain under genetic control are responsible for perception, cognition and memory and are not permanently fixed. They are modified in the course of learning (Chapter 15). The broader theoretical issues relate to the degree of innateness of sensory and perceptual processes and how change is brought about.

Up through most of the latter century, the concept of reflexes was held as the key to understanding behavior. In his great book on reflexes of the nervous system, Sherrington wrote:

> The reflex-arc is the unit mechanism of the nervous system when that system is regarded in its integrative function. *The unit reaction in nervous integration is the reflex*, because every reflex is an integrative action and no nervous action short of a reflex is a complete act of integration. The nervous system of an individual

---

[1] The technique used to show this was to place labeled amino acid precursor at a site on one hemisphere and then allow time for transport of labeled proteins into the fibers terminating at a symmetrical place on the opposite hemisphere from which they were isolated in synaptosomes by differential centrifugation. (Bao, 1972), and (Bao and Ochs, 1972).

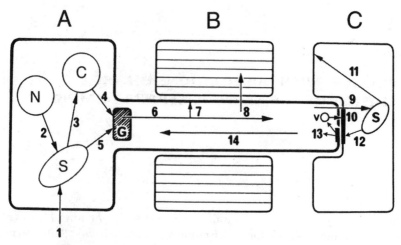

16.1. Diagrammatic representation of a neuron to show the functions controlled by axoplasmic transport. Neuron regions indicated are: **(A)** cell body, **(B)** nerve fiber, and **(C)** terminal region with a synapse on a muscle fiber. **(A)** On uptake of labeled amino acid precursor **(1)**, the nucleus **(N)** controls **(2)** the level and type of synthesis **(S)** of labeled proteins, with the arrow **(3)** showing a compartmentalization **(C)** of some of the synthesized materials in the cell. From the synthetic site and from the compartment, labeled materials move at different times into the axon **(4, 5)**, with **(G)** the Golgi apparatus controlling their egress into the axon. The transport system moving materials down within the nerve fiber is shown by **(6)**. Arrows from it to the membrane **(7)** indicate the insertion of channels, ion pumps, and other components into the axonal membrane. Some components pass to the Schwann cell (horizontal lines) **(8)** to have an effect on the form and function of the myelin sheath. A retrograde axoplasmic transport in the axon directed back to the cell body is indicated by an arrow **(14)**. In the terminal region **(C)**, vesicles **(v)** in the terminal involved in synaptic transmission is shown where the release of their neurotransmitter contents into the synaptic cleft is effected. The transmitters bind to the membrane of the effector cell **(10)** it synapses on (in this example a muscle fiber). Used-up transmitter vesicles and other molecular species taken up by the terminal **(13)** are carried back to the cell bodies for reuse along with molecular signals modulating synthesis. Synthesis of receptor proteins within the muscle is indicated by **(S)** with trophic control from the nerve terminal **(9)**. The synthesis includes the receptor proteins in the muscle membrane **(12)** and components to the rest of the membrane **(11)** responsible for its electrical properties. If the trophic supply is lost, the receptor proteins normally restricted to the receptor area, the end-plate, spreads out over the rest of the fiber membrane. Modified from (Ochs, 1975b), Figure 1.

from what without integration "were a mere aggregation of commensal organs resolves itself into co-ordination by reflex actions."[2]

Sherrington considered that, from the unit reflex, compound reflexes are formed, which are responsible, or at least involved, in higher behavior. He pictured the study of the physiology of the nervous system as being carried

---

[2] (Sherrington, 1947), p. 7.

out from three main points of view: (1) the basic aspect of the neuron as a living cell that requires nutrition and other support functions common to all cells; (2) the special role of nerve to conduct the nerve impulse; and (3) higher integration of neurons, those intracellular and extracellular mechanisms by which the elongated processes of the neuron give speed and "a nicety of adjustment allowing for the coordination of reflex actions."

For the first item, axoplasmic transport is the mechanism that provides proteins and other materials necessary to maintain the structure and viability of the neuron and its fibers (Figure 16.1). For the second item, axoplasmic transport carries down specific key components: the ion channels and ion pumps by which resting and propagated action potentials are generated. For the third item, Sherrington viewed propagated action potentials as the agent of integration, but these are evanescent events; and a permanent change is needed to account for long-lasting learned behavior. Here, too, axoplasmic transport plays a role. By its means, signals carried by the neurons to their terminals and released there are carried by other neurons to their nuclei to form enduring changes in networks in the course of learning and the retention of memories in the brain (Chapter 15).

Reflexes are present from the onset in the embryo as an integrated whole rather than being formed by a sequential chaining of unit reflexes in the course of development (Chapter 14). Additionally, in other species, neural networks similar to those in humans can be traced from our closest evolutionary animal forms, the anthropoid apes, to still earlier animal species and back down to still earlier species. The continuity of the species as they have undergone evolutionary changes is indicated by the similarity of their genetic compositions. With our closest relative, the chimpanzees, we share 98 percent of the DNA in our genetic codes. What happens in the course of evolution appears to be successive modifications of the primitive forms rather than complete de novo transformations. The course of species changes is seen repeated in the embryo as ontology recapitulates phylogeny.

LEARNING PARADIGMS IN THE STUDY OF HIGHER FUNCTIONS

The introspectionists following the lead of Wundt toward the end of the nineteenth century turned to a study of their own conscious states. James elaborated on the characteristics of introspection, with what he termed the *stream of thought*.[3] He dealt with hallucinations, dreams, hypnotism, and religious experiences as well.[4] These phenomena are related to the *terra incognita* of the subconscious, of which the study by Freud is most well known.[5]

---

[3] (James, 1890).   [4] (James, 1963).
[5] (Freud, 1920). Paranormal phenomena was collected by the Psychical Research Society (Myers, 1903). That described in the first volume is of interest, that in the second volume is fantastic.

In a reaction to introspection, the behaviorists[6] devised animal experiments in which problem-solving using mazes and other such devices. In their desire for objectivity, the mind was resolutely left out of the picture. From about the middle of the twentieth century, psychological thinking swung back to the study of the mind by the cognitive psychologists. The intellectual capabilities of chimpanzees were of special interest. They were shown by Köhler to have what had generally been thought of as a prerogative of the human, namely the ability to use tools to attain a goal.[7] To reach food suspended above them out of their direct reach, they set up boxes to reach it, or when given a tube and rod that were each too short to reach the food, they fitted the rod into the tube to extend its length to accomplish their goal, indicating insight in them.[8] This view was further advanced by the remarkable long-term studies of troops of chimpanzees in the African wild of the Gombe National Reserve of Tanzania by Jane Goodall.[9] Her patient, close observations showed the members of the troop to have a rich social life – the individuals having differing personalities and emotional displays. She also saw them make and use tools to achieve a goal. They carefully fashioned a blade of stiff grass and inserted it into an ant mound to pull out termites to eat. In still later studies, she found one community of chimps modeling their tool-making skills on that of another community – evidence indicating intellectual ability. The key differentiation between humans and these closest of our animal relatives is our faculty of speech by which the complexity of individual experiences can be passed on to the social group. It is the accumulated historical experiences that allowed for the evolution of human social groups and, later on to become a civilization.[10]

SELECTIVE CHANGES OF PERCEPTION

How sensation gives rise to perception is an age-old problem.[11] For the most part, sensation is viewed classically as bits of sense-data passed onto

---

[6] (Watson, 1914), (Skinner, 1938), and (Ferster and Skinner, 1957).

[7] (Köhler, 1927).

[8] The German term for their behavior used by Köhler was "Einsicht," translated as both insight and intelligence (Köhler, 1927). Note p. 219.

[9] (Goodall, 2000).

[10] Social Darwinians emphasize the belief that cultures evolve mainly if not wholly by competition, the concept introduced by Spencer in the mid-nineteenth century. It has assumed a cult following today by some who foster the concept that the superior culture will and should win out with the inferior weak and unfit left to die out or be eliminated – a concept that had ruthlessly been acted on by the Nazis. In contrast is the view that diversity and cooperation are also positive values in the evolution of a culture as had been shown by Goodall in her animal studies. That point had been made earlier by Darwin himself (Darwin, 1952), Section on Sociability in the Descent of Man, p. 305 ff.

[11] (Hamlyn, 1961).

primary receptor cortex and then to secondary cortical regions where perception is realized. Recent close clinical studies of sensory and perceptual abnormalities (some aspects of which were discussed in Chapter 15) have thrown more light into the processes that are involved in perception in brain–damaged patients. After loss of a limb, its remaining presence may be felt as a phantom limb showing the ability of the mind to construct conscious awareness. A phantom limb may also appear in addition to that of a normal limb.[12] A curious alteration of perception may occur with the loss of recognition of the face of a parent or spouse, or to consider them as impostores. Such *prosopagnosia* may extend to the inability to recognize one's self in a mirror.[13] The patient tries to carefully piece together the elements of a face and can eventually succeed if there is a distinctive feature that serves as a signal triggering recognition. What is missing in the prosopagnostic is the immediate, emotional grasp of the signs of a threat, fear, or joy.[14] Such instant recognition is of paramount importance for survival of primitive people and animals in the face of friend or foe, prey, or an onrushing predator. Such ready appreciation is a feature that could well have evolved to preserve the species.

This immediacy of recognition is to be contrasted with the generally accepted view that perception is an integration of sensori-motor elements. This was described by Luria.[15] He stated that visual perception "is a complex, active process, consisting of the identification of individual signs...the integration of these signs into groups, and finally, the selection of their meaning from a series of alternations. This is a complicated reflex process requiring the participation of sensory and motor apparatuses, particularly the apparatus of eye movements responsible for performing orienting and exploratory movement." Luria gave a picture of a pretty girl and alongside it the eye movements of a normal observer. The movements are concentrated on the eyes and mouth and to a lesser extent range over the rest of the head. These eye movements are presumed by Luria to be part of the process of a sensori-motor synthesis leading to perception. Such a process seems to be similar to that of the prosopagnostic patient rather than the immediate recognition by a normal person.

---

[12] (McGonigle et al., 2002).
[13] (Sacks, 1970). In the case of the man who mistook his wife for his hat, the patient suffered from a tumor in the visual area. He was a highly accomplished musician who used music to help identify things and places.
[14] (Darwin, 1886). The instant recognition of a person as soon as seen was a feature recognized by Aristotle in his analysis of sensation (Guthrie, 1981), p. 213. An example he gave indicating immediate recognition of a person was that one can directly *see* Socrates.
[15] (Luria, 1962), pp. 134–135.

In recent years the availability of functional magnetic resonance imaging (fMRI) has allowed studies to be made in the human that correlated sensory and perceptual events with neuronal changes in discrete brain regions. Regions outside the striate visual cortex were found that are face-sensitive.[16] The fusiform gyrus on the right side of the cerebrum was found to be significantly most active in this respect.[17] Lesions in the face-sensitive region impairs perception of faces, the defect seen in prosopagnosia patients.[18]

The converse clinical case to the prosopagnostic is the autistic *idiot savant* who shows a hyperawareness and exquisite sensory detail and memory. In the case of idiot savant twins described by Sacks,[19] a box of matches dropped to the floor was seen immediately to consist of 111 matches. The twins then further showed their fantastic number sense by quickly factoring the number into its primes of 37 and 3. They were able to give the prime numbers into numbers of 10 digits and remember events in their life in minute detail down through the years.

SENSORY-PERCEPTUAL PROCESSES

It was Descartes' fundamental view that two substances exist, the thinking immaterial substance of the mind and the extended material substance of the body. Somehow, the sensory impressions carried by the material processes in the nerves to the brain can interact with the immaterial mind, the soul that he placed in the pineal gland that carries within it the inborn ideas of God, morality, and so on (Chapter 5). In opposition, the English empiricists, notably Hume and Locke, held that all that enters into human consciousness comes through the senses, arguing against being born with any innate ideas.[20] In the early part of the twentieth century, studies of Adrian, Mathews, Erlanger, and Gasser showed repetitive action potential discharges in single sensory nerve fibers when their receptors were excited by stimuli appropriate to them.[21] These impulses pass to the brain to terminate in primary sensory regions of the cortex (Chapter 7). The evoked electrical responses in the visual cortex – in response to light flashes in the visual field, touches to the body surface or sounds of different frequencies – are each represented by maps. The somesthetic map representing the body shows variations in the area corresponding to the touch of different parts of the body that depends on the animal species. In animals such as the monkey and human, where the hand is so useful for grasping, the thumb and digits occupy a much greater territory than the feet of four-legged

---

[16] (Puce, et al., 1995).   [17] (Kanwisher, 1997).
[18] (Barton, et al., 2002), and (Barton and Cherkasova, 2003).
[19] (Sacks, 1970), p. 185 ff. A vivid example of such an autistic idiot savant was portrayed by Dustin Hoffman in the movie "Rain Man."
[20] (Hamlyn, 1961).   [21] (Adrian, 1928).

animals (Chapter 15). In the visual system, a map of the visual field had been gained by recording evoked responses in the visual cortex corresponding to the position of a point of light flashed in the visual field. This suggested an analogy to that of a television image formed by a raster movement of the beam of a cathode ray tube over its face. However, the neurons in the visual cortex do not simply respond with a uniform discharge of neurons. Individual neurons in the visual cortex recorded with microelectrodes have shown differences in their responses to light stimulation.[22] Some of the neurons respond preferentially to the onset, others at the termination of a light flash. Others respond to a short bar of light stimulation rather than to a punctate flash of light, with neurons that respond to the bar set at one or another preferred angle in the visual field, or to the movement of a bar in one direction or other. How these various responses are organized to represent an object in the visual field – given its spatial configuration, shading, edge contrast, and color – has been termed the *binding problem*. The various objects of sensory input are presumed to be perceived in secondary sensory cortical regions, after transfer from the primary receptor areas, by a process not well understood.

THE INNATENESS OF PERCEPTIONS

While Locke considered the brain of the newborn to be a clean slate on which sensory input forms ideas, evidence that some perceptions are in fact innate has been noted (Chapter 15). Patients born without an arm or leg may yet perceive its presence as a phantom limb, evidence that the body image is laid down genetically. The experimental observations of ethologists also have shown that certain perceptual expectations of objects in the world are innate (Chapter 14), as in the experimental example given of newly hatched birds that exhibit cowering responses to the specific shape of a model of a predatory falcon passed overhead.

Yet, such experiments touch on only a fraction of the wealth of observations showing that innate propensities are present for certain sensory input that are indicated by animal and human instincts. Darwin[23] dealt with instincts as evolutionarily determined. Just as small variations of body structures can become selected and lead to species differentiation, so do slight spontaneous behavioral deviations in individuals lead to their retention and eventually – if they are advantageous – pass on to the species as new behaviors. The process is exemplified in the domestication of animals, where the various breeds of dogs have been selected to exaggerate one type of behavior or other (e.g., the ferocious pit bull, the docile poodle).

[22] (Hubel and Weisel, 1959).
[23] (Darwin, 1952) in *The Descent of Man*, Chapter 8.

Instincts with their conjoined perceptual powers are not fixed and machine-like in the individual. They show variations in their performance, depending on the contingencies of the circumstance. This has been seen in animals as low in the scale of life as insects.[24] Their ability to respond differently, depending on conditions, meets the criteria of thinking given by Humphrey in his summing up when he states that "thinking...may be provisionally defined as what occurs in experience when an organism, human or animal, meets, recognizes, and solves a problem."[25] The innateness of specific sensory patterns is evident in the response to the display of shape, color, and body designs, as well as ritualized dance-like body movements releasing sexual behavior in the courtship of birds.[26] Such patterns of sexual display are seen extended throughout the range of the animal kingdom.

Facial expressions are of signal importance in human interactions. In his study of the muscles of the human face underlying emotional expression, *G.-B. Duchenne du Boulogne* (1806–1875)[27] used point electrodes to electrically stimulate individual muscles. On stimulating the muscle moving the eyebrows, he was able to evoke an expression of suffering that appeared to include other distant muscles of the face, that of the mouth and nasolabial fold that reinforced the expression of suffering. At first he considered this the result of a reflex spread. However, he determined that the "apparent general of the face was only an illusion...when we see a limited movement and recognize the perfect image of an emotion, it seems to us that the face

[24] (Fabre, 1918), Chapter 10. In Fabre's close study of spiders, the banded Epeira, a small fly or moth landing in her net is rapidly revolved by the spider while she ties silk threads around it from her spinnerets. Then, paralyzed by injection of a venom, the fly's body juices are sucked out. For a larger prey, such as a praying mantis that cannot be so turned, the spider throws a spray of bands or sheets of silk at it until it is overcome and then injected with pacifying venom. But tactics could change from throwing silk bands to revolving its prey when a large beetle, but one not as dangerous as a praying mantis, falls into the net, and silk is thereby conserved. Placing a model of an insect made of red wool on the net and shaking it with a straw to alert the spider, it swaths the model in silk, injects it with venom but then realizes the trick and ejects it from the web. There are more clever spiders who approach the model, explore it with their palpi and after a brief examination eject it from the net without wasting silk and venom on it.

[25] (Humphrey, 1951), p. 311. The relation of neural processes to perception in instinctive behavior is indicated by the effect on the spider's web construction after exposure of spiders to psychotropic drugs. Caffeine, marijuana, benzedrine, and chloral hydrate caused webs to be produced with varied defects, from partial webs with holes, to webs consisting of a few threads strung together at random. Pictures of such defective webs may be viewed on the Internet at www.chaosrealms.net/spiders/html.

[26] (Darwin, 1952), see also (Romanes, 1883).

[27] (Duchenne de Boulogne, 1990), (Duchenne, 1855), and (Duchenne, 1871). Duchenne studied inmates of the Salpêtrière, formerly an arsenal building in Paris which expanded to over 100 buildings serving as an asylum and hospice for indigents and little-treated invalids.

has changed in an overall way." And most significantly he added, "when we experience such illusions, it is by virtue of our own organization, by virtue of a faculty we have possessed since birth."[28] Facial changes are thus part of an innate signaling mechanism for the exhibition and recognition of emotions.

Charles Darwin[29] collected data showing that the recognition of facial expression is generalized, present in infants, the insane, and in races of man in disparate cultures throughout the world, in Australian aborigines, Malay Chinese, Indian tribes in North America, natives of Borneo, Malacca, India, Ceylon, Africa, etc., pointing to its innate basis. The chimpanzee has a similar facial musculature and human-like expressions of pleasure, pain, anger, frowning, astonishment, and terror. Even more subtle expressions of individual personalities were differentiated in them by Jane Goodall that body movements such as patting, kissing, embracing, hugging, or jumping for joy signaled emotions.[30]

Darwin points to the expressions and associated body movements commonly seen in animals: The hostile dog walking upright and stiffly, head upright, tail erect, the hairs bristling along the neck and back, eyes with a fixed stare, its canine teeth exposed. In contrast, is a dog's docile behavior where the body slinks or crouches, the tail lowered and wagged from side to side, the ears depressed, lips loose.[31] The cat threatened arches its back, mouth opened and, more savagely crouches, shows its teeth with claws extended preparatory to an attack. In contrast, the affectionate pet stands upright with back slightly arched, the tail held stiff and perpendicularly upwards, ears erect and pointed, as she rubs against the leg of her master with a purr instead of a growl.[32]

PHILOSOPHICAL ASPECTS OF SENSATION AND PERCEPTION
Perception does not appear to be a simple construct following on passively received sensory reception. Higher processes enter at the most elementary stage of sensation. This concept was introduced by Kant.[33] In his philosophy, an object in space is recognized as such by an a priori formal condition of the mind that is innate for space. This is a framework in which objects

---

[28] (Duchenne de Boulogne, 1990), p. 15.   [29] (Darwin, 1955).
[30] (Goodall, 2000).   [31] (Darwin, 1955), pp. 50–51.   [32] Ibid., pp. 56–57.
[33] Kant's philosophy (Kant, 1961) is notoriously difficult to understand, and an introduction to it is most helpful (Kemp, 1968), as is Kant's own introduction to his work given in his *Prolegomena*, in the preferred English translation of Lucas (Kant, 1953). Kant uses the term *a priori* to indicate innate processes that are independent of previous knowledge from those that are empiric, *a posteriori*, (i.e., given by experience). *Analytic* truths are those derived logically by the mind as distinguished from *synthetic* truths, those that require some input from without. Kant shows that *synthetic* a priori truths are possible, that is, by internal innate categories of space and time, which allows an external object to be intuited and perceived as such.

are recognized as such by the percipient, objects that are given to us as an *intuition*.[34] Kant does not take the position that the objects we experience are wholly constructed by the mind, as in the idealistic philosophy of Berkeley. Kant views objects in the world as real and known to us as *phenomena*, though not as a thing "in itself." Without the innate a priori capability of differentiating an object from the ground of other sense data in the visual field, no perception of sensory objects would be possible. We can consider that the recognition of an object is brought about by those neuronal elements responding to some special aspects of objects in the world. The individual neurons in the primary visual cortex responding to bars of light with certain orientations, colors, contrast at their edges, and so on, must somehow be brought together – allowing an object as such to be sensed in space. By another a priori innate property, that of time, we can recognize an object as being the same when its position in space relative to other objects changes with time.

Tolman found that rats form a cognitive map of their environment – one that appears to have been learned – whereas O'Keefe and Nadel, from their experimental studies, concluded that the cognitive map is innate.[35] They examined single-cell responses in the CA1 field of the hippocampus of rats using implanted electrodes that allowed them to roam freely in a maze. Responses were found to be associated when the animals were in specific places in space.

HIGHER COGNITIVE FUNCTIONS
In addition to an a priori spatial and temporal framework of sensibility, Kant also considered pure faculties of reason present that are "transcendental." Among the categories required for reason to occur is that of cause and effect. As Hume had shown before him, no amount of pairing of one sensory event with another could give rise to the concept of one event causing the other, the philosophical position taken by the associationalists.[36] For Kant, the category of cause and effect must be innately present to

---

[34] The use of the term *intuition* by Kant is of special importance. Its etymology is from the latin *intuere*, "to look into directly." It implies direct apprehension as opposed to deliberation or conscious reflection and it is opposed to instinct, which implies action. The term was used in medieval times when it was associated with an angelic or other higher source that allowed for apprehension by the senses. Kant made use the term to indicate an innate process used when apprehending an object in space (Kant, 1953), p. 36.

[35] (O'Keefe and Nadel, 1978). The authors have an extensive introductory chapter on the philosophical development of the concept and essentially accept the Kantian position.

[36] In British philosophy, James Mill (1773–1836) was a chief proponent of associationism (Peters and Mace, 1967). His son, John Stuart Mill (1806–1873), modified it by introducing the concept of "mental chemistry," considering that, like chemical compounds that exhibit properties not present in their elements, mental concepts

allow for the formation of such an idea. He gave a number of other a priori categories in his critique of reason that were based on logic.[37] The broad current of neurological thinking and experimental psychology has, however, remained Lockian and associationalist for the most part. In the first half of the twentieth century, behaviorism, the approach introduced by Watson,[38] was most popular. Motor performance was considered as essential in behavior as it was in the hands of Skinner and his associates,[39] where the mind was forsaken, replaced by the recording of objective performance. Then, by approximately the middle of the twentieth century, the field of cognitive psychology gained prominence, and the analysis of perceptual functions of mind came under intense study.[40]

In addition to evidence for innate processes for somesthesis and vision, audition, particularly the use and understanding of a language, has similarly been dealt with more recently as having an innate component. Pinker strongly challenged the view of the mind as a blank slate at birth. He gave evidence that the brain has an inborn organization that acts as a generalized grammar, one with internal rules necessary for understanding and learning the use of a language – the concept introduced by Noam Chomsky.[41] This does not mean an inborn ability to learn a specific language such as English or Chinese, but that the various languages formed within a given speech environment are based on an innate neural structure organized for language. Language ability seems to be newly evolved in humans. Attempts to teach chimpanzees a language have failed. They can only imitate sounds or learn specific sounds or sights by conditioning. Missing in them are those structures in the brain needed to form the concepts that lie behind the use of a language. In line with this conception of an innate neural basis for language, Pinker described studies of human birth defects, in which failure to use language was related to specific genetic faults.[42]

GESTALT PSYCHOLOGY

The Gestalt psychologists have pointed to a higher order of integration involved in perceptual processes, a "wholeness" that participates in the

may arise from compilations of sense data, which are then no longer evident as such (Peters and Mace, 1967).

[37] (Kneale and Kneale, 1962). The Kneales make the point that Kant followed the logic of Aristotle, and although this was sufficient for his philosophy, logic has progressed since Aristotle's time, especially so in the nineteenth century with the introduction of mathematical logic. But these arguments are subordinate to Kant's fundamental conception that the mind has innate properties that are required for the primary appreciation of objects. See also (Kant, 1988).

[38] (Watson, 1914).   [39] (Skinner, 1938), and (Ferster and Skinner, 1957).

[40] (Strawson, 1966). Strawson has been influential in leading English philosophy in this direction. See (Warnock, 1958), (Strawson, 1959), and (Baars, 1986).

[41] (Pinker, 1994), and (Pinker, 2002).   [42] (Pinker, 2002), p. 48.

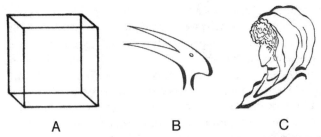

A   B   C

16.2. (A) Necker figure of a box that can appear to be viewed as either from above or from the side. An effort may have to be made to shift from one perception to the other. **(B)** A similar image with alternate possible perceptions, that of either an antelope or a long-billed bird. **(C)** Another figure that may be viewed as either an old woman or a young girl. From (Hanson, 1958), Figures 1, 4, and 2.

process of perception.[43] In the presentation of the picture of an object, such as that of a vase, a face, or an animal with part of it missing, the blank region may be filled in so that the image is seen as complete. This occurs even when the image is stabilized by the use of a cup to fix its position on the eye in order to prevent the missing part from being filled in by eye movements. In another type of experiment showing the power of the higher perceptual processes to participate in the sensory image, special glasses fitted with prisms were used to invert the visual field. After being worn for a few days, the world then appeared to be right side up as it normally does, in spite of the inversion produced by the glasses. On removing the glasses, the world was seen at first to be inverted, but then after a few days reverted to its normal appearance. Another example of the higher integration involved in perception is given by illusionary shifts in the perception of certain figures (Figure 16.2). In (A), the line drawing of a box, it appears as viewed from above or from the side. Only one or the other is perceived at any time. In (B), the figure of what appears to be a bird with a large open beak can also be perceived as a deer with antlers; and in the third figure, (C), an old woman may be seen or appear as a young girl. A concentrated effort may be needed to shift from one perception to the other.[44] Hanson comments that these illusions are not psychological tricks, but are at the frontier of observational science. He analogized the phenomenon to the shift from acceptance of one scientific theory to another as Kuhn has premised.[45] The paradigm for such a conceptual shift is that from Ptolemy's view of the earth as the center around which the sun and planets revolve, to the sun as the center of the solar system in the heliocentric theory of Kepler.

[43] (Köhler, 1929), and (Koffka, 1935).
[44] (Hanson, 1958), Figure 2. The perceptual shift in the wire diagram of a box was first reported in a letter by the Swiss naturalist L. A. Necker. See note 3, pp. 178–179.
[45] (Kuhn, 1962).

Koffka pointed out that while the structure responsible for the gestalt is present, the process is not, a view that he feels, distances Gestalt psychology from the a priori categories of sensation and thought given by Kant.[46] The Gestaltists emphasize the concept of a field, making use of analogy to physical fields of gravity, electricity, and the like. The sensory input then acts to form a trace, which then can allow the neural field to give rise to perceptions of objects in the world. But, rather than some analogical conceptualized field, our present understanding is that neurons are organized into networks in the brain to give rise to perception, cognition, and purposeful action. These networks do not only passively receive sensory information, but also are innately organized to receive selective features in the world. They underlie the recognition of facial displays, body movements signaling threats or pleasant social signals, sexual engagement, and so on, and to act in meaningful ways. On this basis, the world is meaningful because these neural networks have been evolutionarily evolved to match what is meaningful in the world. These innate neuronal networks are not fixed; they are modified by experiences, and the changes are remembered throughout life. In those learning and memory processes, axonal transport and physical changes in neurons are involved, though a sufficient understanding of those processes is yet at its beginnings.

Having come this far, it is of interest to think back to the beginning of our history, to ancient Greece when philosophical inquiries into the nature of the world and man began (Chapter 1). The primitive view that some animating spirit was responsible for consciousness, Plato's concept of an immortal soul, and the theory held by Descartes that the soul is an immaterial substance interacting with the radically different substance of the material body has given way to the present view of the nervous system as an integrated network of neurons responsible for our mind and behavior. If an ancient Greek were transported in time to our day, he could only be dumbfounded at what had evolved since his day, although he might sense some similarity of the ancient thought of the flow of animal spirits in nerve channels to our refined views of action potentials and axonal transport. And, most likely, considering the pace of our advancing knowledge in the neurosciences, if we were transported to a future time, say fifty years hence, we, too, most likely would be similarly astonished. Yet, for all the difference in the accumulated knowledge we have now compared with our ancient forebearers, what remains the same is the spirit of intellectual adventure, the desire to dispel superstition and irrationality – to give a reasoned account of the world in which we exist and account for the nature of our being.

[46] (Koffka, 1935), pp. 548–549.

BIBLIOGRAPHY

Abe, T., Haga, T., and Kurokawa, M. (1973). Rapid transport of phosphatidylcholine occurring simultaneously with protein transport in the frog sciatic nerve. *Biochem. J.*, 136:731–740.

Abe, T., Haga, T., and Kurokawa, M. (1975). Blockage of axoplasmic transport and depolymerisation of reassembled microtubules by methyl mercury. *Brain Res.*, 86:504–508.

Abt, J. P., Essman, W. B., and Jarvik, M. E. (1961). Ether-induced retrograde amnesia for one-trial conditioning in mice. *Science*, 133:1478.

Ackerknecht, E. H., and Vallois, H. V. (1956). *Franz Joseph Gall: Inventor of Phrenology and His Collection*, University of Wisconsin Medical School, Madison.

Adrian, E. D. (1928). *The Basis of Sensation: The Action of the Sense Organs*, Christophers, London.

Adrian, E. D. (1932). *The Mechanism of Nervous Action. Electrical Studies of the Neurone*, University of Pennslyvania Press, Philadelphia.

Agranoff, B. W., Davis, R. E., and Brink, J. J. (1966). Chemical studies on memory fixation in goldfish. *Brain Res.*, 1:303–309.

Aguayo, A. J., Nair, C. P. V., and Midgley, R. (1971). Experimental progressive compression neuropathy in the rabbit: histologic and electrophysiologic studies. *Arch. Neurol.*, 24:358–364.

Alberts, B., Bray, D., Lewis, J., Raff, M., Roberts, K., and Watson, J. D. (1994). *Molecular Biology of the Cell*, 3rd ed., Garland Publishing Co., New York.

Aletta, J. M., and Goldberg, D. J. (1984). Routing of transmitter and other changes in fast axonal transport after transection of one branch of the bifurcate axon on an identified neuron. *J. Neurosci.*, 4:1800–1808.

Alkon, D. L., Vogl, T. P., and Tam, D. (1991). Memory function in neural and artificial networks. In: *Neural Network Models of Conditioning and Action*. M. L. Commons, S. Grossberg, and J. E. R. Staddon, eds. Lawrence Erlbaum Associates, Hillsdale.

Allbutt, T. C. (1921). *Greek Medicine in Rome*. Macmillan, London.

Allen, R. D. (1987). The microtubule as an intracellular engine. *Sci. Am.*, 256:42–49.

Allen, R. D., Allen, N. S., and Travis, J. L. (1981). Video-enhanced contrast, differential interference contrast (AVEC-DIC) microscopy: a new method capable of analyzing microtubule-related motility in the reticulopodial network of *Allogromia laticollaris*. *Cell Mot.*, 1:291–302.

Allen, R. D., Travis, J. L., Hayden, J. H., Allen, N. S., Breuer, A. C., and Lewis, L. J. (1982). Cytoplasmic transport: moving ultrastructural elements common to many cell types revealed by video-enhanced microscopy. In: *Cold Spring Harbor Symposium on Quantitative Biology*, Vol. 46. The Biological Laboratory, Cold Spring Harbor, pp. 85–87.

Allen, R. D., Weiss, D. G., Hayden, J. H., Brown, D. T., Fujiwake, H., and Simpson, M. (1985). Gliding movement of and bidirectional transport along single native microtubules from squid axoplasm: evidence for an active role of microtubules in cytoplasmic transport. *J. Cell Biol.*, 100:1736–1752.

Althaus, J. (1859). *A Treatise on Medical Electricity*. Trubner, London.

Altman, J. (1962). Are new neurons formed in the brains of adult mammals? *Science*, 135:1127–1128.

Altman, J. (1963). Autoradiographic investigation of cell proliferation in the brains of rats and cats. *Anat. Rec.*, 145:573–591.

Altman, J. (1967). Postnatal growth and differentiation of the mammalian brain, with implications for a morphological theory of memory. In: *The Neurosciences, A Study Program*. G. C. Quarton, T. Melnechuk, and F. O. Schmitt, eds. Rockefeller University Press, New York, pp. 723–743.

Alvarez, W. C. (1948). Sixty years of vagotomy: a review of some 200 articles. *Gastroenterology*, 10:413–441.

Amadou, R. (1971). *Le magnétisme animal*. Oeuvres publiées par Robert Amadou avec des commentaires et des notes de F. A. Pattie et J. Vinchon, Payot, Paris.

Ambron, R. T., Schmied, R., Huang, C. C., and Smedman, M. (1992). A signal sequence mediates the retrograde transport of proteins from the axon periphery to the cell body and then into the nucleus. *J. Neurosci.*, 12:2813–2818.

Amos, L. A., Linck, R. W., and Klug, A. (1976). Molecular structure of flagellar microtubules. In: *Cell Motility Book*(C). R. Goldman, T. Pollard, and J. Rosenbaum, eds. Cold Spring Harbor Laboratory, Cold Spring Harbor, pp. 847–867.

Anderson, J. E., Wible, C. G., McCarley, R. W., Jakab, M., Kasai, K., and Shenton, M. E. (2002). An MRI study of temporal lobe abnormalities and negative symptoms in chronic schizophrenia. *Schizophr. Res.*, 58:123–134.

Andriezen, W. (1884). On some of the newer aspects of the pathology of insanity. Brain, 17:548–692.

Antal, M., and Szekely, G. (1987). Phagocytosis of myelin sheath fragments by dendrites. *Exp. Brain Res.*, 66:517–521.

Arcieri, J. P. (1945). *The Circulation of the Blood and Andrea Cesalpina of Arezzo*. S. F. Vanni, New York.

Ariëns-Kappers, C. U. (1929). *The Evolution of the Nervous System in Invertebrates, Vertebrates and Man*. Bohn, Haarlem.

Ariëns-Kappers, C. U., Huber, G. C., and Crosby, E. C. (1936). *The Comparative Anatomy of the Nervous System of Vertebrates Including Man*. 2 Vols. Reprinted in 3 volumes by Hafner, New York, 1967. The Macmillan Company, New York.

Arikuni, T., and Ochs, S. (1973). Slow depolarizing potentials and spike generation in pyramidal tract cells. *J. Neurophysiol.*, 36:1–12.

Aristotle (1910). *The Works of Aristotle*. Edited by Sir David Ross. Vol. IV. *Historia Animalium*. Translation by D'Arcy Wentworth Thompson. Clarendon Press, Oxford.

Aristotle (1964). *On the Soul*. Translated by W. S. Hett, Harvard University Press, Cambridge.

Aristotle (1965). *Aristotle's De Anima in the version of William of Moerbeke* and the commentary of St. Thomas of Aquinas. Translated by K. Foster and S. Humphries, with an introduction by I. Thomas. Yale University Press, New Haven.

Aristotle (1968). *Aristotle's De Anima Books II and III*. Translated with notes by D. W. Hamlyn. Clarendon Press, Oxford.

Arvanitaki, A. (1938). *Propriétés rythmiques de la matière vivante. Variations graduées de la polarisation et rythmicités*, 1st part. Hermann et Cie, Paris.

Arvidson, B. (1992). Inorganic mercury is transported from muscular nerve terminals to spinal and brainstem motoneurons. *Muscle & Nerve*, 15:1089–1094.

Asbury, A. K., and Brown, M. J. (1980). The evolution of structural changes in distal axonopathies. In: *Neurotoxicology*. P. S. Spencer and H. H. Schaumburg, eds. Williams & Wilkins, Baltimore, pp. 179–192.

Axelsson, J., and Thesleff, S. (1959). A study of supersensitivity in denervated mammalian skeletal muscle. *J. Physiol.* (London), 147:178–193.

Baars, B. J. (1986). *The Cognitive Revolution in Psychology*. Guilford Press, New York.

Baglivi, G. (1710). *The Practice of Physic Together with New and Curious Dissertation Particularly of the Tarantula and the Nature of its Poison*. Midwinter et al., London.

Baker, P. F., Blaustein, M. P., Keynes, R. D., Manil, J., Shaw, T. I., and Steinhardt, R. A. (1969). The ouabain-sensitive fluxes of sodium and potassium in squid giant axons. *J. Physiol.* (London), 200:459–496.

Baker, P. F., Hodgkin, A. L., and Shaw, T. I. (1962). The effects of changes in internal ionic concentrations on the electrical properties of perfused giant axons. *J. Physiol.* (London), 164:355–374.

Baker, P. F., Knight, D. E., and Whitaker, M. J. (1980). Calcium and the control of exocytosis. In: *Calcium-Binding Proteins: Structure and Function*. F. L. Siegel, E. Carafoli, R. H. Kretsinger, D. H. MacLennan, and R. H. Wasserman, eds. Elsevier/North Holland, New York, pp. 47–55.

Bamburg, J. R. (1988). The axonal cytoskeleton: stationary or moving matrix? *Trends Neurosci.*, 11:248–249.

Bamburg, J. R., Bray, D., and Chapman, K. (1986). Assembly of microtubules at the tip of growing axons. *Nature*, 321:788–790.

Banks, P., and Mayor, D. (1972). Intra-axonal transport in noradrenergic neurons in the sympathetic nervous system. *Biochem. Soc. Symp.*, 36:133–149.

Banks, P., and Till, R. (1975). A correlation between the effects of anti-mitotic drugs on microtubule assembly in vitro and the inhibition of axonal transport in noradrenergic neurones. *J. Physiol.* (London), 252:283–294.

Bao, D. C. (1972). Protein synthesis in cerebral cortex during spreading depression. Ph.D. thesis submitted to the Department of Physiology, Indiana University School of Medicine.

Bao, D. C., and Ochs, S. (1972). Incorporation of ^{3}H-leucine in rabbit cortex and transport into synaptosomes: effect of spreading depression. *Fed. Proc.*, 31:304.

Bard, P. (1933). Studies on the cerebral cortex. I. Localized control of placing and hopping reactions in the cat and their normal management by small corical remnants. *Arch. Neurol. Psychiatry*, 30:40–74.

Bard. P. (1956). *Medical Physiology*, 10th Ed. C. V. Mosby Co., St. Louis.

Bargmann, W., and Scharrer, E. (1951). The site of origin of the hormones of the posterior pituitary. *Am. Scient.*, 39:255–259.

Barker, L. F. (1899). *The Nervous System and Its Constituent Neurons*, Appleton and Co., New York.

Barnes, C. D., and Worrall, N. (1968). Reinnervation of spinal cord by cholinergic neurons. *J. Neurophysiol.*, 31:689–694.

Barton, J. J., Press, D. Z., Keenan, J. P., O'Connor, M. (2002). Lesions of the fusiform face area impair perception of facial configuration in prosopagnosia. *Neurology.* 58:71–78.

Barton, J. J., Cherkasova, M. (2003). Face imagery and its relation to perception and covert recognition in prosopagnosia. *Neurology.*61:220–225.

Bastholm, E. (1950). *The History of Muscle Physiology from the Natural Philosophers to Albrecht von Haller.* Munksgaard, Copenhagen.

Bates, D. G. (1992). Harvey's account of his "discovery." *Med. Hist.*, 36:361–378.

Bayle, A. L. J., and Thillaye, A. J. (1855). *Biographie Médicale par Ordre Chronologique.* 2 Vols. Adolphe Delahays Libraire, Paris.

Belloni, L. (1968). Die Neuroanatomie von Marcello Malpighi. In: *Steno and Brain Research in the Seventeenth Century.* Analecta Medico-Historica 3. G. Scherz, ed. Pergamon Press, Oxford, pp. 193–206.

Ben-Jonathan, N., Maxson, R. E., and Ochs, S. (1978). Fast axoplasmic transport of noradrenaline and dopamine in mammalian peripheral nerve. *J. Physiol.* (London), 281:315–324.

Benshalom, G., and Reese, T. S. (1985). Ultrastructural observations on the cytoarchitecture of axons processed by rapid-freezing and freeze-substitution. *J. Neurocytol.*, 14:943–960.

Berg, A. (1942). Die lehre von der faser als form- und funktionselement des organismus. *Virchows Arch. Pathol. Anat. Physiol.*, 309:333–460.

Bernard, C. (1844). *Recherches expérimentales sur les fonctions du nerf spinal, étudié spécialement dans ses rapports avec le pneumogastrique.* Labé, Paris.

Bernard, C. (1858). *Leçons sur la Physiologie et la Pathologie du Système Nerveux.* 2 Vols. Baillière et Sons, Paris.

Bernard, C. (1867). *Rapport sur les progrès et la marche de la physiologie générale en France.* Publication faite sous les auspices du Ministère de l'instruction publique. Imprimerie impériale, Paris.

Bernstein, J. (1902). Üntersuchungen zur thermodynamik der bioelektrischen ströme. *Pflügers Arch. Ges. Physiol.*, 92:521–562.

Bernstein, J. (1912). *Elektrobiologie.* Vieweg, Braunschweig.

Berthold, C.-H., and Skoglund, S. (1967). Histochemical and ultrastructural demonstration of mitochondria in the paranodal region of developing feline spinal roots and nerves. *Acta Soc. Med.*, 72:37–70.

Bethe, A. (1904). Der heutige Stand der Neurontheorie. *Deutsche Med. Wschr.*, 1201–1204.

Bible (1970). *The New English Bible with the Apocrypha.* Cambridge University Press, Cambridge.

Bigotte, L., and Olsson, Y. (1987). Degeneration of trigeminal ganglion neurons caused by retrograde axonal transport of doxorubicin. *Neurology*, 37:985–992.

Billings, S. M. (1971). Concepts of nerve fiber development, 1839–1930. *J. Hist. Biol.*, 4:275–305.

Biondi, R. J., Levy, M. J., and Weiss, P. (1972). An engineering study of the peristaltic drive of axonal flow. *Proc. Natl. Acad. Sci. U.S.A.*, 69:1732–1736.

Bisby, M. A. (1976). Orthograde and retrograde axonal transport of labeled protein in motoneurons. *Exp. Neurol.*, 50:628–640.

Bisby, M. A. (1978). Fast axonal transport of labelled protein in sensory axons during regeneration. *Exp. Neurol.*, 61:281–300.

Bisby, M. A. (1980). Retrograde axonal transport. *Adv. Cell Neurobiol.*, 1:69–117.

Bisby, M. A., and Bulger, V. T. (1977). Reversal of axonal transport at a nerve crush. *J. Neurochem.*, 29:313–320.

Bizzi, A., Crane, R. C., Autilio-Gambetti, L. A., and Gambetti, P. (1984). Aluminum effect on slow axonal transport: a novel impairment of neurofilament transport. *J. Neurosci.*, 4:722–731.

Black, M. M., and Lasek, R. J. (1980). Slow components of axonal transport: two cytoskeletal networks. *J. Cell Biol.*, 86:616–623.

Blakeslee, S. (2002). A boy a mother and a rare map of autism's world. *Science Times* (Tuesday, November 19, 2002), D1–D4. *New York Times*, New York.

Bliss, T. V. P., and Gardner-Medwin, A. R. (1973). Long-lasting potentiation of synaptic transmission in the dentate area of the unanaesthetized rabbit following stimulation of the perforant path. *J. Physiol.* (London), 232:357–374.

Bliss, T. V. P., and Lømo, T. (1973). Long-lasting potentiation of synaptic transmission in the dentate area of the anaesthetized rabbit following stimulation of the perforant path. *J. Physiol.* (London), 232:331–356.

Boegman, R. J., and Riopelle, R. J. (1980). Batrachotoxin blocks slow and retrograde axonal transport in vivo. *Neurosci. Lett.*, 18:143–147.

Boerhaave, H. (1743). *Academical Lectures on the Theory of Physic.* W. Innys, London.

Boerhaave, H. (1959). *Hermanni Boerhaave Praelectiones de Morbis Nervorum 1730–1735.* Edited by Benedictus Petrus Maria Schulte. In Dutch and Latin. E. J. Brill, Leiden.

Bonnefoy, J.-B. (1782). *De l'application de l'électricité a l'art de guérir.* Dissertation inaugurale pour son agrégation au College royal de chirurgie. Aimé de la Roche, Lyon.

Bono, J. J. (1984). Medical spirits and the medieval language of life. *Traditio*, 40:91–130.

Borelli, G. A. (1989). *On the Movement of Animals.* Translated by Paul Maquet. Springer-Verlag, Berlin.

Borisy, G. G., Marcum, J. M., Olmsted, J. B., Murphy, D. B., and Johnson, K. A. (1975). Purification of tubulin and associated high molecular weight proteins from porcine brain and characterization of microtubule assembly in vitro. *Ann. N. Y. Acad. Sci.*, 253:107–132.

Bouillaud, J.-B. (1825). *Traité clinique et physiologique de l'encéphalite ou Inflammation du cerveau, et ses suites, etc.* Bailliere, Paris.

Bouldin, T., and Cavanagh, J. (1979a). Organophosphorous neuropathy. I. A teased-fiber study of the spatio-temporal spread of axonal degeneration. *Am. J. Pathol.*, 94:241–252.

Bouldin, T., and Cavanagh, J. (1979b). Organophosphorous neuropathy. II. A fine-structural study of the early stages of axonal degeneration. *Am. J. Pathol.*, 94:253–270.

Bozler, E. (1926). Untersuchungen über das Nervensystem der Coelenteraten. I Teil. Kontinuitat oder kontakt zwischen den Nervenzellen. *Zeit.Mikroskop.Anat.*, 5:244–262.

Bozler, E. (1927). Untersuchungen über das Nervensystem der Coelenteraten. II. *Zeit.Vergl. Physiol.*, 6:255–263.

Bradley, W. G., and Williams, M. H. (1973). Axoplasmic flow in axonal neuropathies. I. Axoplasmic flow in cats with toxic neuropathies. *Brain*, 96:235–246.

Brady, S. T., Chrothers, S. D., Nosal, C., and McClure, W. O. (1980). Fast axonal transport in the presence of high Ca^{2+}: evidence that microtubules are not required. *Proc. Natl. Acad. Sci. U.S.A.*, 77:5909–5913.

Bray, D. (1973). Model for membrane movements in the neural growth cone. *Nature*, 244:93–95.

Brandner, M. D., Buncke, H. J., and Campagna-Pinto, D. (1989). Experimental treatment of neuromas in the rat by retrograde axoplasmic transport of ricin with selective destruction of ganglion cells. *J. Hand Surg.*, 14:710–714.

Brattgård, S. O., Edström, J. E., and Hydén, H. (1957). The chemical changes in regenerating neurons. *J. Neurochem.*, 1:316–325.

Bray, D., and Bunge, M. B. (1981). Serial analysis of microtubules in cultured rat sensory axons. *J. Neurocytol.*, 10:589–605.

Bray, D., and Gilbert, D. A. (1981). Cytoskeletal elements in neurons. *Ann. Rev. Neurosci.*, 4:505–523.

Bray, G. M., and Friedman, H. C. (1999). Schwann cell and axon relationships. In: *Encycopedia of Neuroscience*, 2 Vols., 2nd ed. G. Adelman, and B. H. Smith, eds. Elsevier, Amsterdam, pp. 1823–1825.

Bray, G. M., Rasminsky, M., and Aguayo, A. J. (1981). Interactions between axons and their sheath cells. *Ann. Rev. Neurosci.*, 4:127–162.

Brazier, M. A. B. (1984). *A History of Neurophysiology in the 17th and 18th Centuries*. Raven Press, New York.

Brazier, M. A. B. (1988). *A History of Neurophysiology in the 19th Century*, Raven Press, New York.

Breuer, A. C., Lynn, M. P., Atkinson, M. B., Chou, S.-M., Wilbourn, A. J., Marks, K. E., Culver, J. E., and Fleeger, E. J. (1987). Fast axonal transport in amyotrophic lateral sclerosis: an intra-axonal organelle traffic analysis. *Neurology*, 37:738–748.

Bréhier, É. (1965). *The Middle Ages and the Renaissance*. Translated by W. Baskin. University of Chicago Press, Chicago.

Brimijoin, S. (1979). Axonal transport and subcellular distribution of molecular forms of acetylcholinesterase in rabbit sciatic nerve. *Mol. Pharmacol.*, 15:641–648.

Brimijoin, S. (1982). Axonal transport in autonomic nerves: views on its kinetics. In: *Trends in Autonomic Pharmacology*, Vol. 2. S. Kalsner, ed. Urban and Schwarzenberg, Baltimore, pp. 17–42.

Brimijoin, S. (1975). Stop-flow: A new technique for measuring axonal transport and its application to the transport of dopamine-β-hydroxylase. *J. Neurobiol.*, 6:379–394.

Brimijoin, S., and Hammond, P. I. (1985). Acrylamide neuropathy in the rat: effects on energy metabolism in sciatic nerve. *Mayo Clin. Proc.*, 60:3–8.

Brimijoin, S., and Helland, L. (1976). Rapid retrograde transport of dopamine-β-hydroxylase as examined by the stop-flow technique. Brain Res., 102:217–228.

Brimijoin, S., Olsen, J., and Rosenson, R. (1979). Comparison of the temperature-dependence of rapid axonal transport and microtubules in nerves of the rabbit and bullfrog. *J. Physiol.* (London), 287:303–314.

Brimijoin, S., and Wiermaa, M. J. (1977a). Direct comparison of the rapid axonal transport of norepinephrine and dopamine-β-hydroxylase activity. *J. Neurobiol.*, 8:239–250.

Brimijoin, S., and Wiermaa, M. J. (1977b). Rapid axonal transport of tyrosine hydroxylase in rabbit sciatic nerves. *Brain Res.*, 2:77–96.

Brimijoin, S., and Wiermaa, M. J. (1978). Rapid orthograde and retrograde axonal transport of acetylcholinesterase as characterized by the stop-flow technique. *J. Physiol.* (London), 285:129–142.

Brittanica Staff (1974). *The New Encyclopedia Brittanica.* 30 Vols., 15th ed. The University of Chicago Press, Chicago.

Broca, P. P. (1861). Remarques sur le siege de la faculte du langage articule; suivie d'une observation d'aphemie (perte de la parole). *Bull. Soc. Anthrop.*, 36:330–357.

Brooks, V. B., Curtis, D. R., and Eccles, J. C. (1957). The action of tetanus toxin on the inhibition of motoneurones. *J. Physiol.* (London), 35:655–672.

Brooks, B. R. (1991). The role of axonal transport in neurodegenerative disease spread: a meta-analysis of experimental and clinical poliomyelitis compares with amyotrophic lateral sclerosis. *Canad. J. Neurol. Sci.*, 18:435–438.

Brown, A., and Lasek, R. J. (1993). Neurofilaments move apart freely when released from the circumferential constraint of the axonal plasma membrane. *Cell Mot. Cytoskel.*, 26:313–324.

Brown, G. L. (1937). The actions of acetylcholine on denervated mammalian and frog's muscle. *J. Physiol.* (London), 89:438–46.

Brown, J. O., and McCouch, G. P. (1947). Abortive regeneration in the transected spinal cord. *J. Comp. Neurol.*, 87:131–137.

Brown, M. C., Perry, V. H., Lunn, E. R., Gordon, S., and Heumann, R. (1991). Macrophage dependence of peripheral sensory nerve regeneration: possible involvement of nerve growth factor. *Neuron*, 6:359–370.

Brown, S. S. (1999). Cooperation between microtubule-and actin-based motor proteins. *Annu. Rev. Cell Dev. Biol.*, 15:63–80.

Bruch, C. (1855). Über die regeneration durchschnittenen nerven. *Zeit. Wiss Zool.*, 6:135–138.

Brunetti, M., Di Giamberardino, L., Porcellati, G., and Droz, B. (1981). Contribution of axonal transport to the renewal of myelin phospholipids in peripheral nerves. II. Biochemical study. *Brain Res.*, 219:73–84.

Buller, A. J., Eccles, J. C., and Eccles, R. M. (1960a). Differentiation of fast and slow muscles in the cat hind limb. *J. Physiol.* (London), 50:399–416.

Buller, A. J., Eccles, J. C., and Eccles, R. M. (1960b). Interactions between motoneurones and muscles in respect of the characteristic speeds of their responses. *J. Physiol.* (London), 50:417–439.

Bunge, M. B., Bunge, R. P., and Ris, H. (1961). Ultrastructural study of remyelination in an experimental lesion in adult cat spinal cord. *J. Biophys. Biochem. Cytol.*, 10:67–94.

Burdach, E. (1838). Beitrag zur microscopischen Anatomie der Nerven. *Ann. des Sci. Naturelles Zool.* (2nd Series), 9:96–270.

Bureš, J., Burešová, O., and Křivánek, J. (1974). *The Mechanism and Applications of Leão's Spreading Depression of Electroencephalographic Activity.* Academia, Prague.

Burnet, J. (1948). *Early Greek Philosophy.* Adam and Charles Black, London.

Burnside, B. (1975). The form and arrangement of microtubules: an historical, primarily morphological review. *N. Y. Acad. Sci.*, 253:14–26.

Burton, P. R. (1987). Microtubules of frog olfactory axons: their length and number/axon. *Brain Res.*, 409:71–78.

Byers, M. R. (1974). Structural correlates of rapid axonal transport: evidence that microtubules may not be directly involved. *Brain Res.*, 75:97–113.

Byers, M. R., Fink, B. R., Kennedy, R. D., Middaugh, M. E., and Hendrickson,
A. E. (1973). Effects of lidocaine on axonal morphology microtubules and rapid
transport in rabbit vagus nerve in vitro. *J. Neurobiol.*, 4:25–43.
Byers, T. J., and Branton, D. (1985). Visualization of the protein associations in the
erythrocyte membrane skeleton. *Proc. Natl. Acad. Sci. U.S.A.*, 82:6153–6157.
Caelius Aurelianus (1950). *On Acute Diseases and on Chronic Diseases.* Edited and
translated by I. E. Drablin from the work of Soranus. University of Chicago Press,
Chicago.
Caley, E. R., and Richards, J. C. (1956). *Theophrastus on Stones. A Modern Edition with
Greek Text, Translation, Introduction and Commentary*, The Ohio State University,
Columbus, Ohio.
Campbell, K. (1967). Materialism. In: *The Encyclopedia of Philosophy.* Vol. 5 of 8.
P. Edwards, ed. Macmillan Publications, New York, pp. 179–188.
Campenot, R. B. (1977). Local control of neurite development by nerve growth factor.
Proc. Natl. Acad. Sci. U.S.A., 74:4516–4519.
Canguilhem, G. (1955). *La Formation du Concept de Reflexe aux xvii et xviii Siecles.*
Presses Universitaires de France, Paris.
Cannon, W. B., and Rosenblueth, A. (1949). *The Supersensitivity of Denervated Struc-
tures.* MacMillan, New York.
Carafoli, E., and Crompton, M. (1976). Calcium ions and mitochondria. *Symp. Soc.
Exp. Biol.*, 30:89–115.
Card, D. J. (1977). Denervation: sequence of neuromuscular degenerative changes in
rats and the effect of stimulation. *Exp. Neurol.*, 54:25–265.
Carmichael, A. G. (1997). Tarantism: the toxic dance. In: *Plague, Pox and Pestilence.*
K. F. Kiple, ed. Weidenfeld and Nicolson, London.
Castelli, B. (1761). *Lexicon Medicum Graeco-Latinum.* 4th ed. Valentine Azzolini,
Naples.
Castiglioni, A. (1947). *A History of Medicine*, 2nd Ed. Translated from the Italian and
edited by E. B. Krumbhaar. Knopf, New York.
Cavanagh, J., Chen, F., Kyu, M., and Ridley, A. (1968). The experimental neuropathy
in rats caused by p-bromophenylacetylurea. *J. Neurol. Neurosurg. Psychiatry*, 31:471–
478.
Cavanagh, J. B. (1964). The significance of the "dying back" process in experimental
and human neurological disease. *Int. Rev. Exp. Pathol.*, 3:219–267.
Cavanagh, J. B. (1985). Peripheral nervous system toxicity: a morphological approach. In: *Neurotoxicology.* K. Blum and L. Manzo, eds. Marcel Dekker, New York,
pp. 1–44.
Cave, A. J. E. (1937). *The Human Crania from New Guinea.* Collected by Lord Moyne's
Expedition, Kingswood.
Chalfie, M., and Thomson, J. N. (1979). Organization of neuronal microtubules in
the nematode *Caenorhabditis elegans. J. Cell Biol.*, 82:278–289.
Chan, K. Y., and Bunt, A. H. (1978). An association between mitochondria and microtubules in synaptosomes and axon terminals of cerebral cortex. *J. Neurocytol.*,
7:37–43.
Chan, S. Y., and Iqbal, Z. (1986). Analytical techniques for the study of axoplasmic
transport. In: *Axoplasmic Transport.* Z. Iqbal, ed. CRC Press, Boca Raton, pp. 9–20.
Chan, S. Y., Ochs, S., and Jersild, R. A., Jr. (1979). Calcium localization in mammalian
nerve fibers in relation to its regulation and axoplasmic transport. *Soc. Neurosci.
Abstr.*, 5:59.

Chan, S. Y., Ochs, S., and Jersild, R. A., Jr. (1984). Localization of calcium in nerve fibers. *J. Neurobiol.*, 15:141–155.

Chan, S. Y., Ochs, S., and Worth, R. M. (1980a). The requirement for calcium ions and the effect of other ions on axoplasmic transport in mammalian nerve. *J. Physiol.* (London), 301:477–504.

Chan, S. Y., Worth, R., and Ochs, S. (1980b). Block of axoplasmic transport in vitro by vinca alkaloids. *J. Neurobiol.*, 11:251–264.

Chang, H.-T. (1951). Dendritic potential of cortical neurons produced by direct electrical stimulation of the cerebral cortex. *J. Neurophysiol.*, 14:1–23.

Charcot, J.-M. (1878). *Clinical Lectures on Diseases of the Nervous System*, with an introduction by Ruth Harris. New Sydenham Society, London.

Cherniss, H. (1962). *Aristotle's Criticism of Plato and the Academy*. Russell and Russell, Inc., New York.

Cho, E., Schaumburg, H. H., and Spencer, P. S. (1977). Adriamycin produces ganglioradiculopathy in rats. *J. Neuropathol. Exp. Neurol.*, 36:907–915.

Chou, S.-M., and Hartmann, H. A. (1965). Electron microscopy of focal neuroaxonal lesions produced by β,β'-iminodipropionitrile (IDPN) in rats. *Acta Neuropathol.*, 4:590–603.

Churchland, P. S., and Sejnowski, T. J. (1992). *The Computational Brain*. MIT, Cambridge.

Cicero (1972). *The Nature of the Gods*. Translated by H. C. P. McGregor, with an introduction by J. M. Ross. Penguin Books, Harmondsworth.

Clarke, E. S. (1968). The doctrine of the hollow nerve in the seventeenth and eighteenth centuries. In: *Medicine, Science and Culture*. L. G. Stevenson and R. P. Multhauf, eds. Johns Hopkins Press, Baltimore, pp. 123–141.

Clarke, E. S. (1978). The neural circulation: the use of analogy in medicine. *Med. Hist.*, 22:291–307.

Clarke, E. S., and Bearn, J. B. (1968). The brain "glands" of Malpighi elucidated by practical history. *J. Hist. Med. Allied Sci.*, 23:309–330.

Clarke, E. S., and Bearn, J. B. (1972). The spiral nerve bands of Fontana. *Brain*, 95:1–20.

Clarke, E. S., and Dewhurst, K. (1972). *An Illustrated History of Brain Function*. Sandford Publications, Oxford.

Clarke, E. S., and Jacyna, L. S. (1987). *Nineteenth Century Origins of Neuroscientific Concepts*. University of California Press, Berkeley.

Clarke, E. S., and O'Malley, D. D. (1968). *The Human Brain and Spinal Cord*. University of California Press, Berkeley.

Clodd, E. (1905). *Animism: The Seed of Religion*. Archibald Constable and Co., London.

Codellas, P. S. (1932). Alcmaeon of Croton: his life, work, and fragments. *Proc. Roy. Soc. Med.*, 25:1041–1046.

Coghill, G. E. (1929). *Anatomy and the Problem of Behavior*. Cambridge University Press, Cambridge.

Cohen, I. B. (1957). The germ of an idea or what put Harvey on the scent? *J. Hist. Med.*, 12:102–105.

Cohn, S. A., Ingold, A. L., and Scholey, J. M. (1987). Correlation between the ATPase and microtubule translocating activities of sea urchin egg kinesin. *Nature*, 328:160–163.

Cooper, P. D., and Smith, R. S. (1974). The movement of optically detectable organelles in myelinated axons of *Xenopus laevis*. *J. Physiol.* (London), 242:77–97.

Cornford, F. M. (1935). *Plato's Theory of Knowledge*. The *Theaetetus* and the *Sophist* of Plato translated wtih a running commentary. Routledge and Kegan Paul Ltd., London.

Cornford, F. M. (1937). *Plato's Cosmology*. The *Timaeus* of Plato translated with a running commentary. Routledge and Kegan Paul Ltd., London.

Cotugno, D. F. A. (1764). *De Ischiade nervosa commentarius*. Fratres Simonii, Naples.

Cowan, W. M., and Cuenod, M. (1975). *The Use of Axonal Transport for Studies of Neuronal Connectivity*. Elsevier, New York.

Cranefield, P. F. (1974). *The Way In and the Way Out, François Magendie, Charles Bell and the Roots of the Spinal Nerve*. Futura Publishing Co., Mount Kisco.

Crick, F. (1982). Do dendritic spines twitch? *Trends Neurosci.*, Vol. 5. 44–46.

Critchley, H. (2003). Emotion and its disorders. *Br. Med. Bull.*; 65:35–47.

Cronly-Dillon, J., Carden, D., and Birks, C. (1974). The possible involvement of brain microtubules in memory fixation. *J. Exp. Biol.*, 61:443–454.

Cross, R., and Scholey, J. (1999). Kinesin: the tail unfolds. *Nature Cell Biol.*, 1:E119–E121.

Cruikshank, W. (1779). *Experiments on the Insensible Perspiration of the Human Body: Showing Its Affinity to Respiration*. G. Nichol, London.

Cruikshank, W. (1795). Experiments on the nerves, particularly on their reproduction, and on the spinal marrow of living animals. *Phil. Trans. Roy. Soc.* (abridged version), 85:512–519.

Cruz-Höfling, M. A., Love, S., Brook, G., and Duchen, L. W. (1985). Effects of *Phoneutria Nigriventer* spider venom on mouse peripheral nerve. *Q. J. Exp. Physiol.*, 70:623–640.

Cuenod, M., and Schonbach, J. (1971). Synaptic proteins and axonal flow in the pigeon visual pathway. *J. Neurochem.*, 18:809–816.

Curtis, H. H., and Cole, K. S. (1940). Membrane action potentials from the squid giant axon. *J. Cell. Comp. Physiol.*, 15:147–157.

Dahlström, A. B., and Haggendal, J. (1966) Studies on the transport and life-span of amine storage granules in a peripheral adrenergic neuron system. *Acta Physiol. Scand.*, 67:278–288.

Dahlström, A. B. (1968). Effect of colchicine on transport of amine storage granules in sympathetic nerves of rat. *Eur. J. Pharmacol.*, 5:111–113.

Dahlström, A. B. (1971). Axoplasmic transport (with particular respect to adrenergic neurons). *Phil. Trans. Roy. Soc. B*, 261:325–358.

Dahlström, A. B. (1973). Aminergic transmission. Introduction and short review. *Brain Res.*, 62:441–460.

Dahlström, A. B., Evans, C. A. N., Haggendal, J., Heiwall, P. O., and Saunders, N. R. (1974). Rapid transport of acetylcholine in rat sciatic nerve proximal and distal to a lesion. *J. Neural Transm.*, 35:1–11.

Dahlström, A. B. (1999). Adrenergic nervous system. In: *Encyclopedia of Neuroscience*, 2 Vols., 2nd Ed. G. Adelman and B. H. Smith, eds. Elsevier, Amsterdam, pp. 27–31.

Dahlström, A. B., and Fuxe, K. (1964). A method for the demonstration of adrenergic nerve fibres in peripheral nerves. *Z. Zellforsch. Mikr. Anat.*, 62:602–607.

Darwin, C. (1952). *The Origin of Species by Means of Natural Selection* and *The Descent of Man and Selection in Relation to Sex* (1859 and 1871)., Encyclopedia Britannica, Inc., Chicago.

Darwin, C. (1955). *The Expression of the Emotions in Man and Animals*. First published 1872. With a preface by Margaret Mead, Philosophical Library, New York.

Darwin, C. (1886). *The Expression of the Emotions in Man and Animals*, D. Appelton & Co., New York.

Davis, H. P., and Squire, L. R. (1984). Protein synthesis and memory: a review. *Psychol. Bull.*, 96:518–559.

Dawson, V. P. (1987). *Nature's Enigma: The Problem of the Polyp in Letters of Bonnet, Trembley and Réaumur*. American Philosophical Society, Philadelphia.

DeBrabander, M., Geuens, G., Nuydens, R., Willebrords, R., and DeMey, J. (1982). Microtubule stability and assembly in living cells: The influence of metabolic inhibitors taxol and pH. *Cold Spring Harbor Symp. Quant. Biol.*, 46:227–240.

De Lacy, P. (1972). Galen's platonism. *Am. J. Philol.*, 93:27–39.

De Lacy, P. (1978). *Galen: On the Doctrines of Hippocrates and Plato*, 2 Vols., Akademie Verlag, Berlin.

de La Roche, F. G. (1778). *Analyse des fonctions du système nerveux, pour servir d'introduction à un examen pratique des maux de nerfs*, 2 Vols. du Villard and Nouffer, Geneva.

De Reuck, A. V. S., and Knight, J. (1966). *Touch, Heat and Pain*. Little Brown and Co., Boston.

De Robertis, E. (1958). Submicroscopic morphology and function of the synapse. In: *Experimental Cell Research, Supplement 5*. Academic Press, New York, pp. 347–369.

DeFelipe, J., and Jones, E. G. (1988). *Cajal on the Cerebral Cortex*. An annotated translation of the complete writings, Oxford University Press, New York.

DeFelipe, J., and Fariñas, I. (1992). The pyramidal neuron of the cerebral cortex: Morphological and chemical characteristics of the synaptic inputs. *Prog. Neurobiol.*, 39:563–607.

Deisseroth, K., Bito, H., and Tsien, R. W. (1996). Signaling from synapse to nucleus: postsynaptic CREB phosphorylation during multiple forms of hippocampal synaptic plasticity. *Neuron*, 16:89–101.

Deiters, O. (1865). *Unterschungen über gehirn und ruckenmark des menschen und der saugethiere*. Vieweg, Braunschweig.

Demoor, J. (1898). Le mécanisme et la signification de l'état moniliforme des neurones. *Ann. Soc. Roy. Med. Nat. Bruxelles*. 7:205–250.

Denny-Brown, D. (1970). Augustus Volney Waller (1816–1870). In: *The Founders of Neurology*, 2nd ed. W. Haymaker, Ed. Chas. C Thomas, Springfield, pp. 88–91.

Descartes, R. (1664). *L'Homme*. Angot, Paris.

Descartes, R. (1955). *The Philosophical Works of Descartes*. Rendered into English by E. S. Haldane and G. R. T. Ross., 2 Vols. Originally published as the corrected edition of 1931 by Cambridge University Press. Dover Publications, Inc., New York.

Descartes, R. (1972). *Treatise of Man by Rene Descartes*. With translation and commentary by T. S. Hall, Harvard University Press, Cambridge.

Descartes, R. (1662). *De Homine, figuris, et latinitate donatus a Florentio Schuyl*. F. Moyardum and P. Leffen, Leiden.

Devarajan, P., and Morrow, J. S. (1996). The spectrin cytoskeleton and organization of polarized epithelial cell membranes. In: *Current Topics in Membranes*. W. J. Nelson, ed. Academic Press, New York, pp. 97–128.

Dewitt, N. W. (1967). *Epicurus and His Philosophy*. Meridian, World Publishing Co., Cleveland.

Diamond, M. C., Rosenzweig, M. R., Bennett, E. L., Lindner, B., and Lyon, L. (1972). Effects of environmental enrichment and impoverishment on rat cerebral cortex. *J. Neurobiol.*, 3:47–64.

Diogenes Laertius (1966). *Lives of Eminent Philosophers*, 2 Vols. Translated by R. D. Hicks, Harvard University Press, Cambridge.

Diringer, D. (1982). *The Book Before Printing. Ancient, Medieval and Oriental*. Republication of the work originally published in 1953 under the title *The Hand-Produced Book*, Dover Publications, Inc., New York.

Disraeli, I. (1866). *Curiosities of Literature*, 3 Vols. Edited, with memoir and notes by his son Benjamin Disraeli. Frederick Warne and Co., London.

Dobson, J. F. (1925). Herophilus of Alexandria. *Proc. Roy. Soc. Med. Sect. Hist. Med*, 18:19–32.

Dobson, J. F. (1927). Erasistratus. *Proc. Roy. Soc. Med. Sect. Hist. Med.*, 20:825–832.

Dodds, E. R. (1951). *The Greeks and the Irrational*. University of California Press, Berkeley.

Donat, J. R., and Wisniewski, H. M. (1973). The spatio-temporal pattern of Wallerian degeneration in mammalian peripheral nerves. *Brain Res.*, 53:41–53.

Donoso, J. A., Illanes, J. P., and Samson, F. (1977). Dimethylsulfoxide action on fast axoplasmic transport and ultrastructure of vagal axons. *Brain Res.*, 20:287–301.

Douglas, R. M., and Goddard, G. (1975). Long-term potentiation of the perforant path-granule cell synapse in the rat hippocampus. *Brain Res.*, 86:205–215.

Downing, A. C., Gerard, R. W., and Hill, A. V. (1926). The heat production of nerve. *Proc. Roy. Soc. Lond. B.*, 100:223–251.

Dragstedt, L. R. (1945). Vagotomy for gastroduodenal ulcer. *Ann. Surg.*, 122:973.

Droz, B., Brunetti, M., Di Giamberardino, L., Koenig, H. L., and Porcellati, G. (1979). Transfer of phospholipid constituents to glia during axonal transport. In: *Society for Neuroscience Symposia*, Vol. 4. F. A. Ferrendelli, ed. pp. 344–360.

Droz, B., and Leblond, C. P. (1962). Migration of proteins along the axons of the sciatic nerve. *Science*, 137:1047–1048.

Droz, B., and Leblond, C. P. (1963). Axonal migration of proteins in the central nervous system and peripheral nerves as shown by radioautography. *J. Comp. Neurol.*, 121:325–337.

Droz, B., Giamberardino, L. D., and Koenig, H. L. (1981). Contribution of axonal transport to the renewal of myelin phospholipids in peripheral nerves. I. Quantitative radioautographic study. *Brain Res.*, 219:57–71.

Droz, B., Rambourg, A., and Koenig, H. L. (1975). The smooth endoplasmic reticulum: structure and role in the renewal of axonal membrane and synaptic vesicles by fast axonal transport. *Brain Res.*, 93:1–13.

Duchenne de Boulogne, G.-B. A. (1990). *The Mechanism of Human Facial Expression*. Translation by R. Andrew Cuthbertson of *Mècanisme de la Physionomie humaine or Analyse Électro- physiologique de l'expression des Passions applicable a la pratique des arts plastique*, originally published 1862, Cambridge University Press, Cambridge.

Duchenne de Boulogne, G.-B. A. (1871). *A Treatise on Localized Electrization*. Translated from the 3rd Ed. by Herbert Tibbits, Lindsay and Blakiston, Philadelphia.

Duchenne de Boulogne, G.-B. A. (1855). *De la therapeutique de l'electrisation localisee et de son application a la pathologie*. Bailliere, Paris.

Ductoray de Blainville, H. M. (1808). *Extraites d'un essai sur la respiration, suivies de quelques experiénces sur l'influence de la huitième paire de nerfs dans la respiration*. Didot, Paris.

Dulak, L., and Crist, R. D. (1974). Microtubule properties in dimethylsulfoxide. *J. Cell Biol.*, 63:90–98.

Dunglison, R. (1833). *New Dictionary of Medical Science*, 2 Vols. Charles Bowen, Boston.

Durling, R. J. (1993). *A Dictionary of Medical Terms in Galen.* E. J. Brill, Leiden.

Dustin, P. (1984). *Microtubules.* Springer-Verlag, New York.

Dyck, P. J., Gianni, C., and Lais, A. (1993a). Pathologic alterations of nerves. In: *Peripheral Neuropathy,* Vol. 1, 3rd Ed. P. J. Dyck, P. K. Thomas, J. W. Griffin, P. A. Low, and J. F. Poduslo, eds. W. B. Saunders, Philadelphia, pp. 514–595.

Dyck, P. J., and Hopkins, A. P. (1972). Electron microscopic observations on degeneration and regeneration of unmyelinated fibres. *Brain,* 95:223–234.

Dyck, P. J., and Lofgren, E. P. (1968). Nerve biopsy. Choice of nerve, method, symptoms and usefulness. *Med. Clin. North Am.,* 52:885–893.

Dyck, P. J., Thomas, P. K., Griffin, J. W., Low, P. A., and Poduslo, J. F. (1993b). *Peripheral Neuropathy,* 3rd ed. Saunders, Philadelphia.

Eastwood, B. S. (1982). *The Elements of Vision: The Micro-Cosmology of Galenic Visual Theory According to Hunayn Ibn Ishaq.* The American Philosophical Society, Philadelphia.

Eccles, J. C. (1941a). Changes in muscle produced by nerve degeneration. *Med. J. Aust.,* 1:573–575.

Eccles, J. C. (1941b). Disuse atrophy of skeletal muscle. *Med. J. Aust.,* 2:160–164.

Eccles, J. C. (1953). *The Neurophysiological Basis of Mind.* Clarendon Press, Oxford.

Eccles, J. C. (1964). *The Physiology of Synapses.* Academic Press, New York.

Edström, A., and Hanson, M. (1973). Temperature effects on fast axonal transport of proteins in vitro in frog sciatic nerves. *Brain Res.,* 58:345–354.

Edström, A., and Mattsson, H. (1975). Small amounts of zinc stimulate rapid axonal transport in vitro. *Brain Res.,* 86:162–167.

Edström, A., and Mattsson, H. (1976). Inhibition and stimulation of rapid axonal transport in vitro by sulfhydryl blockers. *Brain Res.,* 108:381–395.

Ehrenberg, C. G. (1838). Observations on the structure hitherto unknown of the nervous system in man and animals. Translated by D. Craigie, with additions and notes from an address read to the Academy of Sciences at Berlin, October 24, 1833. In: *Essays on Physiology and Hygiene.* Anonymous, ed. Haswell, Barrington and Haswell, Philadelphia, pp. 67–112.

Eisenstein, E. L. (1979). *The Printing Press as an Agent of Change. Communications and Cultural Transformations in Early-Modern Europe,* 2 Vols. complete in one. Cambridge University Press, Cambridge.

Ekehorn, G. (1949). *Sherrington's "Endeavor of Jean Fernel" and "Man on His Nature."* Comments by G. Ekehorn. Publication VII: Sherrington's Approach to the Special Mental Integers. Kung Boktryckeriet, P. A. Norstedt and Söner, Stockholm.

Ekström, P. (1987). *Calmodulin and Calmodulin-Binding Proteins in Axonal Transport.* Thesis, pp. 1–126. University of Lund, Sweden.

Ekström, P., Kanje, M., and McLean, W. G. (1987). The effects of trifluoperazine on fast and slow axonal transport in the rabbit vagus nerve. *J. Neurobiol.,* 18:283–293.

Elfvin, L.-G. (1963). The ulstrastructure of the plasma membrane and myelin sheath of peripheral nerve fibers after fixation by freeze-drying. *J. Ultrastruct. Res.,* 8:283–304.

Ellisman, M. H. (1982). A hypothesis for rapid axoplasmic transport based upon focal interactions between axonal membrane systems and the microtrabecular crossbridges of the axoplasmic matrix. In: *Axoplasmic Transport.* D. G. Weiss, ed. Springer-Verlag, Berlin, pp. 390–396.

Ellisman, M. H., and Porter, K. R. (1980). Microtrabecular structure of the axoplasmic matrix: visualization of cross-linking structures and their distribution. *J. Cell Biol.,* 87:464–479.

Erb, W. H. (1883). *Handbook of Electro-Therapeutics*. Translated from the German by L. Putzel. William Wood, New York.

Erdmann, G., Wiegand, H., and Wellhoner, H. H. (1975). Intraaxonal and extraaxonal transport of ^{125}I-tetanus toxin in early local tetanus. *Naunyn-Schmiedeberg's. Arch. Pharmacol.*, 290:357–373.

Eriksson, P. S., Perfilieva, E., Bjork-Eriksson, T., Alborn, A. M., Nordborg, C., Peterson, D. A., and Gage, F. H. (1998). Neurogenesis in the adult human hippocampus. *Nature Med.*, 4:1313–1317.

Eriksson, R. (1959). *Andreas Vesalius' First Public Anatomy at Bologna 1540*. An eye-witness report by Baldasar Heseler medicanae scholaris together with his notes on Matthaeus Curtius' lectures on Anatomia Mundini. Edited with an introduction, translation into English and notes by Ruben Eriksson. Almqvist and Wiksells, Uppsala.

Erlanger, J., and Gasser, H. S. (1924). The compound nature of the action current of nerve as disclosed by the cathode ray oscillograph. *Am. J. Physiol.*, 70:624–666.

Erlanger, J., and Gasser, H. S. (1937). *Electrical Signs of Nervous Activity*. University of Pennslyvania Press, Philadelphia.

Esche, S. (1954). *Leonardo da Vinci. Das anatomische Werk*, Holbein-Verlag, Basel.

Esquerro, E., Garcia, A. G., and Sanchez-Garcia, P. (1980). The effects of the calcium ionophore A23187 on the axoplasmic transport of dopamine-β-hydroxylase. *Br. J. Pharmacol.*, 70:375–381.

Fabre, J. H. (1918). The Wonders of Instinct. Chapters in the *Psychology of Instincts*. Translated from the French by A. Teixeira de Mattos and B. Miall. T. Fisher Unwin Ltd., London.

Fairbrother, R. W., and Hurst, E. W. (1930). The pathogenesis of, and propagation of the virus in experimental poliomyelitis. Part I. *J. Pathol. Bact.*, 33:17–45.

Faller, A. (1968). Die Hirnschnitt-Zeichnungen in Stensens Discours sur l'anatomie du cerveau. In: *Steno and Brain Research in the Seventeenth Century*. Analecta Medico-Historica 3. G. Scherz, ed. Pergamon Press, Oxford, pp. 115–145.

Falck, B. (1962). Observations on the possibilities of the cellular localization of monoamines by a fluorescence method. *Acta Physiol. Scand.*, Suppl. 197:1–26.

Fawcett, C. P., Powell, A. E., and Sachs, H. (1968). Biosynthesis and release of neurophysin. *Endocrinology*, 83:299–310.

Fearing, F. (1930). *Reflex Action. A Study in the History of Physiological Psychology*. Williams & Wilkins, Baltimore.

Feigin, I., Geller, E. H., and Wolf, A. (1951). Absence of regeneration in the spinal cord of the young rat. *J. Neuropath. Exp. Neurol.*, 10:420–425.

Feldberg, W., and Vogt, M. (1948). Acetylcholine synthesis in different regions of the central nervous systema. *J. Physiol.* (London), 7:372–338.

Fernel, J. (2003). *The Physiologia of Jean Fernel* (1567). Translated and annotated by J. M. Forrester., American Physiological Society, Philadelphia.

Fernandez, H. L., Duell, M. J., and Festoff, B. W. (1979). Cellular distribution of 16S acetylcholinesterase. *J. Neurochem.*, 32:581–585.

Fernandez, H. L., Huneeus, F. C., and Davison, P. F. (1970). Studies on the mechanism of axoplasmic transport in the crayfish cord. *J. Neurobiol.*, 1:395–409.

Fernández-Morán, H. (1952). The submicroscopic organization of vertebrate nerve fibres. An electron microscope study of myelinated and unmyelinated nerve fibres. *Exp. Cell Res.*, 3:282–259.

Fernández-Morán, H., and Finean, J. B. (1957). Electron microscope and low-angle X-ray diffraction studies of the myelin sheath. *J. Biophys. Biochem. Cytol.*, 3:725–771.

Ferster, C. B., and Skinner, B. F. (1957). *Schedules of Reinforcement.* Appelton-Century-Crofts, Inc., New York.

Fifková, E. (1987). Mechanisms of synaptic plasticity. In: *Neuroplasticity, Learning, and Memory.* N. W. Milgram, C. M. MacLeod, and T. Petit, eds. Alan R. Liss, New York, pp. 61–86.

Fifková, E., and Morales, M. (1992). Actin matrix of dendritic spines, synaptic plasticity, and long-term potentiation. *Int. Rev. Cytol.*, 139:267–307.

Finean, J. B. (1958). X-ray diffraction studies of the myelin sheath in peripheral and central nerve fibres. *Exp. Cell Res., Suppl.* 5:18–32.

Finean, J. B., and Mitchell, R. H. (1981). Isolation, composition and general structure of membranes. In: *Membrane Structure.* J. B. Finean and R. H. Mitchell, eds. Elsevier, New York.

Finger, S. (1994). *Origins of Neuroscience.* Oxford University Press, New York.

Finger, S. (2000). *Minds Behind the Brain: A History of the Pioneers and Their Discoveries.* Oxford University Press, Oxford.

Fishman, H. M., Krause, T. L., Miller, A. L., and Bittner, G. D. (1995). Retardation of the spread of extracellular Ca^{2+} into transected, unsealed squid giant axons. *Biol. Bull.*, 189:208–209.

Flaig, J. V. (1947). Viscosity changes in axoplasm under stimulation. *J. Neurophysiol.*, 10:211–221.

Flourens, P. (1846). *Phrenology Examined.* Translation of the 2nd Ed. Hogan and Thompson, Philadelphia.

Fontana, F. (1787). *Treatise on the Venom of the Viper; on American Poisons; and on the Cherry Laurel and Some Other Vegetable Poisons to Which Are Annexed Observations on the Primitive Structure of the Animal Body; Different Experiments on the Reproduction of the Nerves and a Description of the Nerves; and a New Description of a New Canal of the Eye,* 2 Vols. Translated from the French by J. Skinner, Murray, London.

Forel, A. (1887). Einige hirnanatomische betrachtungen und ergebnisse. *Arch. Psychiat. Nerven.*, 18:162–198.

Forman, D. S., Padjen, A. L., and Siggins, G. R. (1977). Effect of temperature on the rapid retrograde transport of microscopically visible intra-axonal organelles. *Brain Res.*, 136:215–226.

Forman, D. S., and Shain, W. G., Jr. (1981). Batrachotoxin blocks saltatory organelle movement in electrically excitable neuroblastoma cells. *Brain Res.*, 211:242–247.

Franklin, B. (1941). *Benjamin Franklin's Experiments. A New Edition of Franklin's Experiments and Observations on Electricity.* Edited from the last English edition of 1744 with an Historical Introduction by I. B. Cohen. Harvard University Press, Cambridge.

Frazer, J. G. (1922). *The Golden Bough: A Study in Magic and Religion.* The MacMillan Co., New York.

Freeman, K. (1957). *Ancilla to the Pre-Socratic Philosophers: A Complete Translation of the Fragments in Diels, Fragmente der Vorsikrater,* Harvard University Press, Cambridge.

French, R. (1999). *Dissection and Vivisection in the European Renaissance.* Ashgate, Brookfield.

French, R. D. (1970). Some concepts of nerve structure and function in Britain, 1875–1885: background to Sir Charles Sherrington and the synapse concept. *Med. Hist.*, 14:154–165.

French, R. D. (1971). The origins of the sympathetic nervous system from Vesalius to Riolan. *Med. Hist.*, 15:45–54.

French, R. K. (1969). *Robert Wytt, The Soul, and Medicine*. The Wellcome Institute of the History of Medicine, London.

Freud, S. (1882). Über den Bau der Nervenfasern und Nervenzellen beim Flusskrebs. *Sitz. Akad. Wiss. Wien (Math.-Natur.)*, 11:9–45.

Freud, S. (1920). *A General Introduction to Psychoanalysis*. Translated with a preface by G. Stanley Hall. Boni & Liveright, New York.

Frey, O. (1877). *Die pathologischen lungenveränderungen nach lähmung der nervi vagi*. Engelmann, Leipzig.

Friede, R. L. (1966). *Topographic Brain Chemistry*. Academic Press, New York.

Friede, R. L. (1972). Control of myelin formation by axon caliber (with a model of the control mechanism). *J. Comp. Neurol.*, 144:233–252.

Friede, R. L., and Bischausen, R. (1980). The fine structure of stumps of transected nerve fibers in subserial sections. *J. Neurol. Sci.*, 44:181–203.

Friede, R. L., and Bischausen, R. (1982). How are sheath dimensions affected by axon caliber and internode length? *Brain Res.*, 235:335–350.

Friede, R. L., and Bischhausen, R. (1978). The organization of endoneural collagen in peripheral nerves as revealed with the scanning electron microscope. *Neurol. Sci.*, 36:83–88.

Friede, R. L., and Martinez, A. J. (1970). Analysis of axon-sheath relations during early Wallerian degeneration. *Brain Res.*, 19:199–212.

Friede, R. L., and Samorajski, T. (1970). Axon caliber related to neurofilaments and microtubules in sciatic nerve fibers of rats and mice. *Anat. Rec.*, 167:379–387.

Fritsch, G. T., and Hitzig, E. (1963). Über die elektrische Erregbarkeit des Grosshirns. Arch Anat. Physiol. pp. 300–332. English translation by H. Wilkins, "*Neurosurgical Classics xii.*" *J. Neurosurg.*, 20:904–916.

Fulton, J. F. (1951). *Frontal Lobotomy and Affective Behavior*. Norton, New York.

Gainer, H., Sarne, Y., and Brownstein, M. J. (1977). Biosynthesis and axonal transport of rat neurohypophysial proteins and peptides. *J. Cell Biol.*, 73:366–381.

Gajdusek, D. C. (1977). Unconventional viruses and the origin and disappearance of kuru. Nobel Prize acceptance speech. *Science*, 107:943–960.

Gajdusek, D. C. (1985). Hypothesis: interference with axonal transport of neurofilament as a common pathogenetic mechanism in certain diseases of the central nervous system. *N. Engl. J. Med.*, 312:714–719.

Galen (1962). *Galen on Anatomical Procedures. The Later Books*. Translated by W. L. H. Duckworth. Edited by M. C. Lyons and B. Towers. Cambridge University Press, Cambridge.

Galen (1997). *The Soul's Dependence on the Body*, In: *Galen, Selected Works* Translated with an introduction and notes by P. N. Singer, pp. 150–176., Oxford University Press, Oxford.

Galen (1968). *Galen on the Usefulness of the Parts of the Body*, 2 Vols. Translated from the Greek with an introduction by M. T. May. Cornell University Press, Ithaca.

Galen (1984). *Galen on Respiration and the Arteries*. English translation and commentary of *De usu respirationis, an in arteriis natura sanguis contineatur, de usu pulsuum, de causis respirationis* by D. J. Furley and J. S. Wilkie. Princeton University Press, Princeton.

Galen (1997). *Selected Works*. Translated with an introduction by P. N. Singer. Oxford University Press, Oxford.

Gall, F. J., and Spurzheim, G. (1967). *Recherches sur le système nerveux en général et sur celui du cerveau en particulier; mémoire présenté à l'Institut de France, le 14 mars 1808; suivi d'observations sur le rapport qui en a été fait à cette compagnie par ses commissaires.* Reprinted from the 1809 edition. Bonset, Amsterdam.

Gallant, P. E. (1988). Effects of the external ions and metabolic poisoning on the constriction of the squid giant axon after axotomy. *J. Neurosci.*, 8:1479–1484.

Gallant, P. E., Hammar, K., and Reese, T. S. (1996). Cytoplasmic constriction and vesiculation after axotomy in the squid giant axon. *J. Neurocytol.*, 24:943–954.

Galvani, A. L. (1953a). *Commentary on the Effects of Electricity on Muscular Motion*. A translation from the Latin of Luigi Galvani's *de Viribus Electricitatis In Motu Musculari Commentarius* and *a dissertation on animal electricity by G. Aldini*. Translated from the Latin by of 1792 by R. M. Green. With an introduction by G. C. Pupilli. Elizabeth Licht, Cambridge, Massachusetts.

Galvani, A. L. (1953b). *Commentary on the Effects of Electricity on Muscular Motion*. Translated from the Latin of 1791 by Margaret G. Foley, with notes and a critical introduction by I. Bernard Cohn. Burndy Library, Norfolk.

Gambetti, P., Monaco, S., Autilio-Gambetti, L. A., and Sayre, L. M. (1986). Chemical neurotoxins accelerating axonal transport of neurofilaments. In: *The Cytoskeleton: a Target for Toxic Agents*. T. W. Clarkson, P. R. Sager, and T. L. M. Syversen, eds. Plenum Press, New York, pp. 129–142.

Garcia-Ballester, L. (2002). *Galen and Galensism: Theory and Medical Practice from Antiquity to the European Rennaissance*. Collected papers edited by Arrizabalaga, J., Cabré, M. Cifuentes, L., Salmón, F., Ashgate Publishing Co., Burlington.

Garrison, F. H. (1929). *An Introduction to the History of Medicine*, 4th ed. Reprinted 1963. Saunders, Philadelphia.

Gaskell, W. H. (1916). *The Involuntary Nervous System*. Longmans, Green and Co., London.

Gaskin, J. (1995). *The Epicurean Philosophers*. J. Gaskin, ed. Translated by C. Bailey, R. D. Hicks, J. C. A. Gaskin., J. M. Dent, London.

Gasser, H. S. (1958). Comparison of the structure, as revealed with the electron microscope, and the physiology of the unmedullated fibers in the skin nerves and in the olfactory nerves. *Exp. Cell Res. Suppl.*, 5:3–17.

Gaster, T. H. (1969). *Myth, Legend, and Custom in the Old Testament. A Comparative Study with Chapters from Sir James Frazer's Folklore in the Old Testament*. Harper and Row, New York.

Gaziri, L. C. J., (1984). *Fast Axoplasmic Transport in Relation to Energy Metabolism of Peripheral Nerve*. Ph.D. Thesis, Dept. of Physiology, Indianapolis School of Medicine., Indianapolis.

Gaziri, L. C. J., and Ochs, S. (1983). Increased duration of axoplasmic transport in vitro with added glucose or β-hydroxybutyrate. *Soc. Neurosci. Abst.*, 9:150.

Gaziri, L. C. J., and Ochs, S. (1987). Metabolic support of axoplasmic transport in mammalian nerve. *J. Neurochem.*, 1:1.

Geffen, L. B., and Livett, B. G. (1971). Synaptic vesicles in sympathetic neurons. *Physiol. Rev.*, 51:98–157.

Gelfan, S. (1930). Studies of single muscle fibres I. The all-or-none principle. *Am. J. Physiol.*, 93:1–8.

Gelfan, S., and Gerard, R. W. (1930). Studies of single muscle fibres. II. A further analysis of the grading mechanism. *Am. J. Physiol.*, 95:412–416.

Geller, A. I., Keyomarsi, K., Bryan, J., and Pardee, A. (1990). An efficient deletion mutant packaging system for defective herpes simplex virus vectors: potential applications to human gene therapy and neuronal physiology. *Proc. Natl. Acad. Sci. U.S.A.*, 87:8950–8954.

Gerard, R. W. (1932). Nerve metabolism. *Physiol. Rev.*, 12:469–592.

Geren, B. B. (1954). The formation from the Schwann cell surface of myelin in the peripheral nerves of chick embryos. *Exp. Cell Res.*, 7:558–562.

Gerzon, K. (1980). Dimeric catharanthus alkaloids. In: *Anticancer Agents Based on Natural Product Models*. J. M. Cassady, and J. Douros, eds. Academic Press, New York, pp. 271–317.

Ghabriel, M. N., and Allt, G. (1981). Incisures of Schmidt-Lanterman. *Prog. Neurobiol.*, 17:25–58.

Ghetti, B., Alyea, C., Norton, J., and Ochs, S. (1982). Effects of vinblastine on microtubule density in relation to axoplasmic transport. In: *Axoplasmic Transport* D. G. Weiss, ed. Springer-Verlag, New York, pp. 322–327.

Ghetti, B., and Ochs, S. (1978). On the relation between microtubule density and axoplasmic transport in nerves treated with maytansine. In: *Peripheral Neuropathies*. N. Canal, and G. Pozza, eds. Elsevier, Amsterdam, pp. 177–186.

Gilbert, W. (1958). *De magnete*. First published in 1600. Translated by P. F. Mottelay 1893. Dover Publications, Inc., New York.

Glantz, L., and Lewis, D. A. M. (2000). Decreased dendritic spine density on prefrontal cortical pyramidal neurons in schizophrenia. *Arch. Gen. Psychiatry*, 57:65–73.

Glass, J. D., and Griffin, J. W. (1991). Neurofilament redistribution in transected nerves: evidence for bidirectional transport of neurofilaments. *J. Neurosci.*, 11:3146–3154.

Glass, J. D., and Griffin, J. W. (1994). Retrograde transport of radiolabeled cytoskeletal proteins in transected nerves. *J. Neurosci.*, 14:3915–3921.

Goelet, P., Castellucci, V., Schacher, S., and Kandel, E. R. (1986). The long and the short of long-term memory – a molecular framework. *Nature*, 322:419–422.

Gold, B. G., Griffin, J. W., and Price, D. L. (1985). Slow axonal transport in acrylamide neuropathy: different abnormalities produced by single-dose and continuous administration. *J. Neurosci.*, 5:1755–1768.

Goldberg, D. J., Goldman, J. E., and Schwartz, J. H. (1976). Alterations in amount and rates of serotonin transported in an axon of the giant cerebral neurone of Aplysia californica. *J. Physiol.* (London), 259:473–490.

Goldberg, D. J., Schwartz, J. H., and Sherbany, A. A. (1978). Kinetic properties of normal and perturbed axonal transport of serotonin in a single identified axon. *J. Physiol.* (London) , 28:559–579.

Golding, J. P., and Tonge, D. A. (1993). Expression of GAP-43 in normal and regenerating nerves in the frog. *Neuroscience*, 52:415–426.

Goldman, J. E., Kim, K. S., and Schwartz, J. H. (1976). Axonal transport of (^{3}H) serotonin in an identified neuron of *Aplysia californica*. *J. Cell Biol.*, 70:304–318.

Goldman, R., Pollard, T., and Rosenbaum, J. (1976). *Cell Motility*. Books A, B, C. Cold Spring Harbor Laboratory, Cold Spring Harbor.

Goldscheider, A. (1894a). Zur allgemeinen pathologie des nervensystems. II. Ueber neuron- erkrankungen. *Berlin. Klin. Wochenschr.* 31:444–447.

Goldscheider, A. (1894b). Zur allgemeinen pathologie des nervensystems. I. Ueber die lehre von den trophischen centren. *Berlin. Klin. Wochenschr.* 31:421–425.

Goldstein, L. S. B., and Philp, A. V. (1999). The road less traveled: emerging principles of kinesin motor utilization. *Annu. Rev. Cell Dev. Biol.*, 15:141–183.

Golgi, C. (1967). The neuron doctrine – theory and facts 1906. In: *Nobel Lectures Physiology or Medicine 1901–1921*. Nobel Lectures. Elsevier, Amsterdam.

Goodall, J. (2000). *In the Shadow of Man*. Rev. Ed. originally published 1971. Houghton Mifflin, Boston.

Goodpasture, E. W. (1925). The axis-cylinders of peripheral nerves as portals of entry to the central nervous system for the virus of herpes simplex in experimentally infected rabbits. *Am. J. Pathol.*, 1:11–28.

Gould, E., Reeves, A. J., Graziano, M. S. A., and Gross, C. G. (2001). Neurogenesis in the neocortex of adult primates. *Science*, 286:548–552.

Grafstein, B. (1967). Transport of protein by goldfish optic nerve fibers. *Science*, 157:196–198.

Grafstein, B. (1971). Transneuronal transfer of radioactivity in the central nervous system. *Science*, 172:77–79.

Grafstein, B. (1977). Axonal transport: the intracellular traffic of the neuron. In: *Handbook of Physiology*, Section 1. Vol. 1. *The Nervous System*. American Physiological Society, Bethesda, pp. 691–717.

Grafstein, B., and Forman, D. S. (1980). Intracellular transport in neurons. *Physiol. Rev.*, 60:1167–1183.

Graham, J., and Gerard, R. W. (1946). Membrane potentials and excitation of impaled single muscle fibers. *J.Cell Physiol.*, 28:99–117.

Gray, E. G. (1978). Synaptic vesicles and microtubules in frog nerve endplates. *Proc. Roy. Soc. Lond. B.*, 203:219–227.

Greene, L. A., and Shooter, E. (1980). The nerve growth factor: biochemistry synthesis and the mechanism of actions. *Ann. Rev. Neurosci.*, 3:353–402.

Griffin, J., Price, D., Engel, W. K., and Drachman, D. B. (1977). The pathogenesis of reactive axonal swellings: role of axonal transport. *J. Neuropath. Exp. Neurol.*, 36:214–227.

Griffin, J., Price, D., and Spencer, P. S. (1977). Fast axonal transport through giant axonal swellings in hexacarbon neuropathies. *J. Neuropathol. Exp. Neurol.*, 36:603.

Griffin, J. W., Drachman, D. B., and Price, D. L. (1976). Fast axonal transport in motor nerve regeneration. *J. Neurobiol.*, 7:355–370.

Griffin, J. W., Fahnestock, K. E., Price, D. L., and Hoffman, P. N. (1983). Microtubule-neurofilament segregation produced by β,β'-iminodipropionitrile: evidence for the association of fast axonal transport with microtubules. *J. Neurosci.*, 3:557–566.

Griffin, J. W., Hoffman, P. N., Clark, A. W., Carroll, P. T., and Price, D. L. (1978). Slow axonal transport of neurofilament proteins: impairment by β,β'-iminodipropionitrile administration. *Science*, 202:633–635.

Griffin, J. W., Price, D. L., Hoffman, P. N., and Cork, L. C. (1981). The axonal cytoskeleton: alterations of organization and axonal transport in models of neurofibrillary pathology. *J. Neuropath. Exp. Neurol.*, 40:316.

Gross, C. G. (2001). Neurogenesis in the adult brain: death of a dogma. *Nature Rev. Neurosci.*, 1:67–73.

Gross, G. W., and Beidler, L. M. (1973). Fast axonal transport in the C-fibers of the garfish olfactory nerve. *J. Neurobiol.*, 4:413–428.

Gross, G. W., and Beidler, L. M. (1975). A quantitative analysis of isotope concentration profiles and rapid transport velocities in the C-fibers of the garfish olfactory nerve. *J. Neurobiol.*, 6:213–232.

Gross, G. W., and Weiss, D. G. (1982). Theoretical considerations on rapid transport in low viscosity axonal regions. In: *Axoplasmic Transport*. D. G. Weiss, ed. Springer-Verlag, Berlin, pp. 330–341.

Guiard, E. (1930). *La trépanation cranienne chez les néolithiques et chez les primitifs modernes*. Masson et Cie, Paris.

Gundersen, R. W., and Barrett, J. N. (1979). Neuronal chemotaxis: chick dorsal-root axons turn toward high concentrations of nerve growth factor. *Science*, 206:1079–1080.

Gunther, A. F., and Schoen, J. M. A. (1840). Versuche und bemerkungen über regeneration der nerven und abhangigkeit der peripherischen nerven von den central-organen. *Arch. Anat. Physiol. Wiss. Med.*, 270–286.

Guth, L. (1974). Axonal regeneration and functional plasticity in the central nervous system. *Exp. Neurol.*, 45:606–654.

Guth, L. (1975). History of central nervous system regeneration research. *Exp. Neurol.*, 48:2–3.

Guth, L., Bright, D., and Donati, E. J. (1978). Functional deficits and anatomical alterations after high cervical spinal hemisection in the rat. *Exp. Neurol.*, 58:5–520.

Guthrie, W. K. C. (1965). *A History of Greek Philosophy*. Vol. 2. *The Presocratic Tradition from Parmenides to Democritus*. Cambridge University Press, Cambridge.

Guthrie, W. K. C. (1981). *A History of Greek Philosophy. VI. Aristotle: An Encounter*. Cambridge University Press, Cambridge, NY.

Gutmann, E. (1962). ed. *The Denervated Muscle*. Publishing House of the Czechoslovak Academy of Sciences, Prague.

Gutmann, E., Guttmann, L., Medawar, P. B., and Young, J. Z. (1942). The rate of regeneration of nerve. *J. Exp. Biol.*, 9:4–44.

Gutmann, E., and Hnik, P. (1962). Denervation studies in research of neurotrophic relationships. In: *The Denervated Muscle*. E. Gutmann, ed. Publishing House of the Czechoslovakian Academy of Science, Prague.

Gutmann, E., and Hnik, P. (1963). *The Effect of Use and Disuse of Neuromuscular Functions*. Publishing House of the Czechoslovakian Academy of Science, Prague.

Guy de Chauliac (1923). *On Wounds and Fractures*. Translated by W. A. Brennan, Published by translator, Chicago.

Guy de Chauliac (1890). *La Grande Chirurgie de Guy de Chaliac. Composée en l'an 1363. Revue et collationée sur les manuscrits et imprimes Latins et Francais etc. by E. Nicaise*. Bailliere, Paris.

Ha, H. (1970). Axonal bifurcation in the dorsal root ganglion of the cat: a light and electron microscopic study. *J. Comp. Neurol.*, 140:227–240.

Haak, R. A., Kleinhans, F. W., and Ochs, S. (1976). The viscosity of mammalian nerve axoplasm measured by electron spin resonance. *J. Physiol.* (London), 263:5–37.

Haighton, J. (1795). An experimental inquiry concerning the reproduction of nerves. *Phil. Trans. Roy. Soc.* (abridged version), 85:519–525.

Hall, M. (1826). On the nervous circle which connects the voluntary muscles with the brain. *Phil. Trans. R. Soc.*, 116:163–173.

Hall, M. (1833). On the reflex function of the medulla oblongata and medulla spinalis. *Phil. Trans. Roy. Soc.*, 123:635–665.

Hall, S. M. (1972). The effects of injection of potassium cyanide into the sciatic nerve of the adult mouse: in vivo and electron microscopic studies. *J. Neurocytol.*, 1:233–254.

Halpain, S. (2000). Actin and the agile spine: how and why do dendritic spines dance? *Trends Neurosci.*, 23:141–145.

Hamlyn, D. W. (1961). *Sensation and Perception: A History of the Philosophy of Perception.* Routledge and Kegan Paul, London.

Hammerschlag, R., and Brady, S. T. (1989). Axonal transport and the neuronal cytoskeleton. In: *Basic Neurochemistry*, 4th ed. G. Siegel, B. Agranoff, R. W. Albers, and P. Molinoff, eds. Raven Press, New York, pp. 457–478.

Hammerschlag, R., and Stone, G. C. (1986). Prelude to fast axonal transport: sequence of events in the cell body. In: *Axoplasmic Transport.* Z. Iqbal, ed. CRC Press, Boca Raton, pp. 21–34.

Hammond, G. R., and Smith, R. S. (1977). Inhibition of the rapid movement of optically detectable axonal particles by colchicine and vinblastine. *Brain Res.*, 128:227–242.

Handerson, H. E. (1918). *Gilbertus Anglicus (Gilbert of England). A Study of Medicine in the Thirteenth Century.* With a biography of the author. Cleveland Medical Library Association, Cleveland.

Hanson, M., and Edström, A. (1978). Mitosis inhibitors and axonal transport. *Int. Rev. Cytol. Suppl.*, 7:373–402.

Hanson, N. R. (1958). *Patterns of Discovery.* Cambridge University Press, Cambridge.

Harlow, A. F. (1936). *Old Wires and New Waves.* D. Appelton-Century Co., New York.

Harrison, J. E. (1962). *Epilegomena to the Study of Greek Religion.* First printed in 1921. Published with *Themis: A Study of the Social Origins of Greek Religion.* From the 2nd Rev. Ed. of 1927. University Books, New York.

Harrison, R. G. (1910). The outgrowth of the nerve fiber as a mode of protoplasmic movement. *J. Exp. Zool.*, 9:787–846.

Harry, G. J., Goodrum, J. F., Bouldin, T. W., Toews, A. D., and Morell, P. (1989). Acrylamide-induced increases in deposition of axonally transported glycoproteins in rat sciatic nerve. *J. Neurochem.*, 52:1240–1247.

Hartley, D. (1967). *Observations On Man, His Frame, His Duty, and His Expectations.* 2 Vols. Reprinted from the London edition of 1749. Georg Olms, Hildesheim.

Harvey, D. M. R. (1981). Freeze-substitution. *J. Microscopy*, 127:209–221.

Harvey, W. (1959). *William Harvey's De motu locali animalium 1627.* Edited, translated and introduced by Gwenneth Whitteridge. Cambridge University Press, Cambridge.

Harvey, W. (1970). *Anatomical Studies on the Motion of the Heart and Blood.* Translated by C. D. Leake. Charles C Thomas, Springfield.

Haskins, C. H. (1965). *Studies in Mediaeval Culture.* Reprinted from the first edtion, 1929. Fredrik Unger, New York.

Haygarth, J. (1801). Of the imagination, as a cause and as a cure of disorders of the body: exemplified by fictitious tractors, and epidemical convulsions. Read to the Literary and Philosophical Society of Bath by John Haygarth, a new editor with additional remarks. R. Cruttwell, Strand, London.

Haymaker, W., and Schiller, F. (1970). *The Founders of Neurology. One Hundred and Forty-Six Biographical Sketches by Eight-Nine Authors.* 2nd. Ed. Charles C Thomas, Springfield.

Heath, J., Ueda, S., Bornstein, M. B., Daves, G., and Raine, C. S. (1982). Buckthorn neuropathy in vitro: evidence for a primary neuronal effect. *J. Neuropath. Exp. Neurol.*, 2:204–220.

Hebb, C. O., and Silver, A. (1961). Gradient of choline acetylase activity. *Nature*, 89:23–25.

Hebb, D. O. (1949). *The Organization of Behavior.* Wiley, New York.

Hecker, J. F. C. (1884). *The Epidemics of the Middle Ages*. Translated from the German by B. G. Babington. The Sydenham Society, London.

Heidemann, S. R. (1996). Cytoplasmic mechanisms of axonal and dendritic growth in neurons. *Int. Rev. Cytol.*, 165:235–296.

Heilbron, J. L. (1999). *Electricity in the 17th and 18th Centuries: A Study in Early Modern Physics*. Dover Publications, Inc., Mineola.

Heiwall, P.-O. (1978). *Studies on the Intra-axonal Transport of Acetylcholine and Cholinergic Enzymes in Rat Sciatic Nerve*. Ph.D. thesis. Institute of Neurobiology, University of Göteborg, Sweden.

Helmholtz, A. (1842). *De Fabrica Systematis Nervosi Evertebratorum*. Dissertation, Nietackianis, Berolini (Berlin).

Hendry, I. A., Stockel, K., Thoenen, H., and Iversen, L. L. (1974). The retrograde axonal transport of nerve growth factor. *Brain Res.*, 68:103–121.

Herken, H., and Hucho, F. (1992). eds. *Selective Neurotoxicity*. Vol. 102, Handbook of Experimental Pharmacology, Springer-Verlag, Berlin.

Hermier, D., Forgez, P., and Chapman, M. J. (1985). A density gradient study of the lipoprotein distribution in the chicken, *Gallus domesticus*. *Biochem. Biophys. Acta*, 836:105–118.

Heslop, J. P. (1975). Axonal flow and fast transport in nerves. *Adv. Comp. Physiol. Biochem.*, 6:75–163.

Heslop, J. P., and Howes, E. A. (1972). Temperature and inhibitor effects on fast axonal transport in a molluscan nerve. *J. Neurochem.*, 9:1709–1716.

Hild, W. (1951). Experimentell-morphologische untersuchungen über das verhalten der neurosekretorischen bahn nach hypophysenstieldurchtrennungen eingriffen in den wasserhaushalt und belastung der osmoregulation. *Virchow's Archiv.*, 319:526–546.

Hill, A. V., and Howarth, J. V. (1958). The initial heat production of stimulated nerve. *Proc. Roy. Soc. Lond. B.*, 149:167–175.

Hippocrates (1849). *The Genuine Works of Hippocrates*. Translated from the Greek with a Preliminary Discourse and Annotations by Francis Adams. Two volumes in one. Sydenham Society. William Wood and Co., New York.

Hippocrates (1923). *The Sacred Disease*. Translated by W. H. M. Jones. In: *Hippocrates*, Vol. II. Harvard University Press, Cambridge, pp. 127–183.

Hirokawa, N. (1982). Cross-linker system between neurofilaments, microtubules, and membranous organelles in frog axons revealed by the quick-freeze, deep-etching method. *J. Cell Biol.*, 94:129–142.

Hirokawa, N., Pfister, K. K., Yorifuji, H., Wagner, M. C., Brady, S. T., and Bloom, G. S. (1989). Submolecular domains of bovine brain kinesin identified by electron microscopy and monoclonal antibody decoration. *Cell*, 56:867–878.

Hirokawa, N. (1991). Molecular architecture and dynamics of the neuronal cytoskeleton. In: *The Neuronal Cytoskeleton*. R. D. Burgoyne, ed. Wiley-Liss, New York, pp. 5–74.

Hirokawa, N., Terada, S., Funakoshi, T., and Takeda, S. (1997). Slow axonal transport: the subunit transport model. *Cell Biol.*, 7:384–388.

Hirokawa, N. (1998). Kinesin and dynein superfamily proteins and the mechanism of organelle transport. *Science*, 279:519–526.

Hirokawa, N., and Okabe, S. (1992). Microtubules on the move? *Curr. Biol.*, 2:193–195.

Hirokawa, N., Sato-Yoshitake, R., Kobayashi, N., Pfister, K. K., Bloom, G. S., and Brady, S. T. (1991). Kinesin associates with anterogradely transported membranous organelles in vivo. *J. Cell Biol.*, 114:295–302.

His, W. (1887). Zur Geschichte des Menschlichen Ruckenmarkes und der Nerven-wurzeln. *Abh. Sachs. Ges. Wisen.* (Math-Phys. Klasse), 13:477–514.

Hiscoe, H. B. (1947). Distribution of nodes and incisures in normal and regenerated nerve fibers. *Anat. Rec.*, 99:447–475.

Hodgkin, A. L., and Huxley, A. F. (1939). Action potentials recorded from inside a nerve fibre. *Nature*, 144:710–711.

Hodgkin, A. L., and Huxley, A. F. (1952). A quantitative description of membrane current and its application to conduction and excitation in nerve. *J. Physiol.* (London), 117:500–544.

Hodgkin, A. L., Huxley, A. F., and Katz, B. (1952). Measurements of current-voltage relations in the membrane of the giant axon of *Loligo*. *J. Physiol.*, 116:424–448.

Hoff, H. E. (1936). Galvani and the pre-Galvanian electrophysiologists. *Ann. Sci.*, 1:157–172.

Hoff, H. E. (1959). A classic of microscopy: an early, if not the first, observation on the fluidity of the axoplasm, micromanipulation, and the use of the cover-slip. *Bull. Hist. Med.*, 33:375–379.

Hoff, H. E., and Geddes, L. A. (1957). The rheotome and its pre-history. *Bull. Hist. Med.*, 31: 212–234, 327–347.

Hoffman, P. N., and Griffin, J. W. (1993). The control of axonal caliber. In: *Peripheral Neuropathy*. 2 Vols., 2nd Ed. P. J. Dyck, P. K. Thomas, J. W. Griffin, P. A. Low, and J. F. Poduslo, eds. Saunders, Philadelphia, pp. 389–402.

Hoffman, P. N., Griffin, J. W., and Price, D. L. (1981). Changes in the axonal transport of the cytoskeleton during development aging and regeneration. *Soc. Neurosci. Abstr.*, 7:743.

Hoffman, P. N., and Lasek, R. J. (1975). The slow component of axonal transport. Identification of major structural polypeptides of the axon and their generality among mammalian neurons. *J. Cell Biol.*, 66:351–366.

Hoffman, P. N., and Lasek, R. J. (1980). Axonal transport of the cytoskeleton in regenerating motor neurons: constancy and change. *Brain Res.*, 202:317–333.

Hokfelt, T. (1969). Distribution of noradrenaline storing particles in peripheral adrenergic neurons as revealed by electron microscopy. *Acta Physiol. Scand.*, 76:427–440.

Hollenbeck, P. J. (1989). The transport and assembly of the axonal cytoskeleton. *J. Cell Biol.*, 108:223–227.

Hollenbeck, P. J. (1990). Cell biology. Cytoskeleton on the move. *Nature*, 343:408–409.

Homer (1997). *The Iliad.* Translated by Stanley Lombardo. Hackett Publishing Co., Indianapolis.

Hooker, D. (1952). *The Prenatal Origin of Behavior.* University of Kansas Press, Lawrence.

Horie, H., Takenaka, T., and Inomata, K. (1981). Effects of antimitotic drugs on axoplasmic transport in tissue cultured nerve cells. *Soc. Neurosci. Abstr.*, 7:485.

Horie, H., Takenaka, T., Ito, S., and Kim, U. (1987). Taxol counteracts colchicine blockade of axonal transport in neurites of cultured dorsal root ganglion cells. *Brain Res.*, 420:144–146.

Horwitz, S. B., Lothstein, L., Manfredi, J. J., Mellado, W., Parness, J., Roy, S., Schiff, P. B., Sorbara, L., and Zeheb, R. (1986). Taxol: mechanisms of action and resistance. *Ann. N. Y. Acad. Sci.*, 466:733–744.

Hoy, R. R., Bittner, G. D., and Kennedy, D. (1967). Regeneration in crustacean motoneurons: evidence for axonal fusion. *Science*, 56:25–252.

Hrdlička, A. (1939). Trepanation among prehistoric people, especially in America. *Ciba Symp.*, 1:170–176.

Hubel, D. H., and Weisel, T. N. (1959). Receptive fields of single neurones in the cat's striate cortex. *J. Physiol.*, 148:106–154.

Humphrey, G. (1951). *Thinking: An Introduction to Its Experimental Psychology.* Wiley, New York.

Hunter, R., and MacAlpine, I. (1963a). *De Catarrhis 1627.* Translated with a bibliographical analysis. Dawson of Pall Mall, London.

Hunter, R., and MacAlpine, I. (1963b). *Three Hundred Years of Psychiatry. 1535–1860: A History Presented in Selected English Texts.* Oxford University Press, London.

Hussey, E. (1972). *The Presocratics.* Duckworth, London.

Hyrtl, J. (1970). *Onomatologia Anatomica.* Georg Olms, Hildesheim.

Ingber, D., Karp, S., Plopper, G., Hansen, L., and Mooney, D. (1993). Mechanochemical transduction across extracellular matrix and through the cytoskeleton. In: *Physical Forces and the Mammalian Cell.* J. A. Frangros, ed. Academic Press, New York, pp. 61–79.

Ingber, D. E. (1993). Cellular tensegrity: defining new rules of biological design that govern the cytoskeleton. *J. Cell Sci.*, 104:613–627.

Ingelfinger, F. J. (1958). Esophageal motility. *Physiol. Rev*, 38:532–584.

Inoue, S., and Sato, H. (1967). Cell mobility by labile association of molecules. The nature of mitotic spindle fibers and their role in chromosome movement. *J. Gen. Physiol.*, 50:259–292.

Iqbal, Z. (1986). ed. *Axoplasmic Transport.* CRC Press, Boca Raton.

Iqbal, Z. (1986). Calmodulin and its role in axoplasmic transport. In: *Axoplasmic Transport.* Z. Iqbal, ed. CRC Press, Boca Raton, pp. 45–55.

Iqbal, Z., and Ochs, S. (1980). Calmodulin in mammalian nerve. *J. Neurobiol.*, 11:311–318.

Jackson, J. H. (1931). *Selected Writings of John Hughlings Jackson.* 2 Vols. J. Taylor ed. Hodder and Stoughton, London.

Jackson, S. W. (1970). Force and kindred notions in 18th century neurophysiology and medical psychology. *Bull. Hist. Med.*, 44:397–410 and 539–554.

Jacobson, M. (1978). *Developmental Neurobiology*, 2nd ed. Plenum Press, New York.

Jakobsen, J., Brimijoin, S., and Sidenius, P. P. (1983). Axonal transport in neuropathy. *Muscle & Nerve*, 6:164–166.

Jakobsen, J., Brimijoin, S., Skau, K., and Sidenius, P. P. (1981). Retrograde axonal transport of transmitter enzymes, fucose labelled proteins and nerve growth factor in streptozotocin-diabetic rats. *Diabetes*, 30:797–803.

James, K. A. C., and Austin, L. (1969). The binding in vitro of colchicine to axoplasmic proteins from chicken sciatic nerve. *Biochem. J.*, 117:773–777.

James, W. (1890). *The Principles of Psychology.* Reprinted 1950. Dover Publications, New York.

James, W. (1963). *The Varieties of Religious Experience; A Study in Human Nature, being the Gifford lectures on natural religion delivered at Edinburgh in 1901–1902.* Enlarged edition. Introduction by J. Ratner. University Books, New Hyde Park.

Jefferson, G. (1975). *Marshall Hall, The Grasp Reflex and the Diastaltic Spinal Cord*, Vol. 2. In: *Science, Medicine and History.* 2 Vols. Reprinted from Oxford University Press, London (1953). K. A. Underwood, ed. Arno Press, New York, pp. 303–320.

Jennings, H. S. (1962). *Behavior of the Lower Organisms.* With a new introduction by D. D. Jensen. Originally published in 1906. Indiana University Press, Bloomington.

Jersild, R. A. Jr., and Ochs, S. (1982). Variations in cation content of pyroantimonate precipitates within individual subcellular compartments determined by X-ray microanalysis. *Anat. Rec.*, 202:89a.

Johnson, S. (1827). Dictionary of the English Language. 2nd Ed. With corrections and additions by Rev. A. J. Todd. In 3 Vol., Longman, Rees, Orme, Brown and Green, London.

Johnson, J. L. (1970). Changes in acetylcholinesterase, acid phosphatase, and beta glucuronidase proximal to a nerve crush. *Brain Res.*, 18:427–440.

Johnson, M. (1975). Organophosphorus esters causing delayed neurotoxic effects. *Arch. Toxicol.*, 34:259–288.

Johnstone, J. (1764). Essay on the use of the ganglions of the nerves. *Phil. Trans. Roy. Soc.* (London), 54:177.

Johnstone, J. (1771). *An Essay on the Use of the Ganglions of the Nerves.* J. Eddowes, London.

Johnstone, J. (1795). *Medical Essays and Observations with Disquisitions Relating to the Nervous System.* Evesham, London.

Jones, E. G., and Pons, T. P. (1998). Thalamic and brainstem contributions to large-scale plasticity of primate somatosensory cortex. *Science*, 282:1121–1125.

Jones, R. F. (1965). *Ancients and Moderns.* University of California Press, Berkeley.

Joseph, B. S. (1973). Somatofugal events in Wallerian degeneration: a conceptual overview. *Brain Res.*, 59:1–18.

Kandel, E. R., Schwartz, J. H., and Jessell, T. M. (2000). *Principles of Neural Science*, 4th ed. McGraw-Hill, New York.

Kanje, M., Edström, A., and Ekström, P. (1982). Divalent cations and fast axonal transport in chemically desheathed (Triton X-treated) frog sciatic nerve. *Brain*, 241:67–74.

Kanje, M., Edström, A., and Hanson, M. (1981). Inhibition of rapid axonal transport in vitro by the ionophores X-537 A and A23187. *Brain Res.*, 204:43–50.

Kant, I. (1953). *Prolegomena to Any Future Metaphysics.* Translated by P. G. Lucas. Manchester University Press, Manchester.

Kant, I. (1961). *Immanuel Kant's Critique of Pure Reason.* Translated by Norman Kemp Smith. MacMillan, London.

Kant, I. (1988). *Logic.* Translated, with an Introduction, by Robert S. Hartman and Wolfgang Schwarz and a Preface by Anna Schwarz., Dover Publications, Inc., New York.

Kanwisher, N., McDermott, J, Chun, M. M. (1997). The fusiform face area: a module in human extrastriate cortex specialized for face perception. *J. Neurosci.*, 17:4302–4311.

Kao, C. C., Chang, L. W., and Bloodworth, J. M. B., Jr. (1977). Axonal regeneration across transected mammalian spinal cords: an electron microscopic study of delayed microsurgical nerve grafting. *Exp. Neurol.*, 54:591–615.

Kapeller, K., and Mayor, D. (1969). An electron microscopic study of the early changes distal to a constriction in sympathetic nerves. *Proc. Roy. Soc. Lond. B.*, 172:53–63.

Kaplan, M. S. (1985). Formation and turnover of neurons in young and senescent animals: an electronmicroscopic and morphometric analysis. In: *Hope for a New Neurology.* F. Nottebohm, ed. New York Academy of Science, New York, pp. 173–192.

Kaplan, M. S., and Bell, D. H. (1984). Mitotic neuroblasts in the 9-day-old and 11-month-old rodent hippocampus. *J. Neurosci.*, 4:1429–1441.

Kardel, T. (1994). *Steno on Muscles*. American Philosophical Society, Philadelphia.

Karlsson, J.-O., and Sjöstrand, J. (1971a). Rapid intracellular transport of fucose-containing glycoproteins in retinal ganglion cells. *J. Neurochem.*, 18:2209–2216.

Karlsson, J.-O., and Sjöstrand, J. (1971b). Synthesis migration and turnover of protein in retinal ganglion cells. *J. Neurochem.*, 18:749–767.

Karlsson, U. (1966). Comparison of the myelin period of peripheral and central origin by electron microscopy. *J. Ultrastruct. Res.*, 15:451–468.

Karnovsky, M. J. (1965). A formaldehyde-gluteraldehyde fixative of high osmolality for use in electron microscopy. *J. Cell Biol.*, 27:137A–138A.

Kasai, H., Matsuzaki, M., Noguchi, J., Yasumatsu, N., and Nakahara, H. Structure-stability-function relationships of dendritic spines. (2003). *Trends Neurosci.*, 26:360–368.

Kater, S. B., and Nicholson, C. (1973). *Intracellular Staining in Neurobiology*. Springer-Verlag. New York.

Kates, M., and Manson, L. A. (1984). *Membrane Fluidity*. Plenum, New York.

Kato, S., Yamamoto, S., Iwasaki, Y., Niizuma, H., Nakamura, T., and Suzuki, J. (1988). Experimental retrograde adriamycin trigeminal sensory ganglionectomy. *J. Neurosurg.*, 69:760–765.

Katz, B. (1966). *Nerve, Muscle, and Synapse*. McGraw-Hill, New York.

Katz, B., and Miledi, R. (1965). The measurement of synaptic delay: the time course of acetylcholine release at the neuromuscular junction. *Proc. Roy. Soc. Lond. B.*, 161:483–495.

Keele, K. D. (1952). *Leonardo da Vinci On Movement of the Heart and Blood*. With a forward by Charles Singer. Harvey and Blythe, London.

Keele, K. D. (1957). *Anatomies of Pain*. Charles C Thomas, Springfield.

Keele, K. D. (1963). Leonardo da Vinci's research on the central nervous system. In: *Essays on the History of Italian Neurology*. L. Belloni, ed. Elli and Pagani, Milan, pp. 15–30.

Keele, K. D. (1979). Leonardo da Vinci's 'Anatomia Naturale.' The inaugural John F. Fulton lecture. *Yale J. Biol. Med.*, 52:369–409.

Kellaway, P. (1946). The part played by electric fish in the early history of bio-electricity and electrotherapy. *Bull. Hist. Med.*, 20:112–137.

Keller, F. S., and Schoenfeld, W. N. (1950). *Principles of Psychology. A Systematic Text in the Science of Behavior*, Appleton-Century-Crofts, Inc., New York.

Kemp, J. (1968). *The Philosophy of Kant*. Oxford University Press, London.

Kennedy, R. (1898). Degeneration and regeneration of nerves: an historical review. *Proc. Roy. Phil. Soc.* (Glasgow), 24:193–229.

Khan, M. A., and Ochs, S. (1975). Slow axoplasmic transport of mitochondria (MAO) and lactic dehydrogenase in mammalian nerve fibers. *Brain Res.*, 96:267–277.

Khan, M. A., Ranish, N., and Ochs, S. (1971). Axoplasmic transport of AChE LDH and MAO in mammalian nerve fibers. *Soc. Neurosci. Abstr.*, 1:144.

Kidwai, A. M., and Ochs, S. (1969). Components of fast and slow phases of axoplasmic flow. *J. Neurochem.*, 16:1105–1112.

Kiernan, J. A. (1978). An explanation of axonal regeneration in peripheral nerves and its failure in the central nervous system. *Med. Hypoth.*, 4:15–26.

Kiernan, J. A. (1979). Hypotheses concerned with axonal regeneration in the mammalian nervous system. *Biol. Rev.*, 54:155–197.

Kim, H., Binder, L. I., and Rosenbaum, J. L. (1979). The periodic association of MAP_2 with brain microtubules in vitro. *J. Cell Biol.*, 80:266–276.

Kirschner, D. A., Ganser, A. L., and Caspar, D. L. D. (1984). Diffraction studies of molecular organization and membrane interactions in myelin. In: *Myelin*. P. Morell, ed. Plenum Press, New York, pp. 51–95.

Kirschner, D. A., and Hollingshead, C. J. (1980). Processing for electron microscopy alters membrane structure and packing in myelin. *J. Ultrastruct. Res.*, 73:211–232.

Kisch, B. (1954). Forgotten leaders of medicine: Valentin, Gruby, Remak, Auerbach. *Trans. Am. Phil. Soc. (NS)*, 44:139–317.

Klatzo, I., Wisniewski, H. M., and Streicher, E. (1965). Experimental production of neurofibrillary degeneration. I. Light microscopic observations. *J. Neuropath. Exp. Neurol.*, 24:187–199.

Klein, I. (1979). *A Guide to Jewish Religious Practice*. Jewish Theological Seminary of America, New York.

Kneale, W., and Kneale, M. (1962). *The Development of Logic*. Clarendon Press, Oxford.

Knowles, R. B., Sabry, J. H., Martone, M. E., Deerinck, T. J., Ellisman, M. H., Bassell, G. J., and Kosik, K. S. (1996). Translocation of RNA granules in living neurons. *J Neurosci*. 16:7812–7820.

Koehnle, T. J., and Brown, A. (1999). Slow axonal transport of neurofilament protein in cultured neurons. *J. Cell Biol.*, 144:447–458.

Koenig, E., and Koelle, G. B. (1961). Mode of regeneration of acetylcholinesterase in cholinergic neurons following irreversible inactivation. *J. Neurochem.*, 8:169–188.

Koffka, K. (1935). *Principles of Gestalt Psychology*. Harcourt, Brace, New York.

Kölliker, A. (1853). *Manual of Human Microscopic Anatomy*. 2 Vols. Translated and edited by G. Busk and T. Huxley. Lippincott, Grambo and Co., Philadelphia.

Konorski, J. (1948). *Conditioned Reflexes and Neuron Organization*. Translated by S. Garry. Cambridge University Press, Cambridge.

Konorski, J., and Lubińska, L. (1946). Mechanical excitability of regenerating nerve-fibres. *Lancet*, 2:609–610.

Köhler, W. (1927). *The Mentality of Apes*. Translated from the 2nd rev. Ed. by Ella Winter. Kegan Paul, Trench, Trubner and Co., New York.

Köhler, W. (1929). *Gestalt Psychology*. Horace Liveright, New York.

Krause, T. L., Fishman, H. M., Ballinger, M. L., and Bittner, G. D. (1994). Extent and mechanism of sealing in transected giant axons of squid and earthworms. *J. Neurosci.*, 14:6638–6651.

Kreiman, G., Koch, C., Fried, I. (2000). Imagery neurons in the human brain. *Nature*. 408:357–361.

Kreutzberg, G. W. (1969). Neuronal dynamics and axonal flow. IV. Blockage of intra-axonal enzyme transport by colchicine. *Proc. Natl. Acad. Sci. U.S.A.*, 62:722–728.

Kreutzberg, G. W., and Gross, G. W. (1977). General morphology and axonal ultra-structure of the olfactory nerve of the pike *Esox lucius*. *Cell Tiss. Res.*, 181:443–457.

Kristensson, K. (1978). Retrograde transport of macromolecules in axons. *Ann. Rev. Pharm. Toxicol.*, 18:97–110.

Kristensson, K., Lycke, E., and Sjöstrand, J. (1971). Spread of herpes simplex virus in peripheral nerves. *Acta Neuropathol.*, 19:44–53.

Kroeber, T. (1976). *Ishi in Two Worlds. A Biography of the Last Wild Indian in North America*. University of California Press, Barkeley.

Kruger, L. (1963). Biographical note on Francois Pourfour du Petit. *Exp. Neurol.*, 7:iii–v.

Kuczmarski, E. R., and Rosenbaum, J. L. (1979). Studies on the organization and localization of actin and myosin in neurons. *J. Cell Biol.*, 80:356–371.

Kuhn, T. S. (1996). *The Structure of Scientific Revolutions*. 3rd Ed. University of Chicago Press, Chicago.

Kuriyama, S. (1995). *Pneuma, qi*, and the problematic of breath. In: *The Comparison Between Concepts of Life-Breath in East and West*. Y. Kawakita, S. Sakai, and Y. Otsuka, eds. Ishiyaku EuroAmerica, Inc., Tokyo, pp. 1–31.

Kuznetsov, S. A., Langford, G. M., and Weiss, D. G. (1992). Actin-dependent organelle movement in squid axoplasm. *Nature*, 356:722–725.

Kuznetsov, S. A., Rivera, D. T., Severin, F. F., Weiss, D. G., and Langford, G. M. (1994). Movement of axoplasmic organelles on actin filaments from skeletal muscle. *Cell Mot. Cytoskel.*, 28:231–242.

Kwan, M. K., Wall, E. J., Massie, J., and Garfin, S. R. (1992). Strain, stress and stretch of peripheral nerve. *Acta Orthop. Scand.*, 63:267–272.

Lajtha, A. (1964). Protein metabolism of the nervous system. Vol. 6. In: *Internat. Rev. Neurobiol*. C. C. Pfeiffer, and J. R. Smythies, eds. Academic Press, New York, pp. 1–98.

La Mettrie, J. O. D. (1912). *Man à Machine; with extracts on The Natural History of the Soul*. Philosophical and historical notes by G. C. Bussey. French-English edition. Translated from French book published in Leyden in 1748 and reprinted. Open Court, La Salle.

Langley, J. N. (1921). *The Autonomic Nervous System*. W. Heffer and Sons, Cambridge.

Lasek, R. J. (1967). Bidirectional transport of radioactively labelled axoplasmic components. *Nature*, 216:1212–1214.

Lasek, R. J., and Black, M. M. (1988). *Intrinsic Determinants of Neuronal Form and Function*. Alan R. Liss, New York.

Lasek, R. J., and Hoffman, P. N. (1976). The neuronal cytoskeleton, axonal transport and axonal growth. In: *Cell Motility Book(C)*. R. Goldman, T. Pollard, and J. Rosenbaum, eds. Cold Spring Harbor Laboratory, Cold Spring Harbor, pp. 1021–1049.

Lashley, K. S. (1963). *Brain Mechanisms and Intelligence: A Quantitative Study of Injuries to the Brain*. Reprint of 1929 Ed. University of Chicago, Hafner Publishing Co., New York.

Lavail, J. H., and Lavail, M. M. (1974). The retrograde intraaxonal transport of horseradish peroxidase in the chick visual system: a light and electron microscopic study. *J. Comp. Neurol.*, 57:303–358.

Lavail, M. M., and Lavail, J. H. (1975). Retrograde intraaxonal transport of horseradish peroxidase in retinal ganglion cells of the chick. *Brain Res.*, 85:273–280.

Lavoie, P.-A., Bolen, F., and Hammerschlag, R. (1979). Divalent cation specificity of the calcium requirement for fast transport of proteins in axons of desheathed nerves. *J. Neurochem.*, 32:1745–1751.

Lavoie, P.-A., Collier, B., and Tenenhouse, A. (1976). Comparison of alpha-bungarotoxin binding to skeletal muscles after inactivity or denervation. *Nature*, 260:349–350.

Lavoie, P.-A., and Tiberi, M. (1986). Inhibition of fast axonal transport in bullfrog nerves by dibenzazepine and dibenzocycloheptadiene calmodulin inhibitors. *J. Neurobiol.*, 17:681–695.

Leão, A. A. P. (1944). Spreading depression of activity in the cerebral cortex. *J. Neurophysiol.*, 7:359–390.

Lee, F. C. (1929). The regeneration of nervous tissue. *Physiol. Rev.*, 9:575–623.

Lees, G. J. (1985). Inhibition of the retrograde axonal transport of dopamine-beta-hydroxylase antibodies by the calcium ionophore A23187. *Brain Res.*, 345:62–67.

Legallois, J. J. C. (1813). *Experiments on the Principle of Life, and Particularly on the Principle of the Motions of the Heart, and on the Seat of this Principle*. Translated by N. C. and J. G. Nancrede. Thomas, Philadelphia.

Lehninger, A., Nelson, D. L., and Cox, M. L. (2000). *Principles of Biochemistry*, 3rd ed. Worth Publishers, New York.

Lenhoff, S. G., and Lenhoff, H. M. (1986). *Hydra and the Birth of Experimental Biology*. With a translation from the French of Abraham Trembley's Memoirs Concerning the Natural History of a Type of Freshwater Polyp with Arms Shaped like Horns. First published in 1744. Boxwood Press, Pacific Grove.

Lent, E. (1856). Beitrage zur lehre von der regeneration durchschnittenen Nerven. *Zeit. wiss. Zool.*, 7:145–153.

Leone, J., and Ochs, S. (1978). Anoxic block and recovery of axoplasmic transport and electrical excitability of nerve. *J. Neurobiol.*, 9:229–245.

Letourneau, P., Kater, S. B., and Macagno, E. R. (1991). *The Nerve Growth Cone*. Raven Press, New York.

LeVay, S., Stryker, M. P., and Shatz, C. J. (1978). Ocular dominance columns and their development in layer IV of the cat's visual cortex: a quantitative study. *J. Comp. Neurol.*, 179:223–244.

Levi-Montalcini, R. (1976). The nerve growth factor: its role in growth differentiation and function of the sympathetic adrenergic neuron. *Prog. Brain Res.*, 45:235–258.

Levi-Montalcini, R. (1987). The nerve growth factor 35 years later. *Science*, 237:1154–1162.

Levine, J., and Willard, M. (1980). The composition and organization of axonally transported proteins in the retinal ganglion cells of the guinea pig. *Brain Res.*, 194:137–154.

Lewes, G. H. (1864). *Aristotle: A Chapter From the History of Science, Including Analyses of Aristotle's Scientific Writings*. Smith, Elder and Co., London.

Lépine, M. R. (1895). Théorie méchanique de la paralysie hystérique du somnambulisme, du sommeil naturel et de la distraction. *Compt. Rend. Soc. Biol.*, 85–87.

Lépine, R. (1894). Sur un cas d'hystérie a forme particulière. *Rev. Med*, 14:713–728.

Li, J. Y., Pfister, K. K., Brady, S. T., and Dahlström, A. (2000). Cytoplasmic dynein conversion at a crush injury in rat peripheral axons. *J. Neurosci. Res.*, 2:151–161.

Liddell, E. G. T. (1960). *The Discovery of Reflexes*. Oxford University Press, Oxford.

Lillie, R. S. (1925). Factors affecting transmission and recovery in the passive iron nerve model. *J. Gen. Physiol.*, 7:473–507.

Lim, S.-S., Edson, K. J., Letourneau, P. C., and Borisy, G. G. (1990). A test of microtubule translocation during neurite elongation. *J. Cell Biol.*, 111:123–130.

Lim, S.-S., Sammak, P. J., and Borisy, G. G. (1989). Progressive and spatially differentiated stability of microtubules in developing neuronal cells. *J. Cell Biol.*, 109:253–263.

Lindberg, D. C. (1976). *Theories of Vision from Al-Kindi to Kepler*. University of Chicago Press, Chicago.

Ling, G., and Gerard, R. W. (1949). The normal membrane potential of frog sartoriius muscle. *J. Cell Comp. Physiol.*, 34:382–396.

Lisowski, F. P. (1967). Prehistoric and early historic trepanation. In: *Diseases in Antiquity*. D. Brothwell, and A. T. Sandison, eds. Charles C Thomas, Springfield, pp. 651–672.

Liu, H. M., Balkovic, E. S., Sheff, M. F., and Zacks, S. I. (1979). Production in vitro of a neurotropic substance from proliferative neurolemma-like cells. *Exp. Neurol.*, 64:271–283.

Livett, B. G., Geffen, L. B., and Austin, L. (1968). Axoplasmic transport of 14C-noradrenaline and protein in splenic nerves. *Nature*, 217:278–279.

Livett, B. G., Geffen, L. B., and Rush, R. A. (1971). Immunochemical methods for demonstrating macromolecules in sympathetic neurones. *Phil. Trans. Roy. Soc. B*, 261:360–362.

Livingston, W. K. (1943). *Pain Mechanisms*. Macmillan Co., New York.

Liwnicz, B. H., Kristensson, K., Wisniewski, H. M., Shelanski, M. L., and Terry, R. D. (1974). Observations on axoplasmic transport in rabbits with aluminum-induced neurofibrillary tangles. *Brain Res.*, 80:413–420.

Llinás, R. R., Nicholson, C., Freeman, J. A., and Hillman, D. E. (1968). Dendritic spikes and their inhibition in alligator Purkinje cells. *Science* 160:1132–1135.

Llinás, R. R., and Sugimori, M. (1980). Electrophysiological properties of in vitro Purkinje cell dendrites in mammalian cerebellar slices. *J. Physiol.* 305:197–213.

Lloyd, G. E. R. (1973). *Greek Science After Aristotle*. Norton, New York.

Locke, J. (1959). *An Essay Concerning Human Understanding*. Collated and annotated, with prolegomena, biographical, critical, and historical by Alexander Campbell Fraser. Reprinted from the 1690 edition. Dover Publications, New York.

Longet, F.-A. (1842). *Anatomie et physiologie du système nerveux de l'homme et des animaux vertebres*. 2 Vols. Fortin, Masson et Cie, Paris.

Longrigg, J. (1993). *Greek Rational Medicine: Philosophy and Medicine from Alcmaeon to the Alexandrians*. Routledge, London.

Loomis, P. A., and Goldman, R. D. (1999). Neural intermediate filament systems. In: *Encyclopedia of Neuroscience*, 2nd ed., Vol. 2. G. Adelman, and B. H. Smith, eds. Elsevier, Amsterdam, pp. 1313–1316.

Lorenz, K. (1982). *The Foundations of Ethology. The Principal Ideas and Discoveries in Animal Behavior*. Translated by Konrad Z. Lorenz and Robert Warren Kickert. Simon and Schuster, New York.

Lubińska, L. (1956). Outflow from cut ends of nerve fibres. *Exp. Cell Res.*, 10:40–47.

Lubińska, L. (1964). Axoplasmic streaming in regenerating and in normal nerve fibres. *Prog. Brain Res.*, 13:1–71.

Lubińska, L. (1977). Early course of Wallerian degeneration in myelinated fibres of the rat phrenic nerve. *Brain Res.*, 130:47–63.

Lubińska, L., and Niemierko, S. (1971). Velocity and intensity of bidirectional migration of acetylcholinesterase in transected nerves. *Brain Res.*, 27:329–342.

Lubińska, L., Niemierko, S., and Oberfield, B. (1961). Gradient of cholinesterase activity and of choline acetylase activity in nerve fibers. *Nature*, 189:122–123.

Lubińska, L., Niemierko, S., Oderfeld, B., and Szward, L. (1962). Decrease of acetylcholinesterase activity along peripheral nerves. *Science*, 135:368–370.

Luciani, L. (1915). Muscular and nervous systems, In: *Human Physiology*, Vol. 3. Translated from the 4th Italian edition by F. A. Welby. Macmillan and Co. Ltd., London.

Lucretius (1937). *De Rerum Natura*. Translated by R. C. Trevelyan. Cambridge University Press, Cambridge.

Lucretius (1951). *The Nature of the Universe*. Translated with an introduction by R. Latham. Penguin Books, Baltimore.

Lucretius (1969). *The Way Things Are. The De Rerum Natura of Titus Lucretius Carus*. Translated by R. Humphries. Introduction by B. Feldman. Indiana University Press, Bloomington.

Lugaro, E. (1898). Sulle modificazioni morfologiche funzionali dei dendriti delle cellule nervose. *Rivista di patologia nervosa e mentale*, 3:337–359.

Lundborg, G., and Rydevik, B. L. (1973). Effects of stretching the tibial nerve of the rabbit. *J. Bone Joint Surg.*, 55B:390–401.

Lunn, E. R., Perry, V. H., Brown, M. C., Rosen, H., and Gordon, S. (1989). Absence of Wallerian degeneration does not hinder regeneration in peripheral nerve. *Eur. J. Neurosci.*, 1:27–33.

Luria, A. R. (1962). *Higher Cortical Functions in Man.* Basic Books, New York.

Luscombe, D. (1997). *Medieval Thought.* Oxford University Press, Oxford.

Lynch, G. S. (1986). *Synapses, Circuits, and the Beginnings of Memory,* MIT, Cambridge.

Lynch, G. S., and Staubli, U. (1993). A cortical system for studying cortical memory. *Trends Neurosci.*, 16:24–25.

Macchiori, V. D. (1930). *From Orpheus to Paul. A History of Orphism.* Henry Holt, New York.

Magendie, F. J. (1822). Experiences sur les fonctions des racines des nerfs qui naissent de la moelle epiniere. *J. Physiol. Exp. Pathol.*, 2:366–371.

Maimonides, M. (1956). *The Guide for the Perplexed.* Reprint of 2nd revised edition 1904. Translated by M. Friedlander., Dover Publications, Inc., New York.

Maimonides, M. (2000). *Maimonides and the Sciences.* Edited by R. S. Cohen and H. Levine. Boston Studies in the Philosophy of Science. Vol. 211., Kluwer Academic Publishers, London.

Malamud, M. (1970). Ludwig Turck (1810–1868). In: *The Founders of Neurology.* W. Haymaher and F. Schiller eds. 2nd Ed. pp. 85–88. Charles C Thomas Publisher, Springfield.

Malhotra, S. K., and Van Harreveld, A. (1965). Dorsal roots of the rabbit investigated by freeze-substitution. *Anat. Rec.*, 152:283–292.

Malinow, R., and Malenka, R. C. (2002). AMPA receptor trafficking and synaptic plasticity. *Annu. Rev. Neurosci.*, 25:103–126.

Mandelkow, E., and Johnson, K. A. (1998). The structural and mechanochemical cycle of kinesin. *Trends Biochem. Sci.*, 23:429–433.

Manning, K. A., Erichsen, J. T., and Evinger, C. (1990). Retrograde transneuronal transport properties of fragment C of tetanus toxin. *Neuroscience*, 34:251–263.

Mantyh, P. W., DeMaster, E., Malhotra, A., Ghilardi, J. R., Rogers, S. D., Mantyh, C. R., Liu, H., Basbaum, I., Vigna, S. R., Maggio, J. E., and Simone, D. A. (1995). Receptor endocytosis and dendrite reshaping in spinal neurons after somatosensory stimulation. *Science*, 268:1629–1632.

Manzoni, T. (1998). The cerebral ventricles, the animal spirits and the dawn of brain localization of function. *Arch. Ital. Biol.*, 136:103–152.

Marchand J. F., and Hoff, H. E. (1955). Translation of Felice Fontana: the law of irritability. *J. Hist Med.*, 10:197–206, 302–326, 339–420.

Marchesi, V. T. (1985). Stabilizing infrastructure of cell membranes. *Ann. Rev. Cell Biol.*, 1:531–561.

Margetts, E. L. (1967). Trepanation of the skull by the medicine-men of primitive cultures, with particular reference to present-day native east african practice. In: *Diseases in Antiquity.* D. Brothwell and A. T. Sandison, eds. Charles C Thomas, Springfield, pp. 673–701.

Mark, R. F. (1969). Matching muscles and motoneurones. A review of some experiments on motor nerve regeneration. *Brain Res.*, 14:245–254.

Markin, V. S., Tanelian, D. L., Jersild, R. A., Jr., and Ochs, S. (1999). Biomechanics of stretch-induced beading. *Biophys. J.*, 76:2852–2860.

Markowitsch, H. J. (2000). Neuroanatomy of memory. In: The *Oxford Handbook of Memory*. E. Tulving and F. I. M. Craik, eds. Oxford University Press, Oxford, pp. 465–484.

Marmont, G. (1949). Studies on the axon membrane. *J. Cell. Comp. Physiol.*, 34:351–382.

Marotta, C. A. (1983). ed. *Neurofilaments*. University of Minnesota Press, Minneapolis.

Marshal, A. J., and Burton, J. A. G. (1962). *Catalogue of the Pathological Preparations of W. Hunter, W. Macewen, J. H. Teacher, and J. A. G. Burton in the Museum of the Pathology Department, Glasgow Royal Infirmary*. University of Glasgow, Glasgow.

Martinez, A. J., and Friede, R. L. (1970). Accumulation of axoplasmic organelles in swollen nerve fibers. *Brain Res.*, 19:183–198.

Masurovsky, E. B., and Bunge, R. P. (1971). Patterns of myelin degeneration following the rapid death of cells in cultures of peripheral nervous tissue. *J. Neuropath. Exp. Neurol.*, 30:311–324.

Matsumoto, G., Tsukita, S., and Arai, T. (1989). Organization of the axonal cytoskeleton: differentiation of the microtubule and actin filament arrays. In: *Cell Movement*, Vol. 2. F. D. Warner, and J. R. McIntosh, eds. Alan R. Liss, New York, pp. 335–356.

Matteucci, C. (1844). *Traité des phénomènes électro-physiologiques des animaux*. Fortin, Masson et Cie, Paris.

Maximow, A., and Bloom, W. (1930). *A Text-book of Histology*. W. B. Saunders Company, Philadelphia.

Mayow, J. (1907). *Medico-Physical Works*. Being a translation of Tractus Quintus Medico-Physici first published in 1674. The Alembic Club: Simkin et al. London, Edinburgh.

McEwen, B. S. (1992). Steroid hormones: effect on brain development and function. *Horm. Res.*, 37(suppl. 3):1–10.

McEwen, B. S., Forman, D. S., and Grafstein, B. (1971). Components of fast and slow axonal transport in the goldfish optic nerve. *J. Neurobiol.*, 2:361–377.

McEwen, B. S., and Grafstein, B. (1968). Fast and slow components in axonal transport of protein. *J. Cell Biol.*, 38:494–508.

McGaugh, J. L., and Herz, M. J. (1972). *Memory Consolidation*. Albion, San Francisco.

McGonigle, D. J., Hanninen, R., Salenius, S., Hari, R., Frackowiak, R. S., and Frith, C. D. (2002). Whose arm is it anyway? An fMRI case study of supernumerary phantom limb. *Brain*, 125:1265–1274.

McHenry, L. C. (1969). *Garrison's History of Neurology*. Charles C Thomas, Springfield.

McNeley, J. K. (1981). *Holy Wind in Navajo Philosophy*. The University of Arizona Press, Tucson.

McQuarrie, I. G., Brady, S. T., and Lasek, R. J. (1980). Polypeptide composition and kinetics of SCa and SCb in sciatic nerve motor axons and optic axons of the rat. *Soc. Neurosci. Abstr.*, 6:501.

Mellon, M., Rhebun, L., and Rosenbaum, J. (1976). Studies on the accessible sulfhydryls of polymerizable tubulin. In: *Cell Motility (Book C)*. R. Goldman, T. Pollard, and J. Rosenbaum, eds. Cold Spring Harbor Laboratory, Cold Spring Harbor, pp. 1149–1163.

Melzack, R. (1990). Phantom limbs and the concept of a neuromatrix. *Trends Neurosci.*, 13:88–92.

Mendell, J. R., and Sahenk, Z. (1980). Interference of neuronal processing and axoplasmic transport by toxic chemicals. In: *Neurotoxicology*. P. S. Spencer and H. H. Schaumburg, eds. Williams & Wilkins, Baltimore, pp. 139–160.

Mendell, J. R., Sahenk, Z., Saida, K., Weiss, H. S., Savage, R., and Couri, D. (1977). Alterations of fast axoplasmic transport in experimental methyl n-butyl ketone neuropathy. *Brain Res.*, 133:107–118.

Mendelsohn, E. (1964). *Heat and Life. The Development of the Theory of Animal Heat.* Harvard University Press, Cambridge.

Merzenich, M. (2000). Cognitive neuroscience. Seeing in the sound zone. *Nature*, 404:820–821.

Mesulam, M.-M. (1982). *Tracing Neural Connections with Horseradish Peroxidase.* Wiley-Interscience, New York.

Meyer, A. (1967). Marcello Malpighi and the dawn of neurohistology. *J. Neurol. Sci.*, 4:185–193.

Meyer, A. (1971). *Historical Aspects of Cerebral Anatomy.* Oxford University Press, New York.

Mildvan, A. S. (1970). Metals in enzyme catalsis. Vol. II. In: *The Enzymes.* P. D. Boyer, ed. Academic Press, New York, pp. 445–536.

Miledi, R. (1960). Junctional and extra-junctional acetylcholine receptors in skeletal muscle fibres. *J. Physiol.* (London), 151:24–30.

Miledi, R., and Slater, C. R. (1970). On the degeneration of rat neuromuscular junctions after nerve section. *J. Physiol.* (London), 207:507–528.

Miller, M. S., and Spencer, P. S. (1984). Single doses of acrylamide reduce retrograde transport velocity. *J. Neurochem.*, 43:1401–1408.

Milner, B. (1972). Disorders of learning and memory after temporal lobe lesions in man. *Clin. Neurosurg.*, 19:421–446.

Mire, J. J., Hendelman, W. J., and Bunge, R. P. (1970). Observations on a transient phase of focal swelling in degenerating unmyelinated nerve fibers. *J. Cell Biol.*, 45:9–22.

Mitchell, S. W. (1965). *Injuries of Nerves and Their Consequences.* With a new introduction by L. C. McHenry, Jr. Reprinted from the 1st Ed. Published in 1872. Dover, New York.

Mobley, W. C., Server, A. C., Ishii, D. N., Riopelle, R. J., and Shooter, E. M. (1977). Nerve growth factor. *N. Engl. J. Med.*, 297:1096–1104.

Monaco, S., Jacob, J. M., Jenich, H., Patton, A., Autilio-Gambetti, L. A., and Gambetti, P. (1989). Axonal transport of neurofilament is accelerated in peripheral nerve during 2,5-hexanedione intoxication. *Brain Res.*, 491:328–334.

Monard, D., Solomon, F., Rentsch, M., and Gysin, R. (1973). Glia-induced morphological differentiation in neuroblastoma cells. *Proc. Natl. Acad. Sci. U.S.A.*, 70:1894–1897.

Moore, C. H. (1963). *Ancient Beliefs in the Immortality of the Soul with Some Account of Their Influence on Later Beliefs.* Cooper Square Publishers, Inc., New York.

Morat, J. P. (1906). *Physiology of the Nervous System.* Edited and translated by H. W. Syers. Keener, Chicago.

Moretto, A., Lotti, M., Sabri, M. I., and Spencer, P. S. (1987). Progressive deficit of retrograde axonal transport is associated with the pathogenesis of di-*n*-butyl dichlorvos axonopathy. *J. Neurochem.*, 49:1515–1522.

Moretto, A., and Sabri, M. I. (1988). Progressive deficits in retrograde axon transport precede degeneration of motor axons in acrylamide neuropathy. *Brain Res.*, 440:18–24.

Morgan, J. I., and Curran, T. (1989). Stimulus-transcription coupling in neurons: role of cellular immediate-early genes. *Trends Neurosci.*, 12:459–462.

Morgan, J. I., and Curran, T. (1991). Stimulus-transcription coupling in the nervous system: involvement of the inducible proto-oncogenes fos and jun. *Annu. Rev. Neurosci.*, 14:421–451.

Moruzzi, G. (1963). The electrophysiological work of Carlo Matteucci. In: *Essays on the History of Italian Neurology*. L. Belloni, ed. Elli and Pagani, Milano, pp. 139–147.

Müller, J. (1840). *Elements of Physiology*, 2 Vols. Translated from the German with notes by William Baly 2nd Ed. Taylor and Walton, London.

Müller, J. (1845). *Manuel de Physiologie*, 2 Vols. Traduit de l'allemand sur la quatrième édition (1841) avec des annotations par A.-J.-L. Jourdan. J. B. Baillière, Paris.

Muñoz-Martínez, E. J., Massieu, D., and Ochs, S. (1984). Depression of fast axonal transport produced by Tullidora. *J. Neurobiol.*, 15:375–392.

Muñoz-Martínez, E. J., Núñez, R., and Sanderson, A. (1981). Axonal transport: a quantitative study of retained and transported protein fraction in the cat. *J. Neurobiol.*, 12:15–26.

Murray, M. R., and Herrmann, A. (1968). Passive movements of Schmidt-Lantermann clefts during continuous observation in vitro. *J. Cell Biol.*, 39:149a–150a.

Myers, F. W. H. (1903). *Human Personality and Its Survival of Bodily Death*. In two volumes. Longmans, Green, and Co., London.

Nansen, F. (1887). *The Structure and Combination of the Histological Elements of the Central Nervous System*. J. Grieg, Bergen.

Narahashi, T., Albuquerque, E. X., and Deguchi, T. (1971). Effects of batrachotoxin on membrane potential and conductance of squid giant axons. *J. Gen. Physiol.*, 58:54–70.

Nasse, C. F. (1839). Ueber die veranderungen der nervenfasern nach ihrer durchschneidung. *Arch. Anat. Physiol. Wiss. Med.*, 405–419.

Nastuk, W. L., and Hodgkin, A. L. (1950). The electrical activity of single muscle fibers. *J. Cell. Comp. Physiol.*, 35:39–74.

Needham, D. (1971). *Machina Carnis*. Cambridge University Press, Cambridge.

Nemesius (1955). *On the Nature of Man*, pp. 203–453. In: *Cyril of Jerusalem and Nemesius of Emesa*. Translated from the Greek and edited by W. Telfer. Vol IV, The Library of Christian Classics. The Westminster Press, Philadelphia.

Nennesmo, I., and Kristensson, K. (1986). Effects of retrograde axonal transport of *Ricinus communis* agglutinin I on neuroma formation. *Acta Neuropath.* (Berlin), 70:279–283.

Nernst, W. (1908). Zur theorie das elektrischen reizes. *Pflügers Arch. Ges. Physiol.*, 122:275–314.

Neuburger, M. (1981). *The Historical Development of Experimental Brain and Spinal Cord Physiology Before Flourens*. Translated and edited with additional material by Edwin Clarke. Johns Hopkins University Press, Baltimore.

Newton, I. (1952). *Opticks; or, A Treatise of the Reflections, Refractions, Inflections & Colours of Light*. Based on the 4th ed., London, 1730; with a foreword by Albert Einstein, an introduction by Sir Edmund Whittaker, a preface by I. Bernard Cohen, and an analytical table of contents prepared by Duane H. D. Roller. Reprint. Dover, New York.

Nichols, T. R., Smith, R. S., and Snyder, R. E. (1982). The action of puromycin and cycloheximide on the initiation of rapid axonal transport in amphibian dorsal root neurones. *J. Physiol.* (London), 332:441–458.

Niderst, A. (1969). *L'Ame Materielle* (ouvrage anonyme). Avec une introduction et des notes. Université de Rouen, Rouen.

Nissl, F. (1894). Über eine neue untersuchungsmethode des centralorgans specielle zur feststellung der localisation der Nervenzellen. *Neurol. Zbl.*, 13:507.

Nixon, R. A. (1998). Dynamic behavior and organization of cytoskeletal proteins in neurons: reconciling old and new findings. *BioEssays*, 20:798–807.

Nixon, R. A., and Logvinenko, K. B. (1985). Multiple fates of newly synthesized neurofilament proteins: evidence for a stationary neurofilament network distributed nonuniformly along axons of retinal ganglion cell neurons. *J. Cell Biol.*, 102:647–658.

Nordlander, R. H., and Singer, M. (1973). Degeneration and regeneration of severed crayfish sensory fibers: an ultrastructual study. *J. Comp. Neurol.*, 52:75–92.

Nottebohm, F. (1985). Neuronal replacement in adulthood. In: *Hope for a New Neurology*. F. Nottebohm, ed. New York Academy of Science., New York, pp.143–161.

Oaklander, A. L., and Spencer, P. S. (1988). Cold blockade of axonal transport activates premitotic activity of Schwann cells and Wallerian degeneration. *J. Neurochem.*, 50:490–496.

Ochs, S. (1962). The nature of spreading depression in neural networks. *Int. Rev. Neurobiol.*, 4:1–69.

Ochs, S. (1965a). Axoplasmic flow of labeled protein in fluid compartment of nerve fibers. *Proceedings of the 23rd International Physiological Congress*, Tokyo, p. 69.

Ochs, S. (1965b). *Elements of Neurophysiology*. John Wiley & Sons, New York.

Ochs, S. (1966). Axoplasmic flow in neurons. In: *Macromolecules and Behavior*. J. Gaito, ed. Appleton-Century-Crofts, New York.

Ochs, S. (1971). Characteristics and a model for fast axoplasmic transport in nerve. *J. Neurobiol.*, 2:331–345.

Ochs, S. (1972a). Fast transport of materials in mammalian nerve fibers. *Science*, 176:252–260.

Ochs, S. (1972b). Rate of fast axoplasmic transport in mammalian nerve fibres. *J. Physiol.* (London), 227:627–645.

Ochs, S. (1973). Effect of maturation and aging on the rate of fast axoplasmic transport in mammalian nerve. *Prog. Brain Res.*, 40:349–362.

Ochs, S. (1974). Energy metabolism and supply of $\sim$P to the fast axoplasmic transport mechanism in nerve. *Fed. Proc.*, 33:1049–1058.

Ochs, S. (1975). A brief review of material transport in nerve fibers. In: *The Research Status of Spinal Manipulative Therapy-Monograph #15*. U.S. Department of Health, Education, and Welfare, NINCDS., Bethesda, pp. 189–196.

Ochs, S. (1975a). A unitary concept of axoplasmic transport based on the transport filament hypothesis. In: *Third International Congress on Muscle Diseases*. W. G. Bradley, D. Gardner-Medwin, and J. N. Walton, eds. Excerpta Medica, Amsterdam, pp. 189–194.

Ochs, S. (1975b). Retention and redistribution of proteins in mammalian nerve fibres by axoplasmic transport. *J. Physiol.* (London), 253:459–475.

Ochs, S. (1975c). Waller's concept of the trophic dependence of the nerve fiber on the cell body in the light of early neuron theory. *Clio Med.*, 10:253–265.

Ochs, S. (1977). Axoplasmic transport in peripheral nerve and hypothalamo-neurohypophyseal systems. In: *Hypothalamic Peptide Hormones and Pituitary Regulation*. J. C. Porter, ed. Plenum Press, New York, pp. 13–40.

Ochs, S. (1982). *Axoplasmic Transport and Its Relation to Other Nerve Functions*. John Wiley-Interscience, New York.

Ochs, S. (1983). Essay – Review. Max Neuburger, the historical development of experimental brain and spinal cord physiology before flourens. Translated and edited with additional notes by Edwin Clarke. *Trans. Stud. Coll. Phys. Phil.*, 5:131–143.

Ochs, S. (1988). An historical introduction to the trophic regulation of skeletal muscle. In: *Nerve-Muscle Cell Trophic Communication*. H. L. Fernandez and J. A. Donoso, eds. CRC Press, Boca Raton, pp. 7–22.

Ochs, S. (1989). *Book Reviews: A History of Neurophysiology in the 19th Century.* M. A. B. Brazier. Raven Press, New York, 1988; *Nineteenth-Century Origins of Neuroscientific Concepts.* E. Clarke and L. S. Jacyna, eds. University of California Press, Berkeley, 1987. *J. Neurobiol.*, 20:164–168.

Ochs, S. (1992). Kinetic and metabolic disorders of axoplasmic transport induced by neurotoxic agents. In: *Selective Neurotoxicity. Handbook of Experimental Pharmacology*, Vol. 102. H. Herken and F. Hucho, eds. Springer-Verlag, Berlin, pp. 81–110.

Ochs, S. (1997). Silas Weir Mitchell (1829–1914). In: *Doctors, Nurses, and Medical Practitioners: A Bio-Bibliographical Sourcebook.* L. N. Magner, ed. Greenwood Press, Westport, pp. 183–190.

Ochs, S. (1999). Axonal transport. In: *Encyclopedia of Neuroscience.* 2nd Ed. G. Adelman, and B. H. Smith, eds. Elsevier, New York, pp. 169–176.

Ochs, S., and Barnes, C. D. (1969). Regeneration of ventral root fibers into dorsal roots shown by axoplasmic flow. *Brain Res.*, 5:600–603.

Ochs, S., and Booker, H. (1961). Spatial and temporal interaction of direct cortical responses. *Exp. Neurol.*, 4:70–82.

Ochs, S., and Brimijoin, S. (1993). Axonal transport. In: *Peripheral Neuropathy*, 3rd Ed., Vol. 1. P. J. Dyck, P. K. Thomas, J. W. Griffin, P. A. Low, and J. F. Poduslo, eds. Saunders, Baltimore, pp. 331–360.

Ochs, S., and Burger, E. (1958). Movement of substance proximo-distally in nerve axons as studied with spinal cord injections of radioactive phosphorus. *Am. J. Physiol.*, 94:499–506.

Ochs, S., Dalrymple, D. E., and Richards, G. (1962). Axoplasmic flow in ventral root nerve fibers of the cat. *Exp. Neurol.*, 5:349–363.

Ochs, S., Erdman, J., Jersild, R. A., Jr., and McAdoo, V. (1978). Routing of transported materials in the dorsal root and nerve fiber branches of the dorsal root ganglion. *J. Neurobiol.*, 9:465–481.

Ochs, S., Gaziri, L. C. J., and Jersild, R. A., Jr. (1986). Metabolic and ionic properties of axoplasmic transport in relation to its mechanism. In: *Axoplasmic Transport.* Z. Iqbal, ed. CRC Press, Inc., Boca Raton, pp. 57–68.

Ochs, S., and Hollingsworth, D. (1971). Dependence of fast axoplasmic transport in nerve on oxidative metabolism. *J. Neurochem.*, 18:107–114.

Ochs, S., and Iqbal, Z. (1983). The role of calcium in axoplasmic transport in nerve. In: *Calcium and Cell Function*, Vol. 3. W. Y. Cheung, ed. Academic Press, New York, pp. 325–354.

Ochs, S., Iqbal, Z., Chan, S. Y., Worth, R. M., and Jersild, R. A., Jr. (1980). The role of calcium and calmodulin in axoplasmic transport. *Trans. Soc. Neurochem.*, 11:235.

Ochs, S., and Jersild, R. A., Jr. (1974). Fast axoplasmic transport in nonmyelinated mammalian nerve fibers shown by electron microscopic radioautography. *J. Neurobiol.*, 5:373–377.

Ochs, S., and Jersild, R. A., Jr. (1984). Calcium localization in nerve fibers in relation to axoplasmic transport. *Neurochem. Res.*, 9:823–836.

Ochs, S., and Jersild, R. A., Jr. (1985). Beading of nerve fibers and fast axoplasmic transport. *Soc. Neurosci. Abst.*, 11:1135.

Ochs, S. and Jersild, R. A., Jr. (1986). The beaded form of myelinated nerve fibers and fast axoplasmic transport. Abstract, Nerve Transport Congress, Calgary July 10 session.

Ochs, S., and Jersild, R. A., Jr. (1987). Cytoskeletal organelles and myelin structure of beaded nerve fibers. *Neuroscience*, 22:1041–1056.

Ochs, S., and Jersild, R. A., Jr. (1990). Myelin intrusions in beaded nerve fibers. *Neuroscience*, 36:553–567.

Ochs, S., Jersild, R. A., Jr., Breen, T., Mori, K., and McKitrick, L. (1986). The maintenance of axoplasmic transport by strontium and its localization in nerve fibers. *J. Neurobiol.*, 17:55–61.

Ochs, S., Jersild, R. A., Jr., and Li, J.-M. (1989). Slow transport of freely movable cytoskeletal components shown by beading partition of nerve fibers. *Neuroscience*, 33:421–430.

Ochs, S., Jersild, R. A., Jr., Pourmand, R., and Potter, C. G. (1994). The beaded form of myelinated nerve fibers. *Neuroscience*, 61:361–372.

Ochs, S., and Johnson, J. (1969). Fast and slow phases of axoplasmic flow in ventral root nerve fibres. *J. Neurochem.*, 16:845–853.

Ochs, S., Johnson, J., and Ng, M.-H. (1967). Protein incorporation and axoplasmic flow in motoneuron fibres following intra-cord injection of labeled leucine. *J. Neurochem.*, 4:3–7.

Ochs, S., Pourmand, R., and Jersild, R. A., Jr. (1996). Origin of beading constrictions at the axolemma: presence in unmyelinated axons and after β,β'-iminodipropionitrile degradation of the cytoskeleton. *Neuroscience*, 70:1081–1096.

Ochs, S., Pourmand, R., Jersild, R. A., Jr., and Friedman, R. N. (1997). The origin and nature of beading: a reversible transformation of the shape of nerve fibers. *Prog. Neurobiol.*, 52:391–426.

Ochs, S., Pourmand, R., Si, K., and Friedman, R. N. (2000). Stretch of mammalian nerve *in vitro:* effect on compound action potentials. *J. Periph. Nerv. Syst.*, 5:227–235.

Ochs, S., and Ranish, N. (1969). Characteristics of the fast transport system in mammalian nerve fibers. *J. Neurobiol.*, 1:247–261.

Ochs, S., and Ranish, N. (1970). Metabolic dependence of fast axoplasmic transport in nerve. *Science*, 167:878–879.

Ochs, S., Ranish, N., Hollingsworth, D., and Helmer, E. (1970). Dependence of fast axoplasmic transport in mammalian nerve on oxidative metabolism. *Fed. Proc.*, 29:264.

Ochs, S., Sabri, M. I., and Johnson, J. (1969). Fast transport system of materials in mammalian nerve fibers. *Science*, 163:686–687.

Ochs, S., Sabri, M. I., and Ranish, N. (1970). Somal site of synthesis of fast transported materials in mammalian nerve fibers. *J. Neurobiol.*, 1:329–344.

Ochs, S., and Smith, C. B. (1971). Fast axoplasmic transport in mammalian nerve in vitro after block of glycolysis with iodoacetic acid. *J. Neurochem.*, 18:833–843.

Ochs, S., and Smith, C. B. (1975). Low temperature slowing and cold-block of fast axoplasmic transport in mammalian nerves in vitro. *J. Neurobiol.*, 6:85–102.

Ochs, S., and Suzuki, H. (1965). Transmission of direct cortical responses. *Electroencephal. Clin. Neurophysiol.*, 9:230–236.

Ochs, S., and Worth, R. (1975). Batrachotoxin block of fast axoplasmic transport in mammalian nerve fibers. *Science*, 87:87–89.

Ochs, S., Worth, R. M., and Chan, S. Y. (1977). Calcium requirement for axoplasmic transport in mammalian nerve. *Nature*, 270:748–750.

O'Farrell, P. H. (1975). High resolution two-dimensional electrophoresis of proteins. *J. Biol. Chem.*, 250:4007–4021.

Oka, N., and Brimijoin, S. (1990). Premature onset of fast axonal transport in bromophenylacetylurea neuropathy: an electrophoretic analysis of proteins exported into motor nerve. *Brain Res.*, 509:107–110.

Oka, N., and Brimijoin, S. (1992). Tubulomembranous lesions in BPAU reflect local stasis of fast axonal transport: evidence from electron microscope autoradiography. *Mayo Clin. Proc.*, 67:341–348.

Okabe, S., and Hirokawa, N. (1990). Turnover of fluorescently labelled tubulin and actin in the axon. *Nature*, 343:479–482.

O'Keefe, J., and Nadel, L. (1978). *The Hippocampus as a Cognitive Map*. Oxford University Press, New York.

O'Malley, C. D. (1953). *Michael Servetus*. A translation of his geographical, medical and astrological writings with introductions and notes by C. D. O'Malley. American Philosophical Society, Philadelphia.

O'Malley, C. D. (1964). *Andreas Vesalius of Brussels 1514–1564*. University of California Press, Berkeley.

O'Neill, J. H., Jacobs, J. M., Gilliatt, R. W., and Baba, M. (1984). Changes in the compact myelin of single internodes during axonal atrophy. *Acta Neuropath.* (Berl), 63:313–318.

Ornberg, R. L., and Reese, T. S. (1981). Localization of calcium in quick frozen cells by freeze substitution in tetrahydrofuran and low temperature embedding. *J. Cell Biol.*, 91:397.

Oxford (1944). *The Oxford Universal Dictionary*, 3rd ed. Oxford University Press, London.

Oxford (1971). *Oxford Dictionary. The Compact Edition*. Oxford University Press, New York.

Paget, J. (1863). *Lectures on Surgical Pathology Delivered at the Royal College of Surgeons of England*. Vol. 1 rev. and edited by W. Turner. Longman, London.

Pagel, W. (1982). *Paracelsus: An Introduction to Philosophical Medicine in the Era of the Renaissance*. 2nd, revised Ed., Karger, Basel.

Palmer, T. D., Schwartz, P. H., Taupin, P., Kaspar, B., Stein, S. A., and Gage, F. H. (2001). Cell culture. Progenitor cells from human brain after death. *Nature*, 411:42–43.

Pannese, E. (1996). The black reaction. *Brain Res. Bull.*, 41:343–349.

Papasozomenos, S. C., Autilio-Gambetti, L. A., and Gambetti, P. (1981). Reorganization of axoplasmic organelles following β,β'-iminodipropionitrile administration. *J. Cell Biol.*, 91:866–871.

Papasozomenos, S. C., Autilio-Gambetti, L. A., and Gambetti, P. (1982). The IDPN axon: rearrangement of axonal cytoskeleton and organelles following β,β'-iminodipropionitrile (IDPN) intoxication. In: *Axoplasmic Transport*. D. G. Weiss, ed. Springer-Verlag, Berlin, pp. 241–250.

Papasozomenos, S. C., Yoon, M. G., Crane, R., Autilio-Gambetti, L. A., and Gambetti, P. (1982). Redistribution of proteins of fast axonal transport following administration of β,β'-iminodipropionitrile: a quantitative autoradiographic study. *J. Cell Biol.*, 95:672–675.

Paracelsus (1951). *Paracelsus. Selected Writings*. Edited with an Introduction by Jolande Jacobi. Translated by Norbert Guterman. Vol. 28 in the Bollingen Series., Pantheon Books.

Paré, A. (1649). *The Workes of that Famous Chirurgion Ambroise Parey*. Translated by T. Johnson. Cotes and Dugard, London.

Paré, A. (1840). *Oeuvres complètes d'Ambroise Paré, revues et collationees sur toutes les éditions avec les variantes, acccompanée de notes historiques et critique, etc.*, by J. F. Malgaigne. 3 Vols. Bailliere, Paris.

Parker, G. H. (1913–1914). The origin and evolution of the nervous system. *Harvey Lectures Series* 9:72–84.

Parker, G. H. (1915). The origin and evolution of the nervous system. *Harvey Lectures*, 72–84.

Parker, G. H. (1929a). The neurofibril hypothesis. *Q. Rev. Biol.*, 4:55–78.

Parker, G. H. (1929b). What are neurofibrils? *Am. Naturalist*, 63:97–117.

Parker, G. H., and Paine, V. L. (1934). Progressive nerve degeneration and its rate in lateral-line nerve of the catfish. *Am. J. Anat.*, 54:1–25.

Parsons, E. A. (1952). *The Alexandrian Library*. Elsevier, Amsterdam.

Paul of Aegina. (1844). *The Seven Books of Paulus Aegineta*. Translated from the Greek with a commentary etc., by Francis Adams. Sydenham Society, London.

Paulson, J. C., and McClure, W. O. (1975). Inhibition of axoplasmic transport by colchicine, podophyllotoxin and vinblastine: an effect on microtubules. *Ann. N. Y. Acad. Sci.*, 253:517–527.

Pegis, A. C. (1934). *St. Thomas and the Problem of the Soul in the Thirteenth Century*. Pontifical Inst., Mediaeval studies, Toronto.

Pegis, A. C. (1944). The mind of St. Augustine. *Med. Studies*, 6:24.

Perroncito, A. (1905). La rigenerazione delle fibre nervose. *Boll. d. Soc. Med. Chir. di. Pavia.*, 4:434–444.

Peters, A., Palay, S. L., and Webster, H. D. (1991). *The Fine Structure of the Nervous System: Neurons and their Supporting Cells*. Oxford University Press, New York.

Peters, F. E. (1967). *Greek Philosophical Terms: A Historical Lexicon*. New York University Press, New York.

Peters, R. S., and Mace, C. A. (1967). Psychology. In: *The Encyclopedia of Philosophy*, Vol. 7 of 8. P. Edwards, ed. Macmillan, New York, pp. 1–27.

Philipeaux, J. M., and Vulpian, A. (1859). Note sur des expériences démontrant que les nerfs séparés des centers nerveux peuvent, après être altérés completement, se régeneier tout en demeurant isolés de ces centres, et recouvrer leurs proprietés physiologiques. *Compt. Rend. Hebd. Acad. Sci.*(Paris), 59:507–509.

Phillis, J. W. (1970). *The Pharmacology of Synapses*. Pergamon Press, New York.

Phillis, J. W., and Ochs, S. (1971). Occlusive behavior of the negative wave direct cortical response (DCR) and single cells in the cortex. *J. Neurophysiol.*, 34:374–388.

Piccolino, M. (2000). The bicentennial of the voltaic battery (1800–2000): the artificial electric organ. *Trends Neurosci.*, 23:147–151.

Pick, J. (1970). *The Autonomic Nervous System. Morphological, Comparative, Clinical and Surgical Aspects*. J. B. Lippincott Co., Philadelphia.

Pinker, S. (1994). *The Language Instinct: How the Mind Creates Language*. Morrow, New York.

Pinker, S. (2002). *The Blank Slate*, Viking, Penguin Books, New York.

Plateau, J. A. F. (1873). *Statique experimentale et theoretique des liquides soumis aux seules forces moleculaires*. Gauthier-Villars, Paris.

Plato (1920). *The Dialogues of Plato*, 2 Vols. Translated by B. Jowett. Random House, New York.

Plato (1937). *Plato's Cosmology. The Timaeus*. Translated by F. M. Cornford. Routledge and Kegan Paul, London.

Plato (1963). *Phaedo*. Translated by H. Treddennick. In: *Plato; The Collected Dialogues*. E. Hamilton, and H. Cairns, eds. Princeton University Press, Princeton, pp. 40–98.

Pleasure, D. E., Mishler, K. C., and Engel, W. K. (1969). Axonal transport of proteins in experimental neuropathies. *Science*, 66:524–525.

Politis, M. J., and Spencer, P. S. (1983). An in vivo assay of neurotropic activity. *Brain Res.*, 278:229–231.

Pons, T. P., Garraghty, P. E., Ommaya, A. K., Kaas, J. H., Taub, E., and Mishkin, M. (1991). Massive cortical reorganization after sensory deafferentation in adult macaques. *Science*, 252:1857–1860.

Popper, K. R. (1959). The Logic of Scientific Discovery, Basic Books Inc., New York.

Porter, K. R. (1966). Cytoplasmic microtubules and their functions. In: *Principles of Biomolecular Organization*. CIBA Foundation Symposium. G. E. W. Wolstenholme and M. O'Connor, eds. Little Brown Co., Boston, pp. 308–356.

Pourmand, R., Ochs, S., and Jersild, R. A., Jr. (1994). The relation of the beading of myelinated nerve fibers to the bands of Fontana. *Neuroscience*, 61:373–380.

Prevost, N., and Dumas, J. B. (1823). Sur les phénomènes qui accompagnent la contraction de la fibre musculaire. *J. Physiol. Exp.*, 3:301–344.

Price, D. B., and Twombly, N. J. (1972). *The Phantom Limb: An 18th Century Latin Dissertation Text and Translation, with a medical-historical and linguistic commentary*. Languages and Linguistics Working Papers Number 3. Georgetown University Press, Washington, DC.

Price, D. B., and Twombly, N. J. (1978). *The Phantom Limb Phenomenon. A Medical, Folkloric, and Historical Study. Texts and translations of 10th to 20th century accounts of the miraculous restoration of lost body parts*. Georgetown University Press, Washington, DC.

Price, D. L., Griffin, J., Young, A., Peck, K., and Stocks, A. (1975). Tetanus toxin: direct evidence for retrograde intraaxonal transport. *Science*, 88:945–947.

Priestley, J. (1966). *The History and Present State of Electricity: With Original Experiments*, 2 Vols. Reprinted from the 3rd London ed., 1776. Johnson Reprint Corporation, New York.

Puce, A., Allison, T., Gore, J. C., McCarthy, G. (1995). Face-sensitive regions in human extrastriate cortex studied by functional MRI. *J. Neurophysiol.* 74:1192–1199.

Puchala, E., and Windle, W. F. (1977). The possibility of structral and functional restitution after spinal cord injury: a review. *Exp. Neurol.*, 55:1–42.

Purkinje, J. E. (1838). Über die gangliosen korperchen in verschiedenen theilen des gehirns. *Ber. Vers. Deutsch. Nat. Arzte.* (Prague), 15:174–179.

Purpura, D. P. (1974). Dendritic spine "dysgenesis" and mental retardation. *Science*, 186:1126–1128.

Purpura, D. P., Bodick, N., Suzuki, K., Rapin, I., and Wurzelmann, S. (1982). Microtubule disarray in cortical dendrites and neurobehavioral failure. I. Golgi and electron microscopic studies. *Dev. Brain Res.*, 5:287–297.

Purves, D., and Lichtman, J. W. (1980). Elimination of synapses in the developing nervous system. *Science*, 210:153–157.

Putscher, M. (1973). *Pneuma Spiritus Geist*. Franz Steiner Verlag, Wiesbaden.

Quinn, P. J. (1981). The fluidity of cell membranes and its regulation. *Prog. Biophys. Molec. Biol.*, 38:1–104.

Rabl-Rückard (1890). Sind die Ganglienzellen amöboid? Eine Hypothese zur Mechanik psychischer Vorgänge. *Neurol. Centralbl.*, 9:199–200.

Rahman, F. (1952). *Avicenna's Psychology*. An English Translation of Kitab Al-Najab. Book II. Chapter vi with historico-philosophical notes and textual improvements on the Cairo edition. Hyperion Reprint 1981. Oxford University Press, London.

Rakic, P. (1985). DNA synthesis and cell division in the adult primate brain. In: *Hope for a New Neurology*. F. Nottebohm, ed. New York Academy of Science, New York, pp. 193–211.

Rall, W. (1995). *The Theoretical Foundation of Dendritic Function: Selected papers of Wilfrid Rall with commentaries*. Edited by Idan Segev, John Rinzel, and Gordon M. Shepherd., MIT Press, Cambridge.

Rall, W. (1970). Dendritic neuron theory and dendrodendritic synapses in a simple cortical system. In: *The Neurosciences*. Vol. 2. F. O. Schmitt, ed. Rockefeller University Press, New York, pp. 552–565.

Rall, W., and Rinzel, J. (1971). Dendritic spine function and synaptic attenuation calculations. *Soc. Neurosci. Abst.*, 64.

Ramachandran, V. S., and Blakeslee, S. (1998). *Phantoms in the Brain: Probing the Mysteries of the Human Mind*. William Morrow and Co., New York.

Ramón y Cajal, S. (1954). *Neuron Theory or Reticular Theory*. Translated by U. Purkiss and C. A. Fox. Consejo Superior de Investigaciones Científicas Instituto 'Ramon y Cajal,' Madrid.

Ramón y Cajal, S. (1967). The structure and connexions of neurons. In: *Physiology or Medicine Nobel Lectures 1901–1921*. Elsevier, Amsterdam, pp. 220–253.

Ramón y Cajal, S. (1968). *Degeneration and Regeneration of the Nervous System*, 2 Vols. Translated and edited by R. M. May. Originally published by Oxford University Press, 1928. Hafner, New York.

Ramón y Cajal, S. (1971). *Neuroanatomical Research*. Translated by A. D. Lowey. *Perspect. Biol. Med.*, 15:7–36.

Ramón y Cajal, S. (1995). *Histology of the Nervous System of Man and Vertebrates*, 2 Vols. Translated into French from the Spanish edition by L. Azoulay (1952) and from the French into English by N. Swanson and L. W. Swanson. Oxford University Press, New York.

Ranish, N., and Ochs, S. (1972). Fast axoplasmic transport of acetylcholinesterase in mammalian nerve fibres. *J. Neurochem.*, 19:2641–2649.

Ranish, N., Ochs, S., and Barnes, C. D. (1972). Regeneration of ventral root axons into dorsal roots as shown by increased acetylcholinesterase activity. *J. Neurobiol.*, 3:245–257.

Ranvier, L.-A. (1875). Des tubes nerveux en T et de leurs relations avec les cellules ganglionnaires. *Compt. Rend. Acad. Sci.*, 81:1274–1276.

Rashevsky, N. (1938). *Mathematical Biophysics. Physicomathematical Foundations in Biology*. University of Chicago Press, Chicago.

Reed, S. (1886). *Observations on the Growth of the Mind; With Remarks on Some Other Subjects*. Clapp, Boston.

Reid, J. (1838). *An Experimental Investigation into the Functions of the Eight Pair of Nerves, or the Glosso-pharyngeal, Pneumogastric, and Spinal Accessory*. Reprinted by Haswell, Barrington, and Haswell. New Orleans. From, Reid, J. (1840). *Edinb. Med. Surg. J.*, 134:1–68.

Remak, R. (1838). *Observationes Anatomicae et Microscopicae de Systematis Nervosi Structura* (Dissertation). Reimer, Berlin.

Rezai, A. R., and Mogilner, A. (2002). *Introduction to Magnetoencephalography.* http://mens10.med.nyu.edu/research/meg_new/meg_new.html. New York.

Richardson, P. M., McGuinness, U. M., and Aguayo, A. J. (1980). Axons from CNS neurones regenerate into PNS grafts. *Nature,* 284:264–265.

Ris, H. (1985). The cytoplasmic filament system in critical point-dried whole mounts and plastic-embedded sections. *J. Cell Biol.,* 100:1474–1487.

Rivers, W. H. R. (1924). *Medicine, Magic and Religion.* Kegan Paul, Trench, Trubner & Co., Ltd. London.

Robertson, J. D. (1960). The molecular structure and contact relationships of cell membranes. *Prog. Biophys.,* 10:344–418.

Rolls, E. T. (1990). Theoretical and neurophysiological analysis of the functions of the primate hippocampus in memory. In: *Cold Spring Harbor Symposia on Quantitative Biology.* Cold Spring Harbor Laboratory Press, Cold Spring Harbor, pp. 995–1006.

Romanes, G. J. (1883). *Animal Intelligence.* Kegan Paul, Trench, and Co., London.

Roofe, P. G. (1947). Role of the axis cylinder in transport of tetanus toxin. *Science,* 105:180–181.

Roots, B. I. (1983). Neurofilament accumulation induced in synapses by leupeptin. *Science,* 221:971–972.

Rose, P. R. (1991). How chicks make memories: the cellular cascade from c-fos to dendritic remodelling. *Trends Neurosci.,* 14:390–397.

Rosenblueth, A., and Del Pozo, E. C. (1943). The centrifugal course of Wallerian degeneration. *Am. J. Physiol.,* 139:247–254.

Rosenblueth, A., and Dempsey, E. W. (1939). A study of Wallerian degeneration. *Am. J. Physiol.,* 128:19–30.

Ross, W. D. (1923). *Aristotle. A Complete Exposition of His Works and Thought.* Methuen, London.

Rothschuh, K. E. (1958). Vom Spiritus Animalis zum Nervenaktionsstrom. *Ciba Zeitschr.,* 8:2950–2980.

Rothschuh, K. E. (1960). *Alexander von Humboldt et l'histoire de la découverte de l'électricité animale.* Edition du Palais de la découverte, Paris.

Rothschuh, K. E. (1978). *Konzepte der Medizin in Vergangenheit und Gegenwart.* Hippokrates Verlag, Stuttgart.

Röyttä, M., and Raine, C. S. (1985). Taxol-induced neuropathy: further ultrastructural studies of nerve fibre changes in situ. *J. Neurocytol.,* 14:157–175.

Rusche, F. (1933). *Das Seelenpneuma.* Verlag Ferdinand Schoningh, Paderborn.

Russell, B. (1945). *A History of Western Philosophy.* Simon and Schuster, New York.

Russell, I. S., and Ochs, S. (1963). Localization of a memory trace in one cortical hemiphere and transfer to the other hemisphere. *Brain,* 86:37–54.

Russell, J. F. (1979). Tarantism. *Med. Hist.,* 23:404–425.

Rydevik, B. L., Kwan, M. K., Myers, R. R., Brown, R. A., Triggs, K. J., Woo, S. L.-Y., and Garfin, S. R. (1990). An in vitro mechanical and histological study of acute stretching on rabbit tibial nerve. *J. Orthopaed. Res.,* 8:694–701.

Sabatini, D. D., Bensch, K., and Barrnet, R. J. (1963). Cytochemistry and electron microscopy. The preservation of cellular ultrastructure and enzymatic activity by aldehyde fixation. *J. Cell Biol.,* 17:58.

Sabri, M. I., Dairman, W., Fenton, M., Juhasz, L., Ng, T., and Spencer, P. S. (1989). Effect of exogenous pyruvate on acrylamide neuropathy in rats. *Brain Res.,* 483:1–11.

Sabri, M. I., and Ochs, S. (1971). Inhibition of glyceraldehyde-3-phosphate dehydrogenase in mammalian nerve by iodoacetic acid. *J. Neurochem.,* 8:1509–1514.

Sabri, M. I., and Ochs, S. (1972). Relation of ATP and creatine phosphate to fast axoplasmic transport in mammalian nerve. *J. Neurochem.*, 9:2821–2828.

Sabri, M. I., Soiefer, A. I., Moretto, A., Lotti, M., Miller, M. S., and Spencer, P. S. (1987). Early retrograde transport defects induced by primary axonal toxins. In: *Axonal Transport*. R. S. Smith, and M. A. Bisby, eds. Alan Liss, New York, pp. 459–472.

Sachs, H. (1969). Neurosecretion. *Adv. Enzymol.*, 32:327–372.

Sacks, O. (1970). *The Man Who Mistook His Wife for a Hat and Other Clinical Tales.* Summit Books, New York.

Sahenk, Z., and Lasek, R. J. (1988). Inhibition of proteolysis blocks anterograde-retrograde conversion of axonally transported vesicles. *Brain Res.*, 460:199–203.

Sahenk, Z., and Mendell, J. (1979a). Ultrastructural study of zinc pyridinethione-induced peripheral neuropathy. *J. Neuropath. Exp. Neurol.*, 58:532–550.

Sahenk, Z., and Mendell, J. R. (1979b). Analysis of fast axoplasmic transport in nerve ligation and adriamycin-induced neuronal perikaryon lesions. *Brain Res.*, 171:41–53.

Sammak, P. J., and Borisy, G. G. (1988). Direct observation of microtubule dynamics in living cells. *Nature*, 332:724–726.

Sammak, P. J., Gorbsky, G. J., and Borisy, G. G. (1987). Microtubule dynamics in vivo: a test of mechanisms of turnover. *J. Cell Biol.*, 104:395–405.

Samson, F. E. (1976). Pharmacology of drugs that affect intracellular movement. *Annu. Rev. Pharm. Toxicol.*, 16:143–159.

Samuel, S. (1860). *Die Tropischen Nerven*. Wigand, Leipzig.

Samuels, A. J., Boyarsky, L. L., Gerard, R. W., Libet, B., and Brust, M. (1951). Distribution exchange and migration of phosphate compounds in the nervous system. *Am. J. Physiol.*, 164:1–15.

Schadé, J. P., and Baxter, C. F. (1960). Changes during growth in the volume and surface area of cortical neurons in the rabbit. *Exp Neurol.*, 2:158–178.

Scharrer, E., and Scharrer, B. (1940). Secretory cells within the hypothalamus. *Assoc. Res. Nerv. Ment. Dis.*, 20:170–194.

Scheibel, M. E., Lindsay, R. D., Tomiyasu, U., and Scheibel, A. B. (1976). Progressive dendritic changes in the aging human limbic system. *Exp. Neurol.*, 53:420–430.

Scheibel, M. E., and Scheibel, A. B. (1973). Hippocampal pathology in temporal lobe epilepsy. A Golgi study. In: *Epilepsy: Its Phenomena in Man*. M. A. B. Brazier, ed. Academic Press, New York.

Schiff, M. (1854). Sur la regeneration des nerfs et sur les alterations qui surviennent dans des nerfs paralyses. *Compt. Rend. Hebd. Acad. Sci.* (Paris), 38:448–452.

Schiff, M. (1858). *Lehrbuch der physiologie des Menschen. I. Muskel und Nervenphysiologie*. Lahr, Schanenburg.

Schiff, P. B., Fant, J., and Horwitz, S. B. (1979). Promotion of microtubule assembly in vitro by taxol. *Nature*, 277:665–667.

Schlaepfer, W. W. (1974). Calcium-induced degeneration of axoplasm in isolated segments of rat peripheral nerve. *Brain Res.*, 69:203–215.

Schliwa, M. (1986). *The Cytoskeleton – An Introductory Survey*, Springer-Verlag, Wein.

Schmidt, R. E., Grabau, G. G., and Yip, H. K. (1986). Retrograde axonal transport of [^{125}I] nerve growth factor in ileal mesenteric nerves in vitro: effect of streptozotocin diabetes. *Brain Res.*, 378:325–336.

Schmiedebach, H.-P. (1995). *Robert Remak (1815–1865)*. Gustav Fischer Verlag, Stuttgart.

Schmitt, F. O., Bear, R. S., and Palmer, K. J. (1941). X-ray diffraction studies on the structure of the nerve myelin sheath. *J. Cell Comp. Physiol.*, 18:31–42.

Schmitt, F. O., and Samson, F. E. Jr., (1968). Neuronal fibrous proteins. *Neurosci. Res. Prog. Bull.*, 6:113–219.

Schnapp, B. J., Vale, R. D., Sheetz, M. P., and Reese, T. S. (1986). Microtubules and the mechanism of directed organelle movement. *Ann. New York Acad. Sci.*, 466:909–918.

Schonbach, J., and Cuenod, M. (1971). Axoplasmic migration of protein. A light microscopic autoradiographic study in the avian retino-tectal pathway. *Exp. Brain Res.*, 12:275–282.

Schroeder van der Kolk, J. L. C. (1859). *On the Minute Structure and Function of the Spinal Cord and Medulla Oblongata, and on the Proximate Cause and Rational Treatment of Epilepsy*. Translated from the original by W. D. Moore. The New Sydenham Society, London.

Schubert, P., Kreutzberg, G. W., and Lux, H. D. (1972). Neuroplasmic transport in dendrites: effect of colchicine on morphology and physiology of motoneurones in the cat. *Brain Res.*, 47:331–343.

Schulte, B. P. M., and Endtze, L. J. (1977). *A Short History of Neurology in the Netherlands*. Eleventh International Congress of Neurology, September 1977, Amsterdam.

Schultze, M. (1870). The general characters of the structures composing the nervous system. In: *Human Comparative Histology*. S. Stricker, ed. Sydenham Society, London, pp. 108–136.

Schwab, M. E., Agid, Y., Glowinski, J., and Thoenen, H. (1977). Retrograde axonal transport of ^{125}I-tetanus toxin as a tool for tracing fiber connections in the central nervous system; connections of the rostral part of the rat neostriatum. *Brain Res.*, 126:211–224.

Schwab, M. E., Heumann, R., and Thoenen, H. (1982). Communication between target organs and nerve cells: Retrograde axonal transport and site of action of nerve growth factor. In: *Cold Spring Harbor Symposium on Quantitative Biology*, Vol. 46. The Biological Laboratory, Cold Spring Harbor, pp. 125–134.

Schwab, M. E., and Thoenen, H. (1978). Selective binding uptake and retrograde transport of tetanus toxin by nerve terminals in the rat iris. *J. Cell Biol.*, 77:1–13.

Schwann, T. (1969). *Microscopical Researches into the Accordance in the Structure and Growth of Animals and Plants*. Translated from the German by H. Smith. Reprinted from The Sydenham Society publication of 1847. Kraus Co., New York.

Schwartz, J. H., Goldman, J. E., Ambron, R. T., and Goldberg, D. J. (1975). Axonal transport of vesicles carrying (3H)-serotonin in the metacerebral neuron of *Aplysia californica*. In: *Cold Spring Harbor Symposia on Quantitative Biology*, Vol. 40. The Biological Laboratory, Cold Spring Harbor, pp. 83–92.

Scott, F. H. (1905). On the metabolism and action of nerve cells. *Brain*, 28:506–526.

Scott, F. H. (1906). On the relation of nerve cells to fatigue of their nerve fibres. *J. Physiol.*, 34:145–162.

Scoville, W. B., and Milner, B. (2000). Loss of recent memory after bilateral hippocampal lesions. 1957. *J. Neuropsychiatry Clin. Neurosci.*, 12:103–113.

Selemon, L. D., Kleinman, J. E., Herman, M. M., and Goldman-Rakic, P. S. (2002). Smaller frontal gray matter volume in postmortem schizophrenic brains. *Am. J Psychiatry*, 159:1983–1991.

Segev, I. and Rall, W. (1998). Excitable dendrites and spines: earlier theoretical insights elucidate recent direct observations. *Trends Neurosci.* 21:453–460.

Server, A. C., and Shooter, E. M. (1977). Nerve growth factor. *Adv. Prot. Chem.*, 31:339–409.

Shan, J., Munro, T. P., Barbarese, E., Carson, J. H., and Smith, R. (2003). A molecular mechanism for mRNA trafficking in neuronal dendrites. *J. Neurosci.*, 23:8859–8866.

Shawe, G. D. H. (1955). On the number of branches formed by regenerating nerve fibres. *Br. J. Surg.*, 42:474–488.

Sheehan, D. (1936). Discovery of the autonomic nervous system. *Arch. Neurol. Psychol.*, 35:1081–1115.

Sheetz, M. P., Steuer, E. R., and Schroer, T. A. (1989). The mechanism and regulation of fast axonal transport. *Trends Neurosci.*, 12:474–478.

Shelanski, M. L., and Liem, R. K. H. (1979). Neurofilaments. *J. Neurochem.*, 33:5–13.

Shepherd, G. M., Brayton, R. K., Miller, J. P., Segev, I., Rinzel, J., and Rall, W. (1985). Signal enhancement in distal cortical dendrites by means of interactions between active dendritic spines. *Proc. Natl. Acad. Sci. U.S.A*, 82:2192–2195.

Shepherd, G. M. (1991). Modern revisions of the neuron doctrine. In: *Foundations of the Neuron Doctrine*. Oxford University Press, New York, pp. 271–310.

Sherrington, C. S. (1906). *The Integrative Action of the Nervous System*. Yale University Press, New Haven.

Sherrington, C. S. (1941). *Man on His Nature*. Cambridge University Press, Cambridge.

Sherrington, C. S. (1946). *The Endeavour of Jean Fernel*. Cambridge University Press, Cambridge.

Sherrington, C. S. (1947). *The Integrative Action of the Nervous System*. Revised Ed. Originally published 1906. Yale University Press, New Haven.

Shpetner, H. S., Paschal, B. M., and Vallee, R. B. (1988). Characterization of the microtubule-activated ATPase of brain cytoplasmic dynein (MAP 1C). *J. Cell Biol.*, 107:1001–1009.

Sidenius, P. P. (1982). The axonopathy of diabetic neuropathy. *Diabetes*, 31:356–363.

Sidenius, P. P., and Jakobsen, J. (1983). Anterograde axonal transport in rats during intoxication with acrylamide. *J. Neurochem.*, 40:697–704.

Sidenius, P. P., and Jakobsen, J. (1987). Axonal transport in human and experimental diabetes. In: *Diabetic Neuropathy*. P. J. Dyck, P. K. Thomas, A. K. Asbury, A. I. Winegrad, and D. Porte, Jr., eds. Saunders, Philadelphia, pp. 260–265.

Siegel, G. J., Agranoff, B. W., Albers, R. W., Fischer, S. K., and M. D. Uhler (1999). eds. *Basic Neurochemistry: Molecular, Cellular, and Medical Aspects*. 6th Ed., Lippincott-Raven, Philadelphia.

Siegel, R. E. (1968). *Galen's System of Physiology and Medicine*. Karger, New York.

Sigerist, H. E. (1948). The story of tarantism. In: *Music and Medicine*. D. M. Schullian and M. Schoen, eds. New York, pp. 96–116.

Sigstedt, C. O. (1952). *The Swedenborg Epic: The Life and Word of Emanuel Swedenborg*. Bookman Associates, New York.

Singer, C. (1925). *The Evolution of Anatomy*. Kegan Paul, Trench, Trubner & Co., Ltd., London.

Singer, C. (1914). Notes on the early history of microscopy. *Proc. Roy. Soc. Med.*, 7:247–279.

Singer, C. (1952a). *Vesalius on the Human Brain*. Oxford University Press, London.

Singer, C. (1956). *Galen on Anatomical Procedures*. Oxford University Press, Oxford.

Singer, C., Holmyard, E. J., Hall, A., and William, T. I. (1954). *A History of Technology*, 5 Vols. Oxford University Press, New York.

Singer, C., and Rabin, C. (1946). *A Prelude to Modern Science: Being a Discussion of the History, Sources and Circumstances of the 'Tabulae Anatomicae Sex' of Vesalius.* Cambridge University Press, Cambridge.

Singer, M. (1952b). The influence of the nerve in regeneration of the amphibian extremity. *Q. Rev. Biol.*, 27:169–200.

Singh, K. D. (2002). Functional Imaging of the Brain Using Superconducting Magnetometry. http://psyserver.pc.rhbnc.ac.uk/vision/MAGRES paper/final4web.html. Surrey, England.

Skinner, B. F. (1938). *The Behavior of Organisms.* Appelton-Century-Crofts, Inc., New York.

Smith, D. S., Jarlfors, U., and Cameron, B. F. (1975). Morphological evidence for the participation of microtubules in axonal transport. *Ann. N. Y. Acad. Sci.*, 253:472–506.

Smith, F. P., Kato, S., Yamamoto, T., and Iwasaki, Y. (1989). Retrograde transport of adriamycin. *J. Neurosurg.*, 70:819–820.

Smith, R. S., and Bisby, M. A. (1993). Persistence of axonal transport in isolated axons of the mouse. *Eur. J. Neurosci.*, 5:1127–1135.

Snyder, R. E. (1986). The kinematics of turnaround and retrograde axonal transport. *J. Neurobiol.*, 17:637–647.

Sokolansky, G. (1930). Die morphogenese der markscheide der peripherischen nervenfasern bei manchen wirbeltieren und beim menschen. *Anat. Anz.*, 69:161–184.

Sokoloff L Reivich, M., Kennedy, C., Des Rosiers, M. H., Patlak, C. S., Pettigrew, K. D., Sakurada, O., and Shinohara, M. (1977). The ^{14}C deoxyglucose method for the measurement of local cerebral glucose utilization: theory, procedure, normal values in the conscious and anesthetized albino rat. *J. Neurochem.*, 28:897–916.

Solmsen, F. (1961). Greek philosophy and the discovery of the nerves. *Museum Helveticum*, 18:150–197.

Souques, A. (1936). *Étapes de la neurologie dans l'antiquité Grecque (d'Homère a Galien),* Masson et Cie, Paris.

Soury, J. (1898). L'amiboisme des cellules nerveuses: Théories de Wiedersheim, Rabl-Ruckhard, Tanzi et S. Ramón y Cajal. *Rev. Gén. Sci. Pures Appl.*, 9:370–376.

Soury, J. (1899). *Le systeme nerveaux centrale; structure et fonctions; historie critique des theories et des doctrines.* Carre et Naud, Paris.

Southern, R. W. (1953). *The Making of the Middle Ages.* Yale University Press, New Haven.

Spatz (1993). Nissl und die Theoretische Hirnanatomie. *Arch. Psychiatry Nervenkr.* 87:100–125 (1929).

Speidel, C. C. (1941). Adjustments of nerve endings. *Harvey Lect.*, 36:126–158.

Spencer, P. S. (1972). Reappraisal of the model for "bulk axoplasmic flow." *Nature* (New Biol.), 240:283–285.

Spencer, P. S., Couri, D., and Schaumburg, H. H. (1980). n-Hexane and methyl-n-butyl ketone. In: *Experimental and Clinical Neurotoxicology.* P. S. Spencer and H. H. Schaumburg, eds. Williams & Wilkins, Baltimore, pp. 456–475.

Spencer, P. S., Sabri, M. I., Schaumburg, H. H., and Moore, C. L. (1979). Does a defect of energy metabolism in the nerve fiber underlie axonal degeneration in polyneuropathies? *Ann. Neurol.*, 5:501–507.

Spencer, P. S., Schaumburg, H. H., and Ludolph, A. C. (2000). *Experimental and Clinical Neurotoxicology.* 2nd Ed. Oxford University Press, New York.

Sperry R, W. (1961). Cerebral organization and behavior. *Science*, 133:1749–1757.

Sperry, R. W. (1962). Some general aspects of interhemispheric integration. In: *Interhemispheric Relations and Cerebral Dominance*. V. B. Mountcastle, ed. Johns Hopkins Press, Baltimore.

Spillane, J. D. (1981). *The Doctrine of the Nerves*. Oxford University Press, Oxford.

St. Augustine (1964). *The Greatness of the Soul and the Teacher*. Translated by J. M. Colleran. The Newman Press, Westminster.

Stamm, J. S. (1961). Electrical stimulation of frontal cortex in monkeys during learning of an alternating task. *J. Neurophysiol.*, 24:414–426.

Stamm, J. S., and Pribram, K. H. (1960). Effects of epileptogenic lesions in the frontal cortex on learning and retention in monkeys. *J. Neurophysiol.*, 23:552–563.

Stannius, H. F. (1847). Untersuchungen über muskelreizbarkeit. *Arch. Anat. Physiol. Wiss. Med.*, 433–462.

Steele-Russell, I., and Ochs, S. (1961). One trial inter-hemispheric transfer of a learning engram. *Science*, 33:1077–1078.

Stefansson, H., Sarginson, J., Kong, A., Yates, P., Steinthorsdottir, V., Gudfinnsson, E., Gunnarsdottir, S., Walker, N., Petursson, H., Crombie, C., Ingason, A., Gulcher, J. R., Stefansson, K., and Clair, D. S. (2002). Association of Neuregulin 1 with Schizophrenia Confirmed in a Scottish Population. *Am. J. Hum. Genet.*, 72:83–87.

Steida, L. (1899). *Geschichte der Entwickelung der Lehre von den Nervenzellen und Nervenfasern wahrend des XIX Jahrhunderts. I. Teil: Von Sommering bis Deiters*. Gustav Fischer, Jena.

Steinrück, C. O. (1840). *De nervorum regeneratione*. Decker, Berlin. Abstracted in Schmidt's Jahrbüch-in-und Ausland, 26:102–104.

Steno, N. (1965). *Nicolaus Steno's Lecture on the Anatomy of the Brain*. Published in 1669. Translated into English from the French by A. J. Pollack, with an introduction by Gustav Scherz. Nyt Nordisk Forlag. Arnold Busck, Copenhagen, pp. 113–160.

Steward, O., and Levy, W. (1982). Preferential localization of polyribosomes under the base of dendritic spines in granule cells of the dentate gyrus. *J. Neurosci.*, 2:284–291.

Steward, O., and Schuman, E. M. (2001). Protein synthesis at synaptic sites on dendrites. *Annu. Rev. Neurosci.*, 24:299–325.

Stirling, W. (1902). *Some Apostles of Physiology*. Waterlow and Sons, London.

Stjärne, L. (1966). Storage particles in noradrenergic tissues. *Pharmacol. Rev*, 18:425–432.

Stokke, B. T., Mikkelsen, A., and Elgsaeter, A. (1985). Some viscoelastic properties of human erythrocyte spectrin networks end-linked in vitro. *Biochem. Biophys. Acta*, 816:111–121.

Stone, G. C., Wilson, D. L., and Hall, M. E. (1978). Two-dimensional gel electrophoresis of proteins in rapid axoplasmic transport. *Brain Res.*, 144:287–302.

Stratton, G. M. (1917). *Theophrastus and the Greek Physiological Psychology Before Aristotle*, with a translation of his *De sensibus*. Allen and Unwin, London.

Straub, R. E., Jiang, Y., MacLean, C. J., Ma, Y., Webb, B. T., Myakishev, M. V., Harris-Kerr, C., Wormley, B., Sadek, H., Kadambi, B., Cesare, A. J., Gibberman, A., Wang, X., O'Neill, F. A., Walsh, D., and Kendler, K. S. (2002). Genetic variation in the 6p22.3 gene DTNBP1, the human ortholog of the mouse dysbindin gene, is associated with schizophrenia. *Am. J. Hum. Genet.*, 71:337–348.

Strawson, P. F. (1959). *Individuals*. Methuen and Co. Ltd., London.

Strawson, P. F. (1966). *The Bounds of Sense: An Essay on Kant's Critique of Pure Reason*. Methuen and Co. Ltd., London.

Sugar, O., and Gerard, R. W. (1940). Spinal cord regeneration in the rat. *J. Neurophysiol.*, 3:1–19.

Sunderland, S. (1978a). *Nerves and Nerve Injuries*, 2nd ed., Williams & Wilkins, Baltimore.

Sunderland, S. (1978b). Peripheral nerve fibers. In: *Nerves and Nerve Injuries*, 2nd ed. S. Sunderland, ed. Churchhill Livingstone, Edinburgh.

Suzuki, H., and Ochs, S. (1964). Laminar stimulation for direct cortical responses from intact and chronically isolated cortex. *Electroencephal. Clin. Neurophysiol.*, 7:405–413.

Swammerdam, J. (1758). *The Book of Nature.* Translated from the Dutch and Latin by T. Floyd and H. Boerhaave. Seyfert, London.

Swanson, L. W. (1999). Limbic system. In: *Encyclopedia of Neuroscience*, 2nd ed., Vol. 1 of 2. G. Adelman and B. H. Smith, eds. Elsevier, Amsterdam, pp. 1053–1055.

Swedenborg, E. (1883). *The Brain, Considered Anatomically, Physiologically, and Philosophically.* 2 Vols. Edited, translated, and annotated by R. L. Tafel. J. Speirs, London.

Swedenborg, E. (1899). *On tremulation.* Originally written in 1719. Translated by C. Th. Odhner. Massachusetts New-Church Union, Boston.

Swedenborg, E. (1918). *The Economy of the Animal Kingdom.* Translated by A. Acton. Transaction III. The Medullary Fibre of the Brain and Nerve Fibre of the Body. Swedenborg Scientific Association, Philadelphia.

Takacs, J., Saillour, P., Imbert, M., Bogner, M., and Hamori, J. (1992). Effect of dark rearing on the volume of visual cortex (areas 17 and 18) and number of visual cortical cells in young kittens. *J. Neurosci. Res.*, 32:449–459.

Takenaka, T. (1986). Particle movements in axoplasmic transport. In: *Axoplasmic Transport.* Z. Iqbal, ed. CRC Press, Boca Raton, pp. 109–118.

Takenaka, T., Horie, H., and Sugita, T. (1978). New technique for measuring dynamic axonal transport and its application to temperature effects. *J. Neurobiol.*, 9:317–324.

Takenaka, T., Inomata, K., and Horie, H. (1982). Slow axoplasmic transport of labeled protein in sciatic nerves of streptozotocin-diabetic rats and methylcobalamin treated rats. In: *Diabetic Neuropathy.* Y. Goto, A. Horiuchi, and K. Kogure, eds. Excerpta Medica, Amsterdam, pp. 99–103.

Takenaka, T., and Ochs, S. (1980). External detection of axoplasmic transport using ^{32}P-ATP as precursor. *J. Neurobiol.*, 11:571–5769.

Tashiro, S. (1917). *A Chemical Sign of Life.* University of Chicago Press, Chicago.

Temkin, O. (1951). On Galen's pneumatology. *Gesnerus*, 8:180–189.

Temkin, O. (1973). *Galenism: Rise and Decline of a Medical Philosophy.* Cornell University Press, Ithaca.

Tertullian (1947). *De Anima.* Edited with introduction and commentary by J. H. Waszink. Meulenhof, Amsterdam.

Tertullian (1980). *Tertullian. Über die seele (de anima).* With introduction, translation and commentary by J. H. Waszink. Artemis Verlag, Zürich.

Tertullian (1986). *A Treatise on the Soul.* Translated by P. Holmes. The Writings of the Fathers down to A.D. 325. Vol. 3. *Latin Christianity: Its Founder Tertullian.* Edited by A. Roberts and J. Donaldson. Revised with notes by A. C. Coxe. American reprint of the Edinburgh edition 1870. W. B. Eerdmans Publishing Co., Grand Rapids.

Thoenen, H. (1991). The changing scene of neurotrophic factors. *Trends Neurosci.*, 14:165–170.

Thomas, P. K. (1963). The connective tissue of peripheral nerve: an electron microscope study. *J. Anat.*, 97:35–44.

Thomas, P. K., Berthold, C.-H., and Ochoa, J. (1993). Microscopic anatomy of the peripheral nervous system: nerve trunks and spinal roots. In: *Peripheral Neuropathy*, 3rd ed. P. J. Dyck, P. K. Thomas, J. W. Griffin, P. A. Low, and J. F. Poduslo, eds. Saunders, Philadelphia, pp. 28–91.

Thomas, P. K., King, R. H. M., and Phelps, A. C. (1972). Electron microscope observations on the degeneration of unmyelinated axons following nerve section. *J. Pathol.*, 107:154–160.

Thompson, D. (1942). *On Growth and Form*. Cambridge University Press, Cambridge.

Thompson, R. F., and Yu, J. (1987). The neuroanatomy of learning and memory in the rat. In: *Neuroplasticity, Learning and Memory*. N. W. Milgram, C. M. MacLeod, and T. Petit, eds. Alan Liss, New York, pp. 231–263.

Tiberi, M., and Lavoie, P.-A. (1985). Inhibition of the retrograde axonal transport of acetylcholinesterase by the anti-calmodulin agents amitriptyline and desipramine. *J. Neurobiol.*, 16:245–248.

Tieleman, T. (1996). *Galen and Chrysippus on the Soul*. W. J. Brill, Leiden.

Tinbergen, N. (1974). Ethology and stress diseases. *Science*, 185:20–27.

Tinbergen, N. (1989). *The Study of Instinct*. First published 1949. Clarendon Press, Oxford.

Tiruchinapalli, D. M., Oleynikov, Y., Kelic, S., Shenoy, S. M., Hartley, A., Stanton, P. K., Singer, R. H., and Bassell, G. J. (2003). Activity-dependent trafficking and dynamic localization of zipcode binding protein 1 and beta-actin mRNA in dendrites and spines of hippocampal neurons. *J. Neurosci.*, 23:3251–3261.

Tissot, S. A. D. (1840). *L'Onanism. Dissertation sur les maladies produites par la masturbation*. Originally published in 1809. In: *Oevres de Tissot, Encyclopédie des Sciences Médical*. Bureau de l'encyclopèdie, Paris, pp. 479–545.

Todd, R. B. (1847). Nerve. In: *Cyclopaedia of Anatomy and Physiology*. Vol. III. R. B. Todd, ed. Sherwood, Gilbert, and Piper, London, pp. 590–601.

Todd, R. B. (1985). Philosophy and medicine in John Philoponus' commentary on Aristotle's De Anima. In: *Symposium on Byzantine Medicine*. J. Scarborough, ed. Dumbarton Oaks Research Library and Collection, Washington, DC, pp. 103–110.

Toksvig, S. (1948). *Emanuel Swedenborg Scientist and Mystic*. Faber and Faber Ltd., London.

Tome, F. M. S., Tegner, R., and Chevallay, M. (1988). Varicosities in human fetal sciatic nerve fibres. *Neuropathol. Appl. Neurobiol.*, 14:495–504.

Torrey, T. W. (1934). The relation of taste buds to their nerve fibers. *J. Comp. Neurol.*, 59:203–220.

Triarhou, L. C., and Del Cerro, M. (1987). The histologist Sigmund Freud and the biology of intracellular motility. *Biol. Cell*, 61:111–114.

Tsukita, S., and Ishikawa, H. (1981). The cytoskeleton in myelinated axons: serial section study. *Biomed. Res.*, 2:424–437.

Tylor, S. E. B. (1871). *Primitive Culture*. J. Murray, London.

Vale, R. D. (1987). Intracellular transport using microtubule-based motors. *Ann. Rev. Biol.*, 3:347–378.

Vale, R. D., Funatsu, T., Pierce, D. W., Romberg, L., Harqda, Y., and Yanagida, T. (1996). Direct observation of single kinesin molecules moving along microtubules. *Nature*, 380:451–453.

Vale, R. D., and Milligan, R. A. (2000). The way things move: looking under the hood of molecular motor proteins. *Science*, 288:88–95.

Vale, R. D., Schnapp, B. J., Mitchison, T. J., Steuer, E., Reese, T. S., and Sheetz, M. P. (1985). Different axoplasmic proteins generate movement in opposite directions along microtubules in vitro. *Cell*, 43:623–632.

Valentin, G. (1841). Hirn und nervenlehre. In: *Vom baue des menschichen korpers*, Vol. 4. Th. von Sommering, ed. Leipzig.

Valentin, G. (1843). *Traite de Neurologie*. Traduit de l'Allemande par A. J. L Jourdan. J. B. Brailliere, Paris.

Vallee, R. B., and Shpetner, H. S. (1990). Motor proteins of cytoplasmic microtubules. *Annu. Rev. Biochem.*, 59:909–932.

Vallee, R. B., Shpetner, H. S., and Paschal, B. M. (1989). The role of dynein in retrograde axonal transport. *Trends Neurosci.*, 12:66–70.

van Gehuchten, A. (1900). *Anatomie du Système Nerveux de l'Homme*. Imprimerie des Trois Rois, Louvain.

Van Harreveld, A., and Ochs, S. (1956). Cerebral impedance changes after circulatory arrest. *Am. J. Physiol.*, 187:180–192.

van Breemen, V. L. (1958). An attempt to determine the origin of synaptic vesicles. In: *Experimental Cell Research, Supplement 5*. Academic Press, New York, pp. 153–167.

Van der Loos, H. (1967). The history of the neuron. In: *The Neuron*. H. Hydén, ed. Elsevier Publishing Company, Amsterdam, pp. 1–47.

Van Harreveld, A. (1966). *Brain Tissue Electrolytes*. Butterworths, Washington.

Van Harreveld, A., and Crowell, J. (1964). Electron microscopy after rapid freezing on a metal surface and substitution fixation. *Anat. Rec.*, 149:381–385.

Van Harreveld, A., and Fifková, E. (1975). Swelling of dendritic spines in the fascia dentata after stimulation of the perforant fibers as a mechanism of post-tetanic potentiation. *Exp. Neurol.*, 49:736–749.

van Leeuwenhoek, A. (1968). Letter to Abraham van Bleiswyk (March 2 1717) printed in "Epistolae physiologicale super compluribus naturae arcanis" (Beman, Delft, 1719), Epistola XXXII, pp. 310–327. In part translated by E. Clarke and C. D. O'Malley. In: *The Human Brain and Spinal Cord*. E. Clark and C. D. O'Malley, eds. Berkeley, University of California Press, pp. 32–35.

Vartanian, A. (1967). Holbach, Paul-Henri Thiry, Baron D'. In: *The Encyclopedia of Philosophy*, in 8 volumes, Vol. 4. P. Edwards, ed. Macmillan Publishing Co., New York, pp. 49–51.

Vesalius, A. (1543). De humani corporis fabrica libri septem. *J. Oporini*, Basileae.

Vesalius, A. (1950). *The Illustrations from the Works of Andreas Vesalius of Brussels*. Annoted and translations by J. B. deC. M. Saunders and C. D. O'Malley. The World Publishing Co., Cleveland.

Vogel, V. J. (1970). *American Indian Medicine*. University of Oklahoma Press, Norman.

von Euler, U. S. (1956). *Noradrenaline: Chemistry, Physiology, Pharmacology and Clinical Aspects*. Charles C Thomas, Springfield.

von Haller, A. (1762). *Elementa Physiologiae Corporis Humani. Tomus Quartus: Cerebrum. Nervi. Musculi*. Francisci Grasset, Lausanne.

von Haller, A. (1936). A dissertation on the sensible and irritable parts of animals. English translation by O. Temkin. *Bull. Hist. Med.*, 4:651–699.

von Haller, A. (1966). *First Lines of Physiology*. 2 Vols. in one. Translated from correct Latin edition by William Cullen and compared with the edition published by H. A. Wrisberg Reprint of the 1786 edition with a new introduction by Lester S. King. Johnson Reprint Corporation, New York.

von Staden, H. (1989). *Herophilus: The Art of Medicine in Early Alexandria*. Cambridge University Press, Cambridge.

Vulpian, A. (1866). *Leçons sur la physiologie générale et comparee du systéme nerveux.* Baillier, Paris.

Waddell, H. (1934). *The Wandering Scholars.* Constable, London.

Waldeyer, H. W. G. (1891). Über einige neurer forschungen im gebiete der anatomie des central nervensystems. *Deutsche Med. Wschr.,* 17:1213–1218, 1244–1246, 1267–1269, 1287–1289, 1331–1332, 1352–1356.

Walker, A. (1973). *Documents and Dates of Modern Discoveries in the Nervous System.* With notes and historical introduction by P. F. Cranefield. Reprinted from the 1839 edition. Scarecrow Reprint Corporaton, Metuchen.

Walker, W. C. (1937). Animal electricity before Galvani. *Ann. Sci.,* 2:84–113.

Waller, A. V. (1850). Experiments on the section of the glossopharyngeal hypoglossal nerves of the frog, and observations of the alterations produced thereby in the structure of their primitive fibres. *Phil. Trans. Roy. Soc.* (London), 140:423–429.

Waller, A. V. (1852a). A new method for the study of the nervous system. *London J. Med.,* 43:609–625.

Waller, A. V. (1852b). Huitième mémoire sur le système nerveux. *Compt. Rend. Hebd. Seanc.,* 35:561–564.

Waller, A. V. (1852c). Nouvelle méthode anatomique pour l'investigation du systéme nerveux. Georgi, Bonn.

Waller, A. V. (1852d). Septième mémoire sur le système nerveux. *Compt. Rend. Acad. Sci.,* 35:301–306.

Waller, A. V. (1852e). Sur la reproduction des nerfs et sur la structure et les fonctions des ganglions spinaux. *Anat. Physiol. Wiss. Med.* (Berlin), 392–401.

Waller, A. V. (1861). The nutrition and reparation of nerves; being the substance of a lecture delivered at the Royal Institution of Great Britain Friday May 31 1861. Read and Co., London.

Waller, A. V. (1862). On the nutrition and reparation of nerves; being the substance of a lecture delivered at the Royal Institution of Great Britain, Friday, May 31, 1861. *Proc. Roy. Inst.,* 3:378–381.

Waller, A. V. (1870). On the results of the method introduced by the author of investigating the nervous system, more especially as applied to the elucidation of the functions of the pneumogastric and sympathetic nerves. The Croonian Lecture. *Proc. Roy. Soc. Lond. B.,* 18:339–343.

Waller, A. D. (1891). *An Introduction to Human Physiology,* Longmans, Green, and Co., London.

Walsh, J. J. (1923). *The Story of the Cures That Fail.* Appleton and Co., New York.

Walzer, R. (1962). *Greek Into Arabic. Essays on Islamic Philosophy.* University of South Carolina Press, Columbia, South Carolina.

Wang, N., Butler, J. P., and Ingber, D. E. (1993). Mechanotransduction across the cell surface and through the cytoskeleton. *Science,* 260:1124–1127.

Ward, A. A., Jr. (1969). The epileptic neuron: chronic foci in animals and man. In: *Basic Mechanisms of the Epilepsies.* H. H. Jasper, A. A. Ward, Jr., and A. Pope, eds. Little, Brown and Company, Boston, pp. 263–288.

Warnock, G. J. (1958). *English Philosophy Since 1900.* Oxford University Press, London.

Watson, J. B. (1914). *Behavior: An Introduction to Comparative Psychology,* Henry Holt and Co., New York.

Watson, D. F., Glass, J. D., and Griffin, J. W. (1993). Redistribution of cytoskeleton proteins in mammalian axons disconnected from their cell bodies. *J. Neurosci.,* 13:4354–4360.

Watson, D. F., Hoffman, P. N., Fittro, K. P., and Griffin, J. W. (1989). Neurofilament and tubulin transport slows along the course of mature motor axons. *Brain Res.*, 477:225–232.

Webster's (1966). *Webster's Third New International Dictionary of the English Language.* G. and C. Merriam Co. (Merriam-Webster Co.), Springfield, Massachusetts.

Wedl, C. (1855). *Rudiments of Pathological History.* Translated and edited by G. Busk. The Sydenham Society, London.

Weiss, D. G. (1982). ed. *Axoplasmic Transport.* Springer-Verlag, New York

Weiss, D. G., and Gross, G. W. (1982). The microstream hypothesis of *axoplasmic transport.* transport: characteristics, predictions and compatibility with data. In: *Axoplasmic Transport.* D. G. Weiss, ed. Springer-Verlag, Berlin, pp. 362–383.

Weiss, H. D., Walker, M. D., and Wiernik, P. H. (1974). Neurotoxicity of commonly used antineoplastic agents (2nd of two parts). *N. Engl. J. Med.*, 291:127–133.

Weiss, P. (1939). *Principles of Development. A Text in Experimental Embryology.* Henry Holt and Company, New York.

Weiss, P. (1961). The concept of perpetual neuronal growth and proximo-distal substance convection. In: *Regional Neurochemistry.* S. S. Kety and J. Elkes, eds. Pergamon Press, New York, pp. 220–240.

Weiss, P. (1972a). Neuronal dynamics and axonal flow. V. The semisolid state of the moving axonal column. *Proc. Natl. Acad. Sci. U.S.A.*, 69:620.

Weiss, P. A. (1972b). Neuronal dynamics and axonal flow: axonal peristalsis. *Proc. Natl. Acad. Sci. U.S.A.*, 69:1309–1312.

Weiss, P., and Davis, H. (1943). Pressure block in nerves provided with arterial sleeves. *J. Neurophysiol.*, 6:269–286.

Weiss, P., and Hiscoe, H. B. (1948). Experiments on the mechanism of nerve growth. *J. Exp. Zool.*, 107:315–395.

Weiss, P., and Hoag, A. (1946). Competitive reinnervation of rat muscles by their own and foreign nerves. *J. Neurophysiol.*, 9:413–418.

Weiss, P., and Holland, Y. (1967). Neuronal dynamics and axonal flow II. The olfactory nerve as model test object. *Proc. Natl. Acad. Sci. U.S.A.*, 57:258–264.

Weller, R. O., Mitchell, J., and Daves, G. D., Jr. (1980). Buckthorn (Karwinskia Humboldtiana) toxins. In: *Experimental and Clinical Neurotoxicity.* P. S. Spencer and H. H. Schaumburg, eds. Williams & Wilkins, Baltimore, pp. 336–347.

Wellman, K. (1992). *La Mettrie. Medicine, Philosophy, and Enlightenment.* Duke University Press, Durham.

White, W. C. (1966). *Chinese Jews: A Compilation of Matters Relating to the Jews of K'ai-fêng Fu.* Paragon Book Gallery, New York.

White, W. H. (1886). On neurorheuma or nervous energy. *Lancet*, 2:161–162.

Whittaker, E. (1951). *A History of the Theories of Aether and Electricity*, 2 Vols. Philosophical Library, New York.

Whytt, L. L. (1967). Unconscious. In: *The Encyclopedia of Philosophy.* Vol. 8 of 8. P. Edwards, ed. Macmillan Publishing Co., New York, pp. 185–189.

Wiesel, T. N., Hubel, D. H., and Lam, D. M. K. (1974). Autoradiographic demonstration of ocular-dominance columns in the monkey striate cortex by means of transneuronal transport. *Brain Res.*, 79:273.

Wiley, R. G., Blessing, W. W., and Reis, D. J. (1982). Suicide transport: destruction of neurons by retrograde transport of ricin, abrin and modeccin. *Science*, 216:889–890.

Wiley, R. G., and Stirpe, F. (1988). Modeccin and volkensin but not abrin are effective suicide transport agents in rat CNS. *Brain Res.*, 438:145–154.

Willard, M. (1983). Neurofilaments and axonal transport. In: *Neurofilaments*. C. A. Marotta, ed. University of Minnesota Press, Minneapolis, pp. 86–116.

Willard, M., Cowan, W. M., and Vagelos, P. R. (1974). The polypeptide composition of intra-axonally transported proteins: evidence for four transport velocities. *Proc. Natl. Acad. Sci. U.S.A.*, 71:2183–2187.

Willard, M. B., and Hulebak, K. L. (1977). The intra-axonal transport of polypeptide H: evidence for a fifth (very slow) group of transported proteins in the retinal ganglion cells of the rabbit. *Brain Res.*, 36:289–306.

Willey, B. (1940). *The Eighteenth Century Background. Studies on the Idea of Nature in the Thought of the Period.* Columbia University Press, New York.

Willey, B. (1962). *The Seventeenth Century Background: Studies in the Thought of the Age in Relation to Poetry and Religion.* Columbia University Press, New York.

Williams, P. L., and Hall, S. M. (1970). In vivo observations on mature myelinated nerve fibres of the mouse. *J. Anat.*, 107:31–38.

Williams, P. L., and Hall, S. M. (1971a). Chronic Wallerian degeneration – an in vivo and ultrastructural study. *J. Anat.*, 109:487–503.

Williams, P. L., and Hall, S. M. (1971b). Prolonged in vivo observations of normal peripheral nerve fibres and their acute reactions to crush and deliberate trauma. *J. Anat.*, 8:397–408.

Willis, T. (1965a). *The Anatomy of the Brain and Nerves*. 2 Vols. W. Feindel, ed. with a note on Pordage's English translation and a bibliographic survey of Cerebri Anatome. McGill University Press, Montreal.

Willis, T. (1965b). The description and uses of the nerves. In: *The Anatomy of the Brain and Nerves*, 2 Vols. Translated by S. Porridge. Originally published in 1681. W. E. D. Feindel, ed. McGill University Press, Montreal.

Willis, T. (1971a). *The Anatomy of the Brain*. USV Pharmaceutical, Tuckahoe, NY.

Willis, T. (1971b). *Two Discourses Concerning the Soul of Brutes.* Translation by S. Pordage, 1683 of *De Anima Brutorum*. R. Davis, Londini, 1672, Scholars' Facsimiles and Reprints, Gainsville.

Wilson, Kinnier, S. A. (1989). *Neurology*. 3 Vol. Edited by A. N. Bruce. Originally published E. Arnold, London 1940. Special Ed. Classics of Neurology and Neurosurgery Library.

Wilson, D. L., and Stone, G. C. (1979). Axoplasmic transport of proteins. *Ann. Rev. Biophys. Bioeng.*, 8:27–45.

Wilson, L., and Meza, I. (1973). The mechanism of action of colchicine. Colchicine binding properties of sea urchin sperm tail outer doublet tubulin. *J. Cell Biol.*, 58:709.

Wilson, L. G. (1959). Erasistratus, Galen, and the Pneuma. *Bull. Hist. Med.*, 33:293–314.

Wilson, L. G. (1961). William Croone's theory of muscular contraction. Notes and records. *Roy. Soc. London*, 16:158–178.

Windle, W. F. (1955). *Regeneration in the Central Nervous System*. Charles C Thomas, Springfield.

Woodhall, B., and Beebe, G. W. (1956). *Peripheral Nerve Regeneration: A Follow-up Study of 3,656 World War II Injuries*. U.S. Government Printing Office, Washington, DC.

Worth, R. M., and Ochs, S. (1976). The effect of repetitive electrical stimulation on axoplasmic transport. *Soc. Neurosci. Abstr.*, 2:350.

Worth, R. M., and Ochs, S. (1982). Dependence of batrachotoxin block of axoplasmic transport on sodium. *J. Neurobiol.*, 13:537–549.

Wright, G. P. (1955). The neurotoxins of clostridium botulinum and clostridium tetani. *Pharmacol. Rev.*, 7:413–465.

Wyburn-Mason, R. (1950). *Trophic Nerves. Their Role in Physiology and Pathology with Especial Reference to the Aetiology of Malignant, Neurological and Mental Disease and Inflammatory and Atrophic Changes.* Henry Kimpton, London.

Yamada, K. M., Spooner, B. S., and Wessells, N. K. (1971). Ultrastructure and function of growth cones and axons of cultured nerve cells. *J. Cell Biol.*, 49:614–635.

Yamamoto, T., Iwasaki, Y., and Konno, H. (1984). Experimental sensory ganglionectomy by way of suicide axoplasmic transport. *J. Neurosurg.*, 60:108–114.

Yawo, H., and Kuno, M. (1985). Calcium dependence of membrane sealing at the cut end of the cockroach giant axon. *J. Neurosci.*, 5:1626–1632.

Young, J. Z. (1934). Structure of nerve fibres in Sepia. J. Physiol. (London) Proc. 83:27P–28P.

Young, J. (1929). Malpighi's *De Pulmonibus*. Proc. Roy. Soc. Med., 23:1–10.

Young, J. Z. (1936a). Structure of nerve fibres and synapses in some invertebrates. In: *Cold Spring Harbor Symposia on Quantitative Biology*, Vol. 4. *Excitation Phenomena.* The Biological Laboratory, Cold Spring Harbor, pp. 1–6.

Young, J. Z. (1936b). The structure of nerve fibers in cephalopods and crustacea. *Proc. Roy. Soc. Lond. B.*, 121:319–337.

Young, J. Z. (1944a). Contraction, turgor and the cytoskeleton of nerve fibres. *Nature*, 153:333–335.

Young, J. Z. (1944b). Surface tension and the degeneration of nerve fibres. *Nature*, 154:521–522.

Young, J. Z. (1945). The history of the shape of a nerve fiber. In: *Essays on Growth and Form Presented to D'Arcy Wentworth Thompson*. W. E. LeGros Clark and P. B. Medawar, eds. Clarendon Press, Oxford, pp. 41–94.

Young, J. Z. (1949). Factors influencing the regeneration of nerves. *Adv. Surg.*, 1:165–220.

Younkin, S. G., and Younkin, L. H. (1988). Trophic regulation of skeletal muscle. In: *Nerve-Muscle Cell Trophic Communication*. H. L. Fernandez and J. A. Donoso, eds. CRC Press, Boca Raton, pp. 41–59.

Zelena, J. (1968). Bidirectional movement of mitochondria along axons of an isolated nerve segment. *Z. Zellforsch. Mikr. Anat.*, 92:186–196.

Zelena, J., Lubinska, L., and Gutmann, E. (1968). Accumulation of organelles at the ends of interrupted axons. *Z. Zellforsch. Mikr. Anat.*, 91:200–219.

Zeller, E. (1931). *Outlines of the History of Greek Philosophy*. Thirteenth edition revised by Dr. Wilhelm Nestle and translated by L. R. Palmer. Kegan Paul, Trench, Trubner and Co. Ltd., London.

Zenker, W., and Hohberg, E. (1973). Motorische nervenfaser: axonquerschnittsflache von stammfaser und endasten. *Z. Anat. Entwick.-Gesch.*, 139:163–172.

Zola-Morgan, S. (1995). Localization of brain function: the legacy of Franz Joseph Gall (1758–1828). *Ann. Rev. Neurosci.*, 18:359–383.

Zysk, K. G. (1995). Vital Breath (prana) in ancient indian medicine and religion. In: *The Comparison Between Concepts of Life-Breath in East and West.* Y. Kawakita, S. Sakai, and Y. Otsuka, eds. Ishiyaku EuroAmerica, Inc., Tokyo, pp. 33–65.

INDEX

reciprocal innervation, 69
recirculation of activity –
  memory, 326
recognition – an
  integration of sensory elements, 357
recovery without reunion, 199
redundant neurites,
  pruning of, 208
reflex pattern
  laid down as a whole, 313
reflexes, 70
  fossilized intelligence, 310
  integrated, 355
  purposeful, 307
reflexologic school
  reflex chaining during development, 313
regenerating fibers –
  following partial compression, 219
regeneration
  discribed by Waller, 198
  in fibers of CNS, 142
  into their own functional type, 197–198
  of limb nerves, 196–197
  peripheral fibers, 142
Remak, Robert
  early neuron doctrine, 136–138
  form of nerve, 134–135
  3-germ layers, 144, 165
  Jewishness and acceptance of his early
    neuron theory, 143, 165
Remak–Valentin
  controversy, 136–142
resonance
  brain unity, 319
respiration movements
  act on ganglia, 106
  in brain and animal spirits, 101
resting potential
  muscle, 128
rete mirabili, 27, 75
  Galen description of, 27
retrograde transport, 235–236
reunion
  accepted by Waller, 200–201
  analogy to soldering, 200
  contra regeneration, 198–200
reverberatory activity
  trace formation in cell-assembly, 327
rheotome
  form of action potential, 123
ricin, 286
Roger of Salerno, 187
Roland of Parma, 188
rolling vesicle
  model of transport, 266

Roman empire, 37
routing
  in *Aplysia* fibers, 255
  into fiber branches of T-shaped cells,
    233–235
  in neurites, 208
  to neuron spines, 335
Ruysch, Frederick
  injection of blood vessels, 94

Salerno
  City of Hippocrates, 44
saltatory conduction
  myelinated nerve fiber, 147
SCa
  compatibility with regeneration rate, 276
  decrease with distance along nerve, 276
  slow component a, 245
SCb
  slow component b, 245
schizophrenia
  genetic changes in, 351
Schleiden, Matthias Jakob
  plant cells, 143
Schwann, Theodore
  animal cells, 143
Schwann cell
  proliferative spread in Wallerian
    degeneration, 182
Schwann cells
  nerve fibers formed by, 143–144
sciatic nerve
  eating of proscribed by Cherokee Indians,
    3
  eating of proscribed in Bible, 2–3
sciatica
  described by Cotugno, 115, 121
  oily substance obstruction, 121
scriptoria
  manuscript transcriptions, 47
SDS-PAGE protein separation, 245
secretory cells of HNS
  paraventricular nucleus (PVN), 257
  supraophic nucleus (SON), 257
sensation
  a priori formal conditions of (Kant),
    361–362
  involves higher processes, 361–362
sensation to perception, 358
sense data
  primary cortex reception, 356
senses
  vibration of ether, 90
sensible and irritable tissues
  of body, 96

of metabolic – related substances, 216
models of, 263–283
of respiratory ferments, 216
in vitro, 237–238
transport by filaments
hypothesized, 264
transport filament
model, 266–268
transport rate depends on
drop-off kinetics from carriers, 282
trepanation, 6
for spirits, 6–7
medical procedure, 7
tricarboxylic acid cycle, 261
block by fluoracetate, 292
trifluoperazine
anti-calmodulin agent, 291
tripartate soul, 25
tritium (^{3}H) amino acid precursor
protein transport tracer, 224, 226
triton X-100
permeabilizes sheath and axolemma,
288
tRNA (*see* transfer RNA)
trophic influence of cells
on its fibers, 172–173
trophic nerve fibers, 185
tropic influences on
regenerating sciatic nerve fibers, 205
TTX (*see* tetrodotoxin)
tubulin
assembly into microtubules, 211
subunits, α and β, 149
tubulin binding agents, 294
microtubule block, 265
tulidora
neuropathy, 302
turnaround, 236
thiol proteases in, 301
two-dimensional protein separation,
246
tyrosine
to l-DOPA, 254
tyrosine hydroxylase transport of, 254

unconditioned stimulus, 325
unit membrane, 146
unit reaction in
reflex integration, 353
Unitary Hypothesis
for all rates of transport, 282–283
universities
growth of, 48
uptake in nerve terminals
"back-door" for pathogens and toxins and
viruses, 235

vacuum behavior
in hungry coelenterata, 313
in hungry starlings, 313
vagal nerve
innervation of heart, 52–53
regeneration of, 190
vagotomy
esophageal function defect, 196
lethal effects, 191
vagus nerve
regeneration of, 190
Valentin, Gabriel Gustave
fibers and cell separate entities, 136
valves in veins
toward the heart, 60
varicosities (beading), 153
artifacts of preparation, 134
early phase of degeneration, 180–181
normal nerve property, 134
see beading, 153
caused by DFP, 301
in unmyelinated axons, 181
vasa nervorum
blood supply of nerve, 164, 166
vasopressin, 257
ventricles
compression and incision of, 30–31
vermis
valve between brain cells, 29, 41
ventricular fluid flow control, 29
Vesalius, Andreas
anatomy demonstration at Bologna, 55
challenges to Galenic teaching, 55, 57
De corporis humani fabrica, 55
function of laryngeal nerve, 55
nerve sheath not a path for animal spirits,
57
remnants of Galenic teaching, 55
similarity of brain parts in animals and
man, 56
transformation of spirits in brain
substance, 56
vesicle movement
by kinesin, 268–273
vesicles
roll along microtubule, 265–266
vibrations
in solid nerve fibers, 89
vinblastine, 294
vinca alkaloids
iontophoretically administered, 286
vincristine, 294
visual cortex
thinner in kittens raised in darkness, 324
visual field mapped
in visual cortex, 359